Essential Analog Electronics

Essential Analog Electronics

Owen Bishop

Newnes
An imprint of Butterworth-Heinemann
Linacre House, Jordan Hill, Oxford OX2 8DP
A division of Reed Educational and Professional Publishing Ltd

A member of the Reed Elsevier plc group

OXFORD BOSTON JOHANNESBURG
MELBOURNE NEW DELHI SINGAPORE

First published 1997

British Library Cataloguing in Publication Data
Bishop, O. N. (Owen Neville), 1927–
Essential analog electronics
1. Analog electronic systems
I. Title
621.3′815

ISBN 0 7506 2898 7

Library of Congress Cataloguing in Publication Data
Bishop, O. N. (Owen Neville)
Essential analog electronics/Owen Bishop.
p. cm.
Includes index.
ISBN 0 7506 2898 7 (pbk.)
1. Analog electronic systems. I. Title.
TK7867.B52 96–42137
621.381–dc20 CIP

Typeset by Laser Words, Madras, India
Printed and bound in Great Britain by
Biddles Ltd, Guildford and King's Lynn

Contents

Introduction

Analog electronics is a topic in its own right but also relates to most other aspects of electronics. This book is intended to provide the student of any branch of electronics with a concise guide to the essentials of analog electronics. With the increasing use of computer simulations by electronic design engineers, the book is illustrated exclusively with graphs and numerical data obtained from computer analyses. Now that the computer has taken over the burdensome calculations, there is even greater need that the student should fully understand how each circuit works, and in what ways it is expected to behave. For this reason, the emphasis in this book is on understanding analog electronics, rather than on blindly applying equations and formulae.

Acknowledgements

Figures 2.3, 2.5, 2.7, 2.8, 2.13, 2.16, 2.24, 2.25, 2.26, 3.5, 3.7, 3.14, 3.15, 4.2–4.9, 4.14, 4.15, 4.18, 4.19, 4.21, 5.2, 5.3, 5.11–5.14, 5.29, 6.2, 6.3, 6.11, 6.15, 6.18, 6.20, 6.23–6.25, 6.31, 7.3, 7.5, 7.6, 8.1, 8.4, B.1, B.2, B.4, C.5–C.8, C.13, C.14, C.16, C.17 and E.9 are the results of computer simulations using SpiceAge© for Windows™ (published by Those Engineers Ltd).

Figures 2.25, 6.5–6.8, 6.14, 6.19, 6.21 and 6.22 were produced using Nodal (published by Macallan Consulting Inc.) in conjunction with Mathematica® (published by Wolfram Research Inc.). The routines used in the calculations are from *Applied Electronics Engineering with Mathematica®*, by Alfred Riddle and Samuel Dick (published by Addison-Wesley Publishing Co., 1995).

Symbols

The format for voltages:

Quiescent or average values	V_B, V_{IN}
Instantaneous total value of a signal or potential difference	v_{CE}, v_{OUT}
Instantaneous departure of signal from quiescent value	v_{in}, v_b
Fixed supply or bias voltages	V_{CC}, V_{DD}, V_+, V_-
Voltage amplitude of sinusoids	V_0, V_1
Voltages at input terminals of an op amp	v_+, v_-

Formats for currents and impedances: substitute I, i, Z or z in the above.

Abbreviations

→	cause → effect
‖	R1 ‖ R2 = resistance of R1 and R2 in parallel
AF	audio frequency (30 Hz–20 kHz)
BJT	bipolar junction transistor
CB	common base
CC	common collector
CD	common drain
CE	common emitter
CS	common source
emf	electromotive force
FET	field-effect transistor
ic	integrated circuit
JFET	junction field effect transistor
KCL	Kirchhoff's current law (D1.1)
KVL	Kirchhoff's voltage law (D1.2)
NMOS	n-channel MOSFET
MOSFET	Metal oxide semiconductor FET
ntc	negative temperature coefficient
oc	open circuit
PMOS	p-channel MOSFET
pot	potentiometer
ptc	positive temperature coefficient
pd	potential difference
RF	radio frequency (30 kHz–30 GHz)
sc	short circuit
tempco	temperature coefficient

1 Capturing analogs

1.1 Analogs

An analog is something that is similar to or equivalent to something else (the prototype) in one or more significant ways. In analog electronics, the analog is a voltage or current signal in which the duration of the signal, its frequency (or mixture of frequencies) and its amplitude correspond to specified characteristics of the prototype. The correspondence may be direct (e.g. waveform of audio signal corresponding to sound waves in air) or indirect (e.g. frequency of signal corresponding to magnitude of a sensed temperature).

An analog is imperfect if:

- it is non-linear — equal changes in a given quantity in the prototype do not produce equal changes in the analog;
- it has a distorted frequency composition — different frequencies in the prototype are not equally represented in the analog;
- inappropriate quantities affect it — e.g. temperature affects the analog when temperature is not one of the quantities to be represented by it;
- it is noisy — the analog contains signals not present in the prototype.

The aim of analog electronics is to process analogs while minimizing such imperfections.

A digital signal, in which numeric values are coded in binary form, can be said to be an analog (as when the data from an audio compact disc are processed), but such signals are not dealt with in this book.

1.2 Sensors

The first stage in translating a prototype physical event into its electronic analog is to use a *sensor* or a *transducer*. A sensor is a device for detecting a physical quantity. A transducer is a device which converts one form of energy into another form of energy. A quantity that is to be detected or measured by a sensor or transducer is a *measurand.*

Below are listed the sensors and transducers that are most commonly used as interfaces between the real world and analog circuits. Relatively few sensing devices rely on transduction. The only ones listed here being the thermocouple (1.3.3), the pyroelectric sensor (1.3.5) and the electret, electromagnetic and piezo-electric microphones (1.6.3–1.6.5) so we include them with the sensors. Numerical values of properties are based on representative types.

1.2.1 A typical sensor

The electrical resistance of a thermistor varies with temperature (1.3.2). A thermistor displays all the essential attributes of a sensor:

- It is affected by the physical quantity it is designed to detect, the measurand, but is relatively unaffected by other physical quantities. For example, the thermistor is affected by heat but is unaffected by light or non-destructive pressure.
- Any change in the measurand produces a change in a property of the sensor; in a thermistor, increase in temperature → decrease in resistance.
- The amount of change in the property is related directly or inversely to the change in the measurand. In many sensors the relationship is linear, but the thermistor has a complicated exponential relationship (1.3.2).
- A very small change in the measurand may or may not produce a perceptible change in the property. There is a limit on the sensitivity of the sensor. The resistance of a thermistor is, in theory, continuously variable with respect to temperature so its sensitivity is high. But there are practical difficulties, one of which is that it is necessary to pass a current through the thermistor in order to measure its resistance. This current causes the thermistor to become slightly warmer than the ambient temperature.
- The range of response of the sensor is limited. At low temperatures the resistance of a thermistor becomes too high to measure precisely; at high temperatures the thermistor is physically destroyed.
- The response of a sensor may be dependent upon frequency. There is a certain delay in cooling or heating; a thermistor of large physical size does not respond to rapid changes of temperature, particularly rapidly alternating heating and cooling. Its frequency response depends on the physical size, shape, encapsulation (if any), and heat capacity of the thermistor.
- An electronic sensor requires an analog electronic circuit to detect or measure the changes in the sensor's property. A thermistor requires a circuit which responds to changes of resistance.

- The accuracy of the sensor may be affected by age, mistreatment or other circumstances. A thermistor can be damaged by overheating by excess current.

1.3 Thermal sensors

1.3.1 Platinum resistance thermometer

Measurand: Temperature.
Principle: Resistivity of metals increases with increasing temperature (A.3).
Sensory element: coil of platinum wire (wound non-inductively) or a zigzag film of platinum deposited on an insulating base (long narrow track in a small space).
Range: −200°C to 1000°C (lower maximum for platinum film sensors).
Output analog: Increase of resistance with increase of temperature.
Measuring circuit: Bridge (3.3.1).
Advantages: Stable with age, high melting-point of Pt permits measurement of high temperatures.
Disadvantages: Low resistivity and a small temperature coefficient (+0.00385/K for temperatures around 0°C). Non-linear response, acceptable over limited range.
Types: Wide range suited to numerous industrial and scientific applications.

1.3.2 Thermistor

Measurand: Temperature.
Principle: Resistivity of semiconductors decreases with increasing temperature (see Types).
Sensory element: Rod, bead or disc of semiconductor with two leads attached. Semiconductor is a sintered mixture of sulphides, selenides or oxides of Ni, Mn, Co, Cu, or Fe.
Range: −50°C to 400°C.
Output analog: Resistances in the range 100 Ω to 470 kΩ. The tempco of thermistors is appreciably larger than that of metals, usually between −0.01/K and −0.06/K.
Measuring circuit: Bridge (3.3.1) for high precision (low current minimizes self-heating), otherwise various.
Advantages: High resistance, high tempco, small size of element (smallest are 0.5mm dia.; low heat capacity and a rapid response to changes of temperature). Suitable for measuring the temperatures of tiny objects, in minute cavities, subcutaneously, etc.

Disadvantages: Self-heating due to bridge current. Non-linear; resistance R_T of a thermistor at temperature T (in kelvin) is given by:

$$R_T = R_{REF} e^{-\beta(1/T - 1/T_{REF})}$$

where R_{REF} is the resistance at a reference temperature T_{REF}. β depends on the composition of the thermistor material. Thermistors are best used over a limited temperature range.
Types: Above describes negative tempco (ntc) type. Ptc thermistors consist of doped barium titanate, with a high positive tempco (up to +0.7/K) over limited temperature range (−50°C to +200°C), suited to switching applications. Also used as current-limiting resistors in circuits to give temperature stability (high current → heating → resistance increases → reduced current).

1.3.3 Thermocouple

Measurand: Temperature difference.
Principle: Number of free electrons in a metal or alloy depends on temperature and composition. With two junctions between dissimilar metals (Fig. 1.1), there is a pd proportional to the difference of temperature between the hot and cold junctions (*Seebeck effect*).
Sensory element: Probe consisting of two wires of dissimilar metals or alloys welded together.
Range: −200°C to +2500°C (type K).

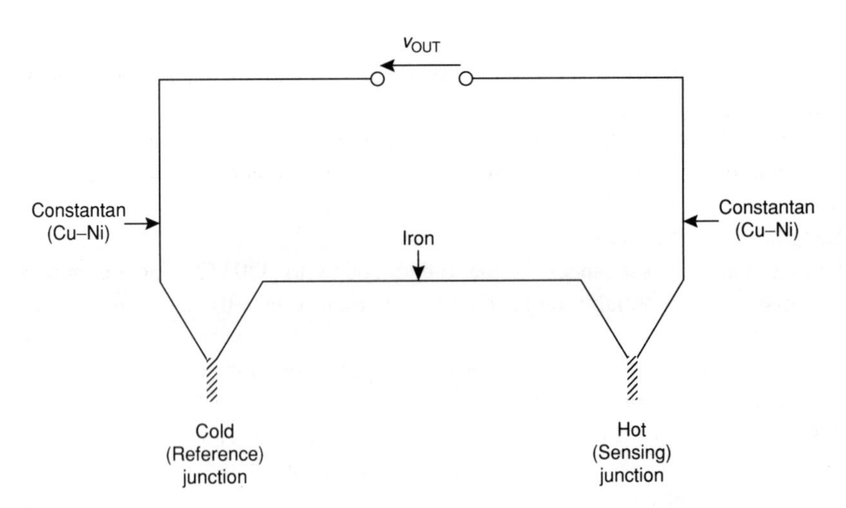

Figure 1.1

Output analog: Pd, approx. 50 μV/°C (type J), or 40 μV/°C (type K); high Z_{OUT}.
Measuring circuit: Voltage amplifier.
Advantages: Linear response over wide range, small junction, rapid response time, cheap.
Disadvantages: For absolute temperature measurement, reference junction must be at a known temperature.
Types: Foil types used for faster (10 μs) response. The thermocouple in Fig. 1.1 is type J. Type K (Ni–Cr for positive arm, Ni–Al for negative arm) is most widely used. Tungsten/rhenium alloys or platinum/rhodium–platinum alloys for highest temperatures. Several thermocouples in series (*thermopile*) produce larger pd.

1.3.4 Band-gap sensor

Measurand: Temperature.
Principle: Current through a pn junction (A.5) proportional to temperature (B.1).
Sensory element: A 3-terminal ic (V_+, 0 V, v_{OUT}).
Range: −40°C to 110°C.
Output analog: v_{OUT} linearly proportional to temperature; in most types proportional to Celsius temperature at 10 mV per degree (e.g. output 280 mV ↔ 28°C). Low Z_{OUT}.
Measuring circuit: Voltmeter for direct reading of temperature, or voltage amplifier.
Advantages: Direct reading, precision approx. ±0.4°C at 25°C. Rapid response.
Disadvantages: Restricted range.
Types: Different ranges and precision.

1.3.5 Pyroelectric sensor

Measurand: Rate of change of thermal (infra-red) radiation.
Principle: Emf generated in sensitive element when incident infra-red changes.
Sensory element: Block of (e.g.) lead–zirconate–titanate heated, then cooled, in a strong magnetic field. Usually mounted behind a plastic Fresnel lens to divide field of view into alternately visible and invisible zones. Warm objects passing through the zones to cause rapid and frequent changes in incident infra-red.
Range: Can detect warmth from human body as far as 40 m away.
Output analog: Pulsating voltage.
Measuring circuit: Differential amplifier to detect varying voltage.
Advantages: Extreme sensitivity, response time < 25 ms.
Disadvantages: Also sensitive to vibration (sound).

1.4 Force sensors

1.4.1 Strain gauge

Measurand: Distortion of metallic or other substrate on which gauge is mounted.
Principle: Extension of filaments of metal foil → reduces area of cross-section → increased resistance. In piezo-resistive strain gauges, atoms become separated under strain → force attracting electrons to the atom increased → forbidden energy gap (A.1) is increased → fewer electrons escape → conductivity reduced.
Sensory element: Metal foil (Fig. 1.2) or piezo-resistive film on adhesive plastic base.
Range: Up to 4% change in length.
Output analog: Change in resistance. Gauge factor $= \frac{\text{change in resistance}}{\text{unstressed resistance}} \times \frac{\text{unstressed length}}{\text{length}}$
Typical gauge factor is 2 to 5.
Measuring circuit: Bridge (3.3.1).
Advantages: Used indirectly to measure large forces (including weight). The force acts on a *load cell* for measuring compression (Fig. 1.3a) or *proving ring* for measuring tension (Fig. 1.3b) in (e.g.) a cable → distortion of cell/ring → distortion of strain gauge(s) → output calibrated in terms of force/weight.
Disadvantages: Small change in resistance, resistance temperature dependent, Seebeck effect at junction with leads (1.3.3), dimensions of substrate may be affected by temperature.
Types: With temperature compensation for mounting on steel or aluminium structures. Multiple gauges, e.g. four gauges connected as a bridge. Piezo-resistive strain gauges have higher gauge factors (50–200) (1.4.2).

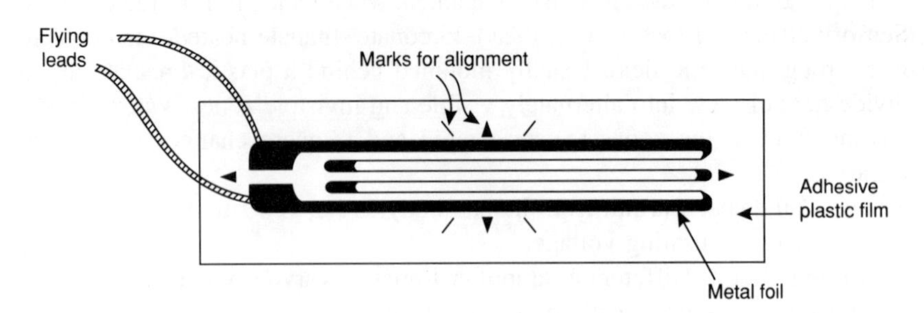

Figure 1.2

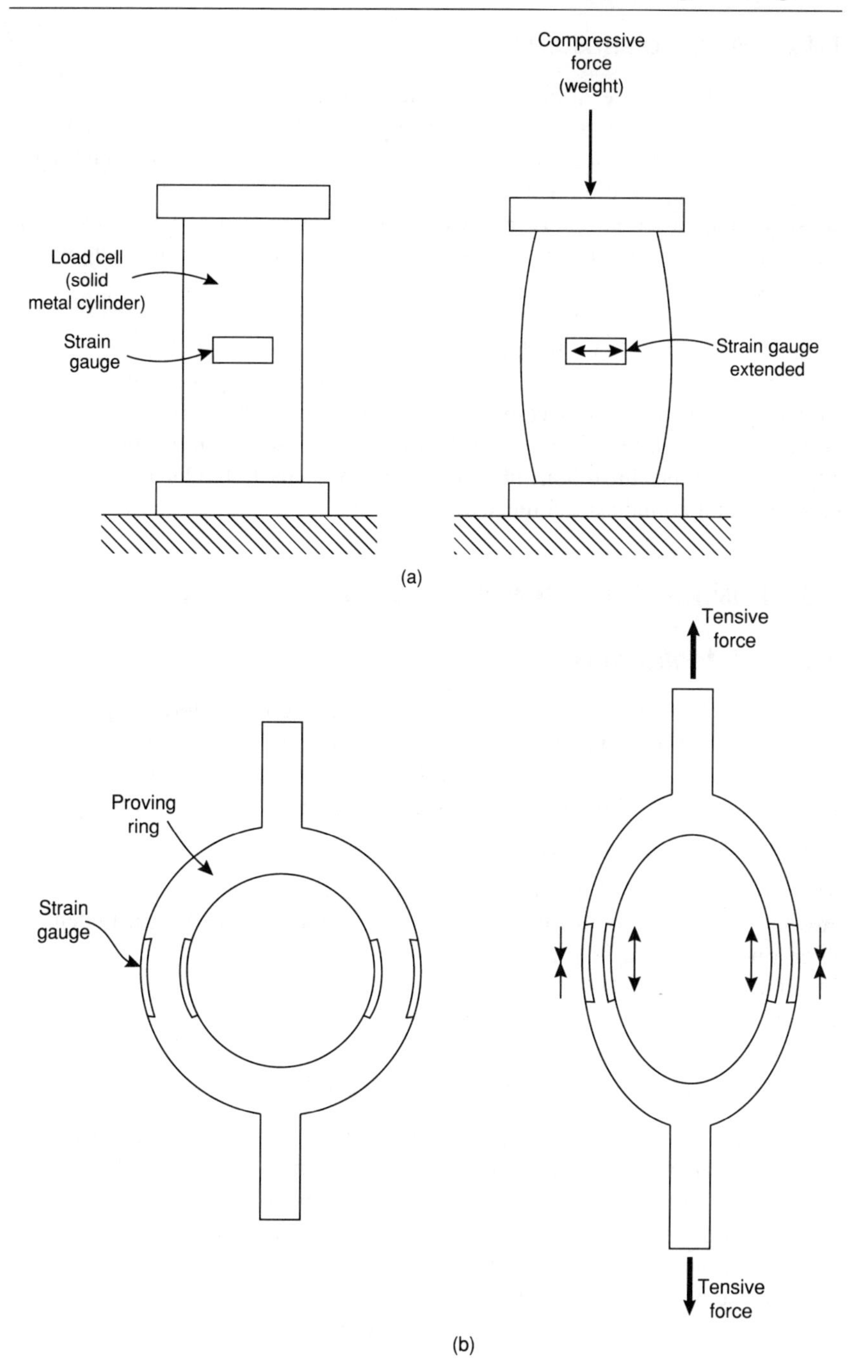
Compressive
force
(weight)
Load cell
(solid
metal cylinder)
Strain
gauge
Strain gauge
extended
(a)
Tensive
force
Proving
ring
Strain
gauge
Tensive
force
(b)

Figure 1.3

1.4.2 Pressure sensor ic

Measurand: Difference in pressures of two volumes of gas or liquid.
Principle: Two chambers separated by a silicon diaphragm contain gases/liquids at different pressures. A pressure difference causes the diaphragm to bulge one way or the other.
Sensory element: Four piezo-resistive strain gauges (1.4.1) connected as a bridge and diffused into the surface of the diaphragm measure bulging of diaphragm.
Range: Typically 0 to 300 kPa ($\approx$ 0 to 3 atmosphere).
Output analog: Pd across bridge. Typically 0.6 mV/kPa. Low Z_{OUT}.
Measuring circuit: Bridge supply voltage and voltage amplifier for output.
Advantages: Almost linear response. Rapid response ($\approx$ 1 ms)
Disadvantages: Output falls by 0.19% for each K rise in temperature.
Types: Some types have built-in temperature compensation. One chamber may be sealed and contain a reference vacuum.

1.5 Position (displacement) sensors

1.5.1 Potentiometer

Measurand: Linear or angular displacement, relative to a reference position.
Principle: Potential division at the wiper of a potentiometer (Fig. 1.4).
Sensory element: Linear or rotary pot linked to displaced object.
Range: Limited by dimensions of linear pot, or turning angle of rotary pot.
Output analog: Pd. Displacement, $d = l \times \frac{v_{OUT}}{v_{IN}}$.
Measuring circuit: Voltmeter or voltage amplifier.
Advantages: Simple construction. Voltage easily measurable. Linearity depends upon linearity of track; cermet or wire-wound pots are best.
Disadvantages: Friction in pot mechanism. Track wear, especially with carbon track.

1.5.2 Linear variable differential transformer

Measurand: Linear or angular displacement, relative to a reference position.
Principle: Compares the reluctance (magnetic coupling) between the primary coil and the two secondary coils (Fig. 1.5). Secondary coils are connected so that their signals are out of phase. At central position, signals from secondary coil cancel out $\rightarrow$ no output signal. To left or right, signal increases in amplitude, phase depends on direction of motion.
Sensory element: Three coils with common nickel–iron core linked to moving object.
Range: Usually a few mm or cm.

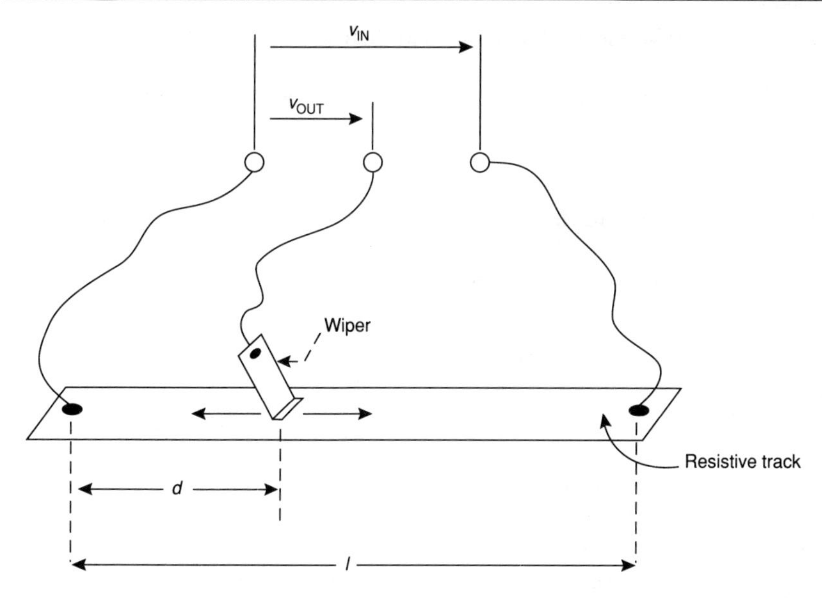

Figure 1.4

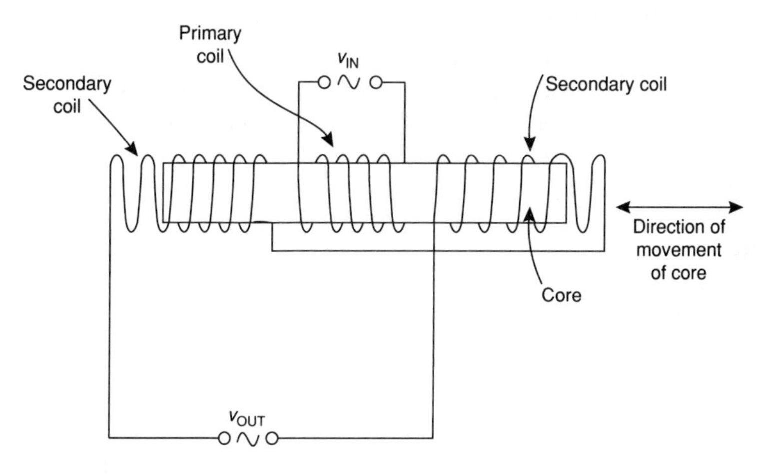

Figure 1.5

Output analog: Sine wave with amplitude linearly proportional to displacement. Accuracy 1% or better. Phase indicates direction of displacement from central (zero) position. High Z_{OUT}.
Measuring circuit: To measure amplitude and detect phase of output signal.
Advantages: High precision, no wear, virtually no friction.

Disadvantages: Circuit more complicated than with most sensors (but special ics available).
Types: Varying in full-scale displacement, and in accuracy. Rotary types (RVDTs) measure angles to minutes or seconds of arc.

1.6 Sound sensors

1.6.1 Carbon microphone

Measurand: Variation of air pressure due to sound waves.
Principle: Resistance of loosely packed carbon granules varies with pressure changes.
Sensory element: Container with metal diaphragm at front and contact at rear, loosely packed with carbon granules.
Output analog: Varying resistance.
Measuring circuit: Voltage source in series with microphone.
Advantages: Large amplitude signal, robust, cheap.
Disadvantages: Resonances distort signal, excessive noise due to random movements of granules.

1.6.2 Capacitor microphone

Measurand: Variation of air pressure due to sound waves.
Principle: Varying capacitance between vibrating and fixed plate.
Sensory element: Metal or metallized plastic diaphragm with metal plate mounted close behind it.
Output analog: Varying capacitance, high Z_{OUT}.
Measuring circuit: Voltage source in series with microphone; signal fed to voltage amplifier.
Advantages: Good linearity → high fidelity.
Disadvantages: Low signal level.

1.6.3 Electret microphone

Measurand: Variation of air pressure due to sound waves.
Principle: Variation in capacitance between vibrating diaphragm and fixed plate.
Sensory element: Similar to capacitor microphone, but with a dielectric material between the diaphragm and plate. In manufacture this is heated, then cooled in strong electric field → permanent electric field in the dielectric.
Range: 50 Hz to 18 kHz.
Sensitivity: Typically −64 dB. (Sensitivity $= 20\log_{10}(V/p)$, where $V =$ rms output and $p =$ effective pressure of sound, usually measured at 1 kHz. Usually $p = 1\,\mu\text{bar}$).

Output analog: Varying voltage. $Z_{OUT} \approx 600\,\Omega$.
Measuring circuit: No external voltage source required; signal fed to voltage amplifier.
Advantages: Good linearity → high fidelity.
Disadvantages: Low signal level.

1.6.4 Electromagnetic microphone

Measurand: Variation of air pressure due to sound waves.
Principle: Motion of a coil in a magnetic field induces a varying emf in the coil (transducer).
Sensory element: Coil attached to diaphragm moves in the field of a permanent magnet (Fig. 1.6).
Range: Response 50 Hz to 15 kHz.
Sensitivity: −60 dB to −80 dB (1.6.3).
Output analog: Varying current. $Z_{OUT} \approx 150\,\Omega$ (low impedance), often switchable to $50\,k\Omega$ (high impedance).
Measuring circuit: Current amplifier.
Advantages: Less subject to hum than moving-iron type, good linearity, compact.
Disadvantages: May resonate within AF range.
Types: This describes the moving-coil (*dynamic*) type. In the moving-iron type, the vibrating metal diaphragm induces a varying emf in a pick-up coil

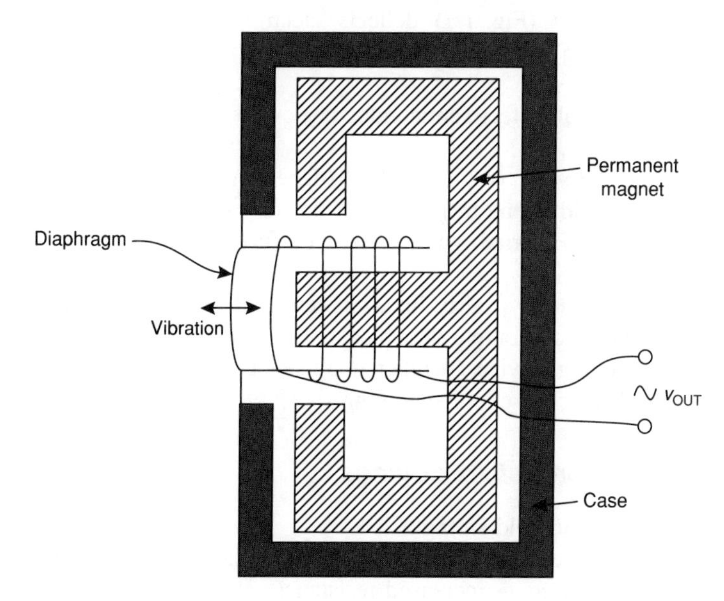

Figure 1.6

wound round a permanent magnet. Moving-iron type is bulky and subject to electromagnetic interference (especially mains hum from nearby equipment).

1.6.5 Piezo-electric microphone

Measurand: Variation of air pressure due to sound waves.
Principle: Distortion of crystal lattice by external pressure causes pd to develop between opposite faces (piezo-electric effect, a transducer) (1.4.1).
Sensory element: Crystal or piezo-electric ceramic element mounted between a vibrating diaphragm and a firm support. Sound vibrates diaphragm → stresses crystal → generation of pd.
Output analog: Small varying emf. $Z_{OUT} \approx 1\,\mathrm{M\Omega}$.
Measuring circuit: Amplifier. FET pre-amplifier in the microphone case if microphone is on a long cable, with battery or external power supply.
Advantages: Cheap.
Disadvantages: Poor linearity.
Types: Ultrasonic microphone with crystal cut to resonate at (e.g.) 40 kHz.

1.7 Magnetic sensors

1.7.1 Linear Hall-effect device

Measurand: Magnetic field strength.
Principle: *Hall effect* (Fig. 1.7) deflects electrons flowing in a block of semiconductor (InSb) in a magnetic field → pd between opposite sides of the block.
Sensory element: Hall-effect semiconductor device with on-chip amplifier.
Range: 0 G to 10 kG.
Output analog: Varying voltage, offset (dependent on supply) ± amount proportionate to field strength (positive or negative).
Measuring circuit: External voltage source; further amplification may be needed.
Advantages: 1% accuracy, fairly linear in centre of range, rapid response.
Types: May have second output which decreases with increasing field strength. Hall-effect switching devices, triggered by level and direction of magnetic field, are useful for contactless switching.

1.7.2 Magnetoresistive devices

Measurand: Magnetic field strength.
Principle: In a magnetic field, the electrons are deflected and follow a longer path through the device → increased resistance.
Sensory element: Block of semiconductor (InSb) with inclusions (NiSb).

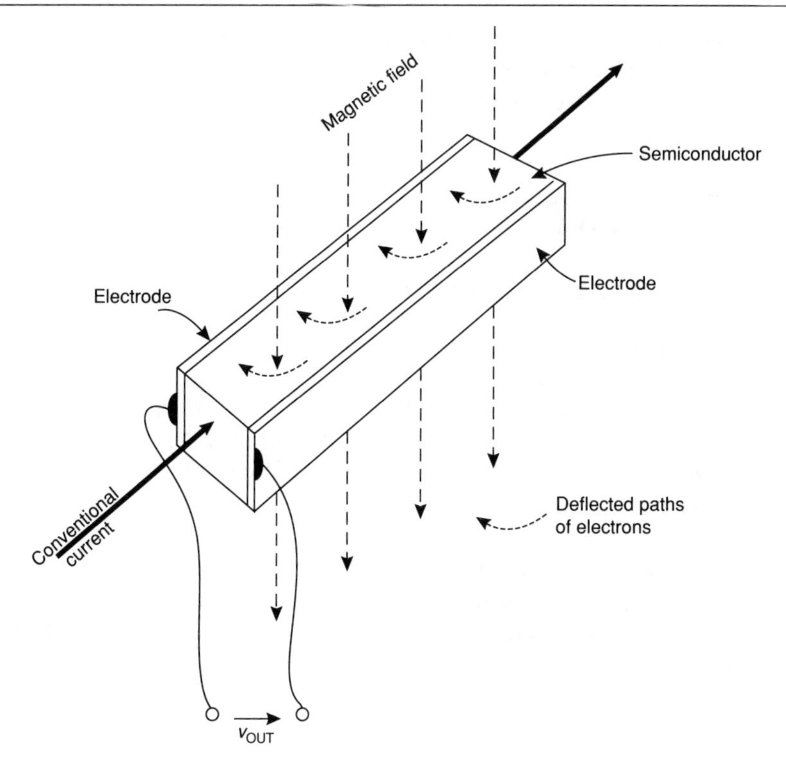

Figure 1.7

Output analog: Change of resistance.
Measuring circuit: Constant current source across magnetoresistor, with amplifier to measure pd produced.
Advantages: Sensitivity sufficient to read data from magnetic tapes, disks and stripes.

1.7.3 Inductive sensor

Measurand: Motion of object made from magnetic material.
Principle: An object of magnetic material in the vicinity of a coil affects the self-inductance of the coil. The effect is greatest for ferromagnetic materials but also occurs with paramagnetic and diamagnetic materials.
Sensory element: Coil of wire with ferromagnetic core.
Range: Frequencies up to 20 kHz.
Output analog: Varying self-inductance.
Measuring circuit: Voltage source and load resistor (100 kΩ) across coil, with voltage amplifier to measure varying pd. If object is a rotating toothed metal

wheel, frequency of alternating pd is proportional to rate of rotation; use a tachometer circuit to measure rotation speed (up to several thousand rpm).
Advantages: Small size, cheap, reliable and rugged (especially in heavy machinery).

1.8 Light sensors

1.8.1 Light-dependent resistor

Measurand: Light intensity.
Principle: Reduced resistance.
Sensory element: Block of semiconductor (CdS or CdSe) with interdigitating conductive strips (electrodes) on exposed surface (Fig. 1.8).
Range: Typically 0 to 10 000 lux; most sensitive to 480 nm (red) to 690 nm (green).
Output analog: Varying resistance (10 MΩ to 80 Ω).
Measuring circuit: Bridge or other.
Advantages: Large change of resistance, sensitive. Can be operated with AC.
Disadvantages: Slow response time (turn-on = 350 ms, turn-off = 75 ms). Non-linear.

1.8.2 Photodiode

Measurand: Light intensity.
Principle: Light liberates minority carriers in semiconductor → increased leakage current.

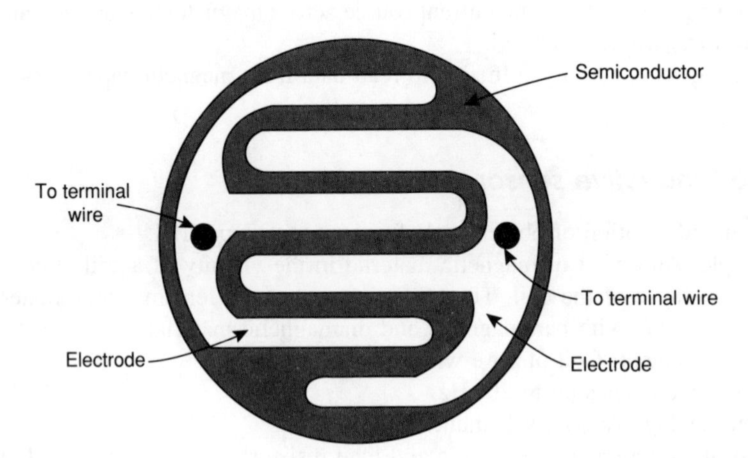

Figure 1.8

Sensory element: Si diode in transparent case, often with focusing lens. Usually a p-i-n diode (B.3) to allow high frequency response.
Range: Typical currents with 10 V reverse voltage: dark, 30 nA; 1000 lux, 60 μA. Wavelength range 325 nm to 1100 nm, peaking at about 800 nm (red).
Output analog: Varying current.
Measuring circuit: Reverse-biased diode in series with resistor; measure pd across resistor.
Advantages: High sensitivity (up to 0.5 A/W); linear response, rapid response time (250 ns, some types much faster down to 0.5 ns).
Disadvantages: Noise from resistor (7.2.1) corrupts signal at low light levels.
Types: Visible light, human-eye response (peak 550 nm), infra-red (remote control, etc.). In infra-red types, case is transparent to infra-red but not to visible light. Large-area types for maximum sensitivity.

1.8.3 Phototransistor

Measurand: Light intensity
Principle: Light generates minority carriers in the base region; equivalent of base current; current amplified to give larger collector current.
Sensory element: BJT (npn) in transparent case, often with focusing lens.
Range: 0 to 1000 lux.
Output analog: Varying collector current.
Measuring circuit: Common emitter connection, with isolated or biased base. Measure pd across collector resistor.
Advantages: More sensitive than photodiode.
Disadvantages: Slower response than photodiode (2 to 15 μs).
Types: Some have amplifier on the same chip.

1.9 Switching sensors

There are many sensors with switching outputs, which change state at a fixed value of the measurand. Examples are proximity switches (inductive, capacitative, Hall-effect), microswitches, tilt switches, light-activated switches, thermal (bimetallic) switches and optical position encoders. Their output is essentially binary and more suited to digital processing, so they are not dealt with in this book.

2 Originating analogs

Most analog signals originate in the outside world (Chapter 1) but some arise within electronic circuits. One group of these includes noise and related signals (Chapter 7). Another major group, the subject of this chapter, includes those which are a mathematical function of time, produced by circuits collectively known as function generators.

2.1 Ramp generator

A ramp generator produces an analog of time equivalent to that produced by an hourglass or a water-clock. Current flows into a capacitor at constant rate $\rightarrow$ pd across capacitor increases at a constant rate $\rightarrow$ pd is a measure of elapsed time. A ramp generator requires a constant current source.

2.1.1 Constant current sources

This term is often used not only for true sources of current but also for circuits that are really current *sinks*. We describe one source applied as a ramp generator and some other sources that could be used (e.g.) to provide constant current flow through a resistive sensor.

2.1.1.1 FET source

The simplest form, often marketed as 'constant current diodes', is a JFET with its gate connected to its source (Fig. 2.1). It has constant current of i_{DSS} (Fig. C.14, curve for $v_{GS} = 0$ V). It is used to control current flowing to a capacitor to make a linear ramp generator (Fig. 2.2). Figure 2.3 shows the early part of the ramp, beginning with C uncharged. Current i_{DSS} is 16.06 mA. If this flows for 5 ms, charge stored = $q = it = 16.06 \times 10^{-3} \times 5 \times 10^{-3} = 8.03 \times 10^{-5} C$. Then $v_{RAMP} = q/C =$

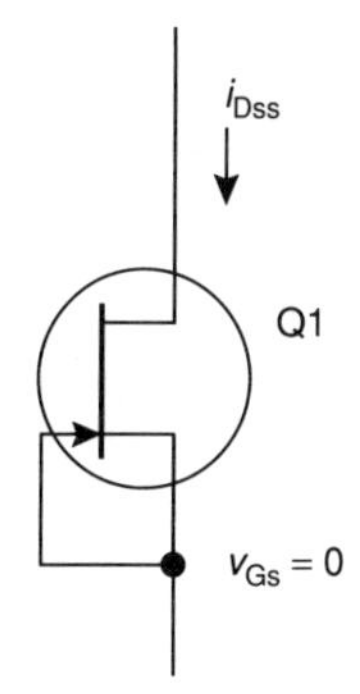

Figure 2.1

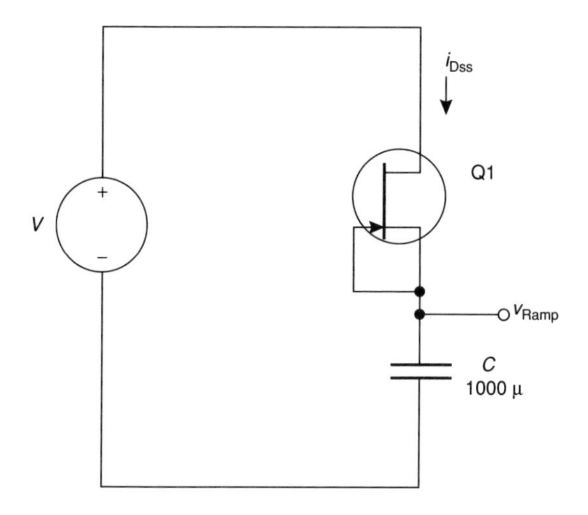

Figure 2.2

$(8.03 \times 10^{-5})/(1000 \times 10^{-6}) = 80.98\,\text{mV}$. Note that i_{DSS} varies widely between transistors of the same type, so it is necessary to select the JFET to produce the required current.

If the current is reduced by biasing the gate negative of the source (Fig. 2.4), Q1 is operating on one of the lower curves of Fig. C.14, where slope is nearer to the horizontal (i.e. current is less affected by changes in v_{DS}). Bias is $-i_D R$. Figure 2.5 shows ramps for $R = 300\,\Omega$ to $1.1\,\text{k}\Omega$, over a longer time scale than Fig. 2.3. Ramps are linear until v_{RAMP} approaches approximately 8 V. Then the compliance limit is reached because $V_{DS} < 2\,\text{V}$. Eventually v_{RAMP} climbs to 10 V. Only a minute current can be drawn from the output (i.e. from the capacitor) without flattening the ramp. Use an op amp voltage follower (Fig. D.3b) as a buffer.

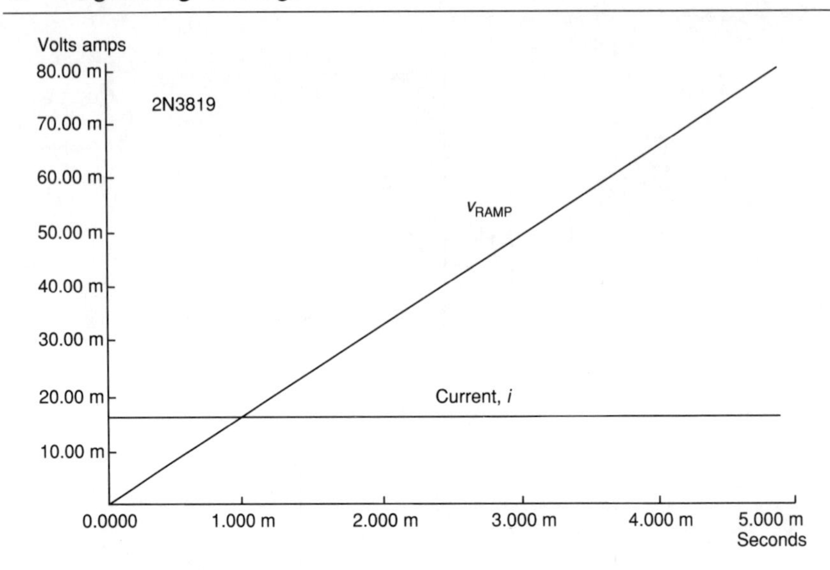

Figure 2.3

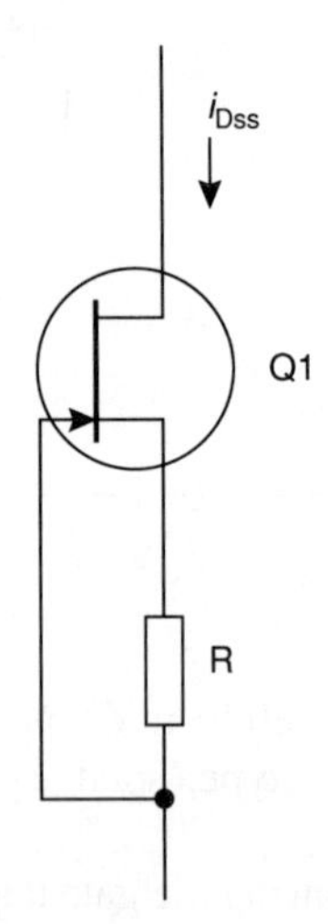

Figure 2.4

2.1.1.2 BJT sources

Figure 2.6 is the essential circuit. The base of Q1 is held at constant voltage V_{REF} by the Zener diode. R1 has the value required to supply sufficient base current. Emitter voltage is constant at V_{BE} ($= 0.6\,V$) below V_{REF}. Then $i_E =$

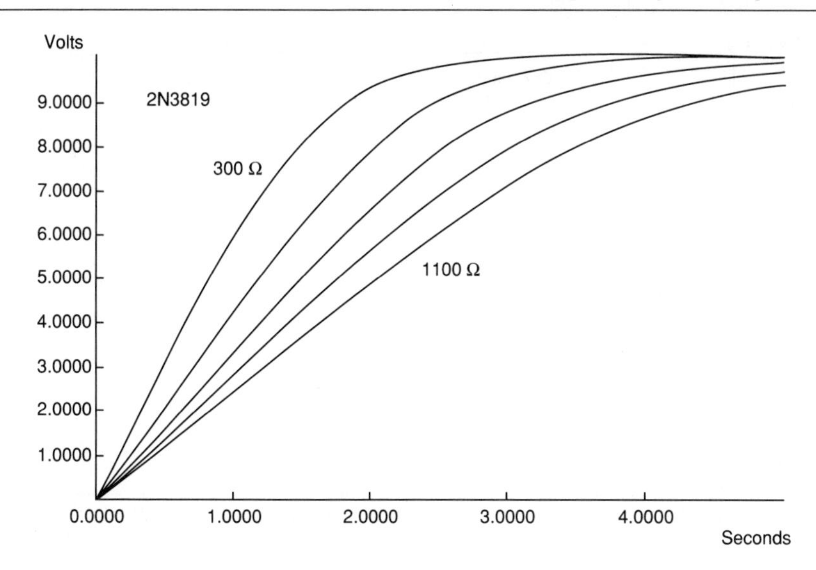

Figure 2.5

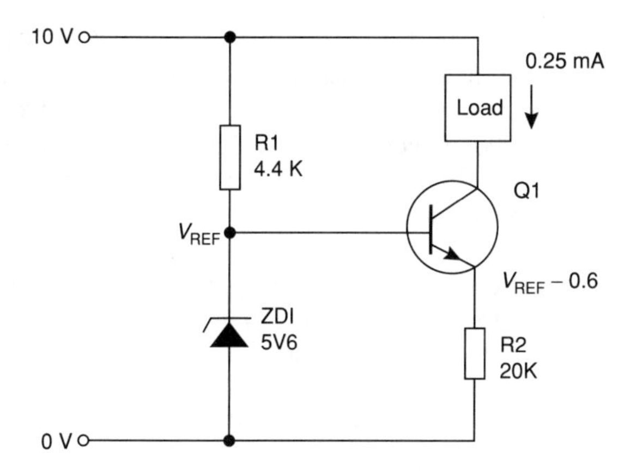

Figure 2.6

$(V_{REF} - 0.6)/R2$. If the resistance of the load increases → pd across the load increases → reduces collector voltage of Q1 → reduced v_{CE}. But this has little effect on i_C (Fig. C.5) provided i_B is fixed (which it is). The same applies if the resistance of the load decreases → constant current with varying load.

Example: In Fig. 2.6, make the current through the Zener = 1 mA. For this, $R1 = (10 - 5.6)/0.001 = 4400\,\Omega$. Decide on value of constant emitter

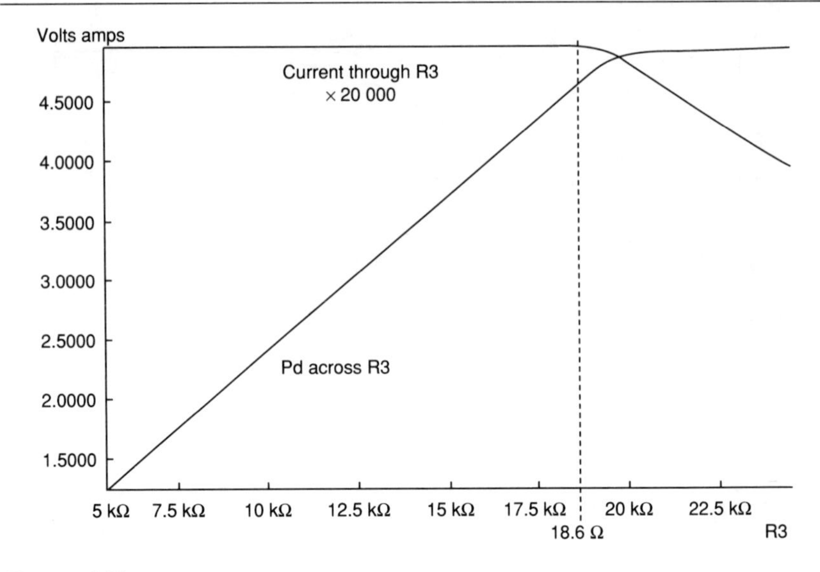

Figure 2.7

current, $i_E = 0.25$ mA. Make R2 $= (V_{REF} - 0.6)/0.00025 = 20\,k\Omega$. Current through the load is slightly less than this because $i_C = i_E - i_B$ (C.2.4.1). Figure 2.7 shows the effect of varying the load (R3) from $5\,k\Omega$ to $25\,k\Omega$. Current (plotted $\times$ 20 000) is constant at 0.25 mA until R3 exceeds $18.6\,\Omega$. The pd across R3 rises in proportion ($v = iR$) until it exceeds 4.58 V, when Q1 is saturated and compliance ceases.

Precision can be increased by (1) making i_E larger (size of i_B and variations in i_B proportionately less), (2) by using a high-gain transistor so that i_B is relatively smaller. But (1) reduces the range of compliance, so the allowable variation in the load is reduced.

Current is temperature dependent because the Zener voltage rises by about 2 mV/°C, V_{BE} falls by about the same amount (C.2.6) and h_{fe} and i_B are also affected. The net result is that the load current falls by about 20 μA between 0°C and 110°C (Fig. 2.8). This is the effect of ambient temperature on the whole circuit. There may also be local effects, (e.g.) because of transistor heating with high load current.

Ambient temperature may be compensated for by a circuit such as Fig. 2.9. The fall in V_{BE} of Q2 is offset by the fall in V_{BE} of Q1. To obtain a load current of (e.g.) 0.25 mA, values of V_{REF} and R2 are as before. R1 must drop 4.4 V when a current of (say) 1 mA passes → R1 $= 4.4\,k\Omega$. Base of Q2 must be at $v_{REF} - 0.6 = 5$ V. Use two equal resistors (R3, R4) to split the supply voltage. The load current is 0.25 mA, the same as in Fig. 2.7, but falls more

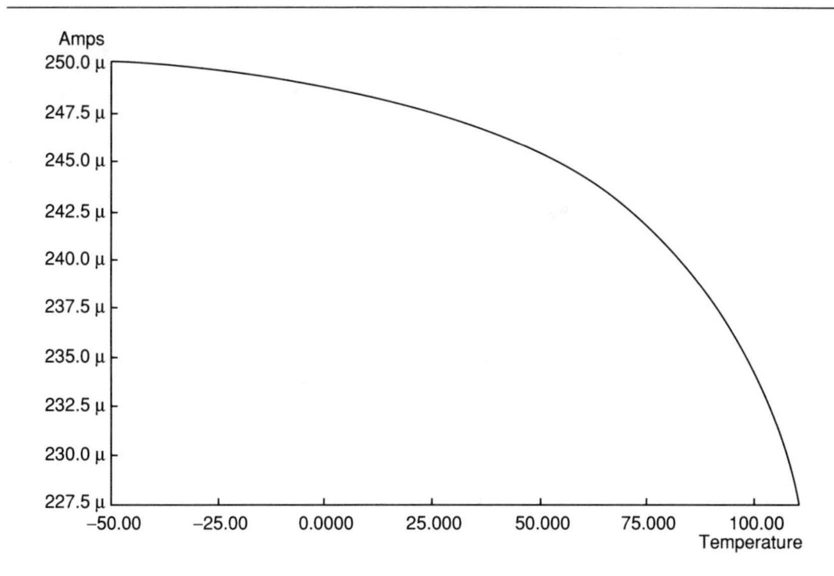

Figure 2.8

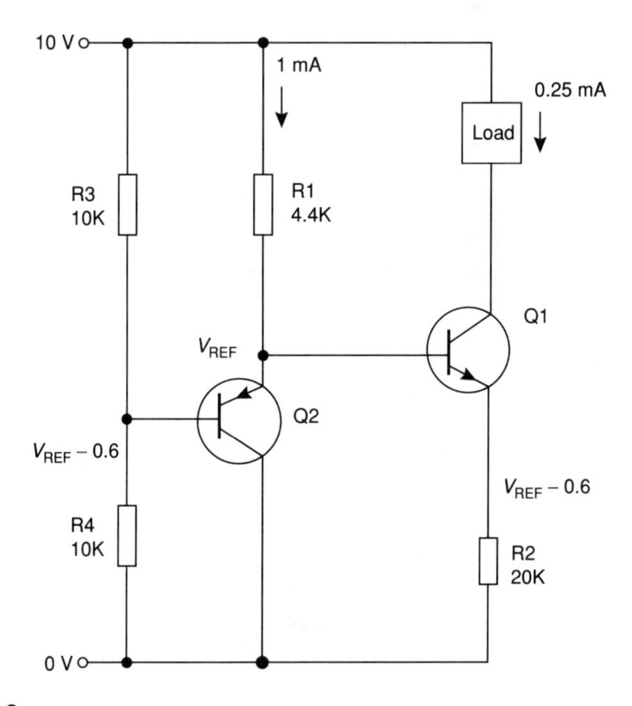

Figure 2.9

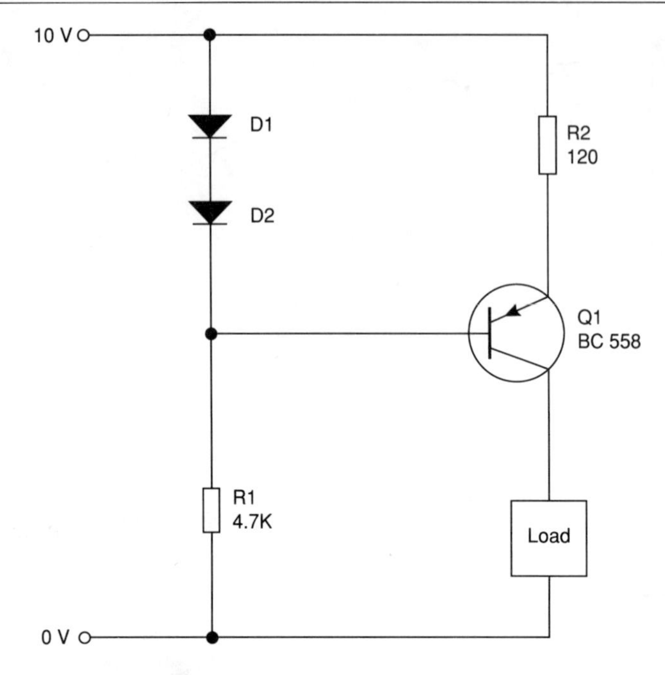

Figure 2.10

steeply above 18.6 Ω. In contrast with Fig. 2.6, between 0°C and 100°C, the load current rises, not falls, and only by 1 μA.

Another source suitable for a ramp generator is Fig. 2.10. Two diodes produce a total drop of 1.2 V, so the base of Q1 is at 8.8 V. Its emitter is at 9.4 V and the current through R2 (and hence the load) is 5 mA. This produces a constant current for any load from 0 Ω to about 1650 Ω, when the pd across the load is over 8 V. Use a capacitor as the load to obtain a voltage ramp.

2.1.2 Current mirrors

This is another technique for obtaining constant current, suitable for ramp generators and other applications. The constant current I_L is programmed by setting the programming current I_P. I_L may be equal to I_P, or any fraction or multiple of I_P. Often used in ics where a single resistor (internal or external) is used to program several different constant currents.

The two (or more) transistors are identical, best fabricated close together on a single chip, so they are at the same temperature. In Fig. 2.11, I_P is set by the value of R_P. The bases of Q1 and Q2 are V_{BE} (≈ 0.6 V) below V_{CC} and:

$$I_P = (V_{CC} - 0.6)/R_P$$

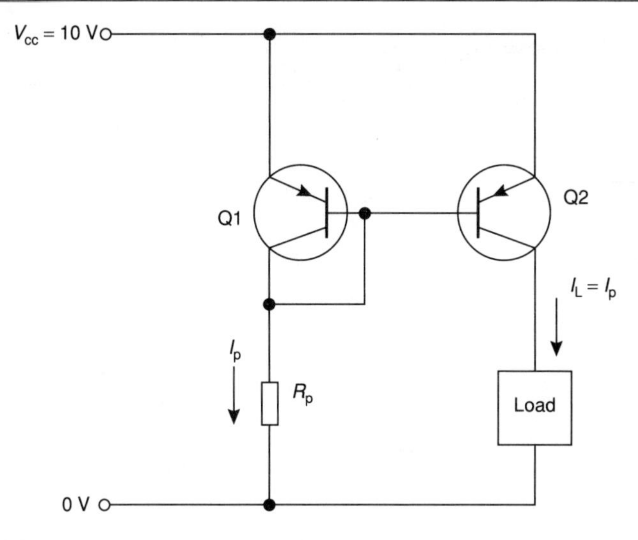

Figure 2.11

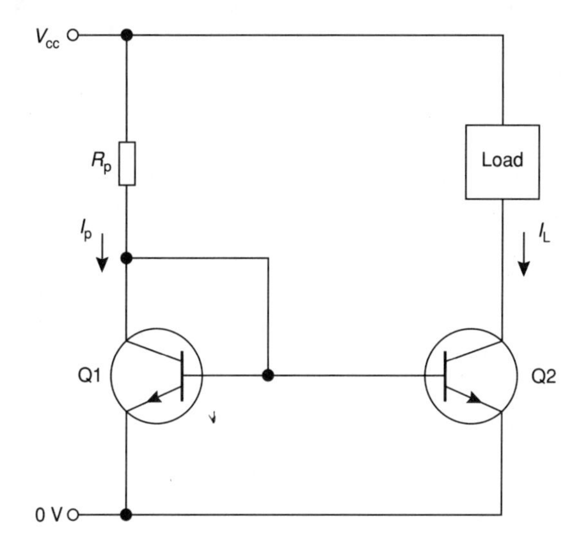

Figure 2.12

In the figure, if $R_P = 376\,\Omega$, then $I_P = 9.4/376 = 25\,\text{mA}$. The exact V_{BE} of Q1 is determined by current I_P, and temperature. V_{BE} for Q2 is identical, so it passes an equal current $\rightarrow I_L = I_P$.

Figure 2.12 uses npn transistors to make a current sink. If the load is a capacitor, a negative ramp is obtained between the collector of Q2 and the 0 V

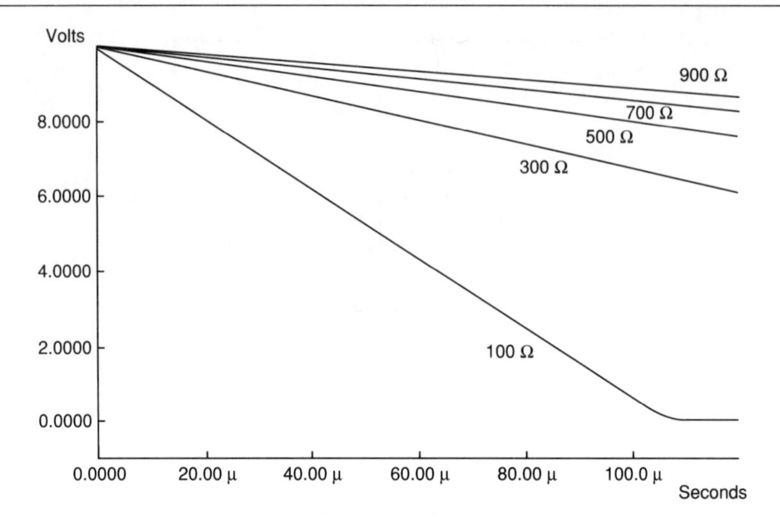

Figure 2.13

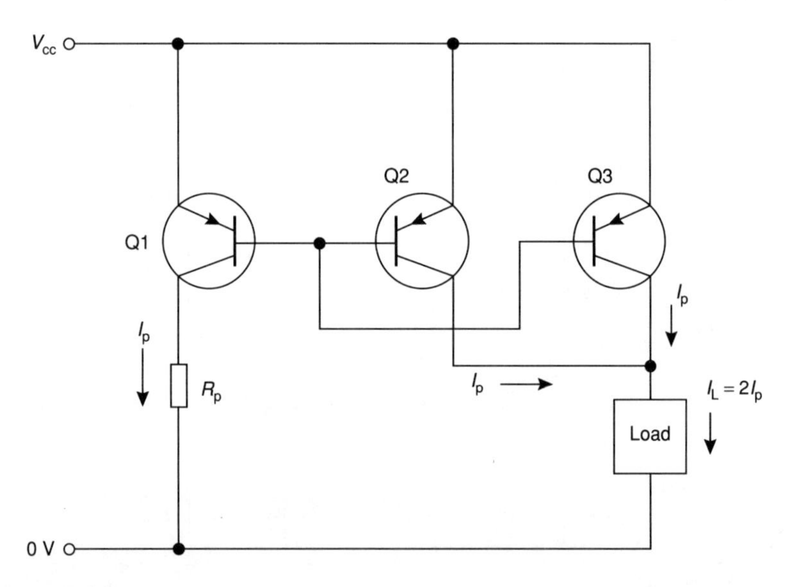

Figure 2.14

rail. Figure 2.13 shows ramps obtained when $V_{CC} = 10\,V$, the load is a $1\,\mu F$ capacitor and R_P has values between $100\,\Omega$ and $900\,\Omega$.

With three transistors, the outputs of two being fed to the load (Fig. 2.14), $I_L = 2I_P$. If Q1 and Q2 feed to R_P and only Q3 feeds to the load, $I_L = I_P/2$. Other integral multiples and fractions are obtained by using more transistors.

Non-integral multiples and fractions are obtained by inserting a resistor(s) between the emitter of one or more transistors and the supply line.

2.2 Oscillators

Output is periodic, with period P (seconds) and frequency f (hertz), where $P = 1/f$. For sine waves, frequency is also expressed as angular velocity or angular frequency ω, where $\omega = 2\pi f$ (rad/s). Main oscillator types are:

- Relaxation oscillator. A capacitor is charged, then discharged. Produces square waves, ramps, sawtooth waves or triangle waves, depending on circuit configuration.
- Resonance oscillator. A resonant circuit (often with inductor and capacitor) is kept in oscillation by amplifying and feeding back part of its output. Produces sine waves.
- Crystal oscillator. Attains stable and highly precise frequencies because of the accurately cut quartz crystal.

2.2.1 Relaxation oscillators

Figure 2.15 is the classic BJT *astable*, consisting of two cross-connected BJT switches, linked by capacitors. Figure 2.16 shows the voltages during two cycles. With Q1 on and Q2 off, v_{B2} ramps up from a low value to 0.6 V → turns on Q2 → v_{C2} falls sharply to 0 V, a drop of 12 V → brings down v_{B1} from 0.6 V to −11.4 V. In one half-cycle v_{B1} rises from −11.4 V as current

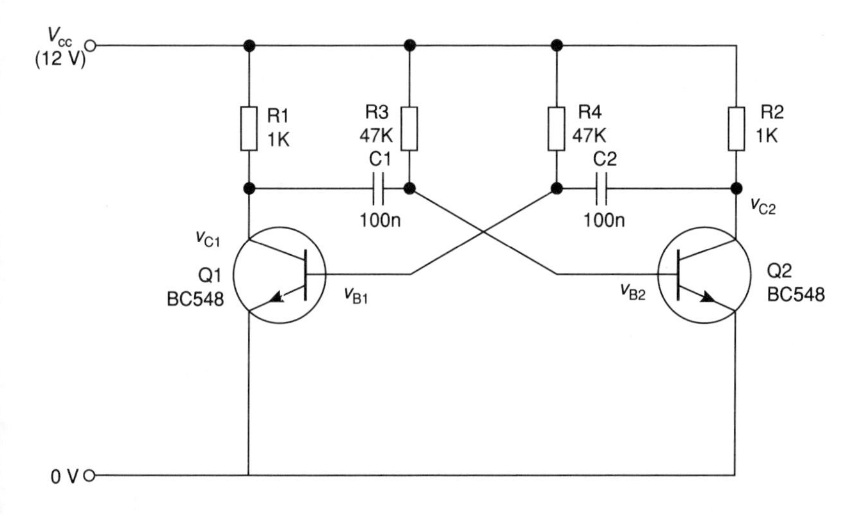

Figure 2.15

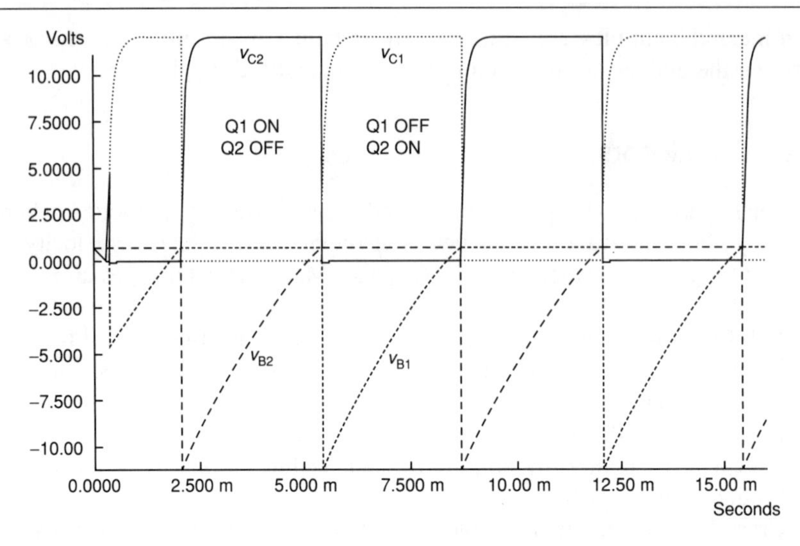

Figure 2.16

flows through R4, eventually reaching 0.6 V, when Q1 is turned on again. But positive end of R4 is connected to V_{CC}(+12 V) so pd across R4 has decreased from 23.4 V to 11.4 V. Mean pd = 17.4 V (this is lower part of exponential charging curve so it is roughly linear and the arithmetic mean is near enough). Mean current through R4 is 17.4/47 000 = 370 μA. As pd across R4 rises by 12 V, the charge on C2 increases by $12 \times 100 \times 10^{-9} = 1.2\,\mu\text{C}$. Time required is $q/i = (1.2 \times 10^{-6})/(370 \times 10^{-6}) = 3.24\,\text{ms}$, for a half-cycle. Period of waveform is 2 × 3.24 ms = 6.48 ms. $f = 1/6.48 = 154\,\text{Hz}$.

From the above:

$$f \approx \frac{3V_{CC} - 1.2}{4V_{CC}R_tC_t}$$

Where $R_3 = R_4 = R_t$ and $C1 = C2 = C_t$.

In real life, slight differences between transistors, capacitors or resistors ensure that the circuit swings completely one way or the other when power is first applied, with one transistor on and the other off. After that, it oscillates. If the circuit is perfectly symmetrical, as it could be in a simulation, it quickly reaches a stable state with neither transistor turned on. It fails to oscillate. In the circuit simulated in Fig. 2.16, R3 is made unequal to R4, to give asymmetry. If R3 = 480 kΩ → C2 charges more quickly than C1 → forces Q1 to turn on first at the start (small peak in Fig. 2.16).

In a practical circuit, the square-wave output is taken from the collector of Q1 or Q2 (or both), using an emitter follower (4.2), or an op amp voltage follower (D.2.2), to minimize loss of current from the oscillator, which would make the waveform and its timing asymmetrical.

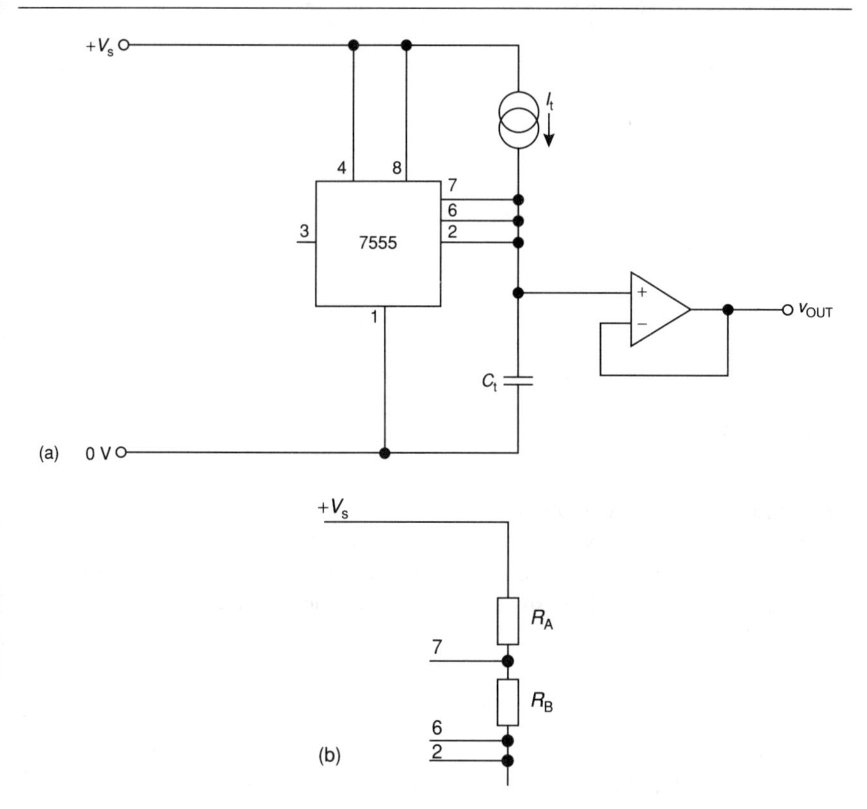

Figure 2.17

Another relaxation oscillator is Fig. 2.17a. A constant current source (2.1.1.1, 2.1.1.2) produces timing current I_t which charges timing capacitor C_t under the control of the timer ic (7555 CMOS version preferred). In each cycle, the capacitor is charged from 1/3 to 2/3 of supply voltage V_S, and is then discharged almost instantly. Produces a series of identical upward-sweeping ramps → sawtooth wave. Output is buffered by a voltage follower (D.2.2). A square-wave output is available from pin 3.

Charge accumulating on C_t during one period is $q = I_t P$ → pd across C_t rises by $I_t P/C_t$. The timer action ensures that the pd across C_t rises by $V_S/3$:

$$I_t P/C_t = V_S/3$$

$$\Rightarrow \quad P = \frac{V_S C_t}{3I_t}$$

$$\Rightarrow \quad f = \frac{3I_t}{V_S C_t} \text{ Hz}$$

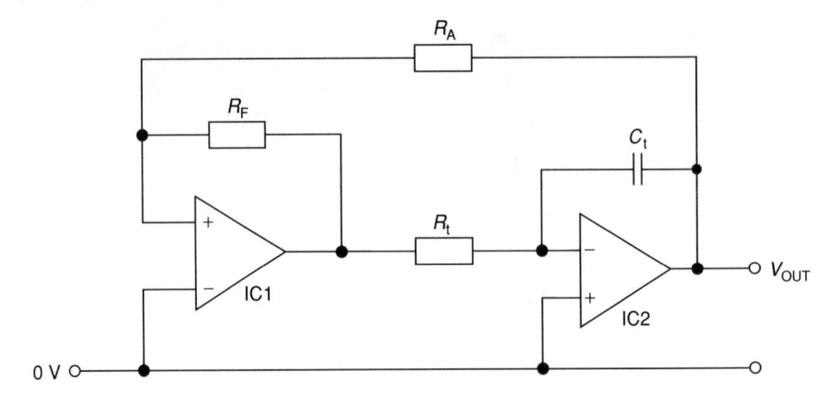

Figure 2.18

The ramp is linear, but frequency depends on constancy of V_S. If the current generator is replaced by two resistors (as in standard 555 astable circuit, Fig. 2.17b), the ramp is exponential (though possibly acceptable since it covers an intermediate charging range) but frequency is independent of supply voltage:

$$f = \frac{1.44}{(R_A + 2R_B)C_t} \text{ Hz}$$

Make $R_A \gg R_B \rightarrow$ slow ramp up (charging) and rapid ramp down (discharging). Output can be fed to an inverter to produce series of downward ramps.

A triangular wave is generated when a ramp generator is switched to alternately charge and discharge a capacitor. In Fig. 2.18, IC1 is a Schmitt trigger (10.1) and IC2 is an integrator (5.2.1). Begin when the output of IC1 has positive saturation $\rightarrow$ current through R_t flows to $C_t \rightarrow v_{OUT}$ ramps down. As v_{OUT} crosses 0 V, IC1 swings to negative saturation $\rightarrow$ current flows from C_t through R_t to IC1 $\rightarrow C_t$ discharged $\rightarrow v_{OUT}$ ramps up $\rightarrow$ crosses 0 V $\rightarrow$ IC1 saturated high $\rightarrow$ repeat. v_{OUT} is a triangular wave, $f = R_F/4R_AR_tC_t$, amplitude $= v_{SAT} \times R_A/R_F$. A square-wave output of the same frequency, amplitude v_{SAT}, can be taken from IC1. This simple design can be refined to allow control of v_{OUT} amplitude and frequency.

2.2.2 Resonance oscillators

A network consisting of a capacitor and inductor (possibly a resistor too) connected in series or in parallel is the analog of a mechanical system such as a pendulum or a mass suspended by a spring. Following a small disturbance of its equilibrium (an electrical pulse, or a mechanical impulse) it oscillates at a fixed frequency. The oscillations are sinusoidal, but decay because energy is lost (electrical resistance, air resistance). In an oscillator we maintain the oscillations by supplying energy periodically, at the same frequency and in phase with the oscillator. This is the principle of the pendulum clock, or the balance wheel of a watch.

The Colpitt's oscillator network consists of two capacitors in series connected in parallel with an inductor (Fig. 2.19). As the network oscillates v_C rises and falls. As v_C falls $\rightarrow$ v_{FB} falls $\rightarrow$ v_E falls $\rightarrow$ v_{BE} exceeds 0.6 V $\rightarrow$ Q1 turned on $\rightarrow$ pulls v_C down further (positive feedback), supplying energy to the network. Recovery of network makes v_C rise $\rightarrow$ v_{BE} falls below 0.6 V and becomes negative $\rightarrow$ Q1 turned off $\rightarrow$ v_C rises further $\rightarrow$ recovers $\rightarrow$ v_C falls $\rightarrow$ repeats. Output may be taken from an emitter follower or voltage follower connected to the collector of Q1. Alternatively, wind a pick-up coil on the same former as L1. Colpitt's oscillators can also be based on an FET or an op amp.

Analysing the network, to find its impedance z_{NET}: complex impedances (D.9) of capacitors are $1/sC_1$ and $1/sC_2$. Complex impedance of inductor is sL.

$$z_{NET} = sL \| (1/sC_1 + 1/sC_2)$$

$$= \frac{sL(1/sC_1 + 1/sC_2)}{sL + 1/sC_1 + 1/sC_2}$$

$$= \frac{sL}{\dfrac{s^2 C_1 C_2 L}{C_1 + C_2} + 1}$$

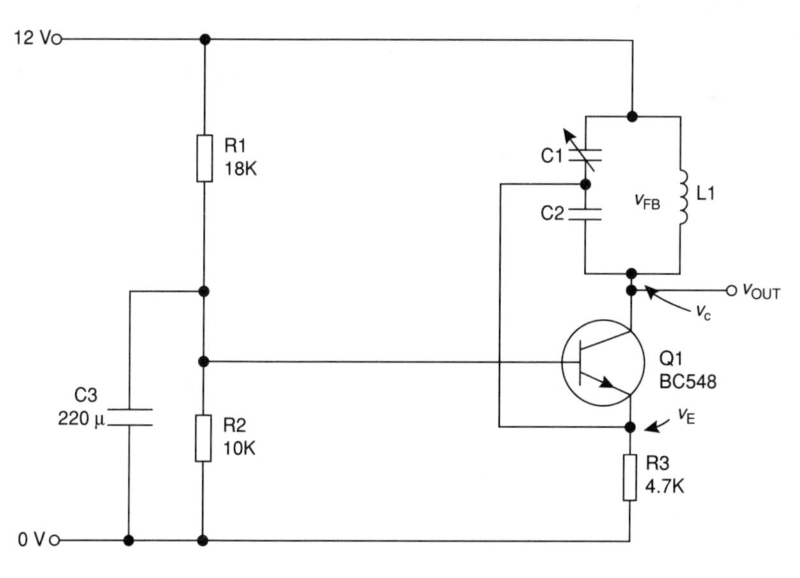

Figure 2.19

z_{NET} rises to infinity when the denominator is zero, i.e. there is a pole (6.1.2) when:

$$\frac{s^2 C_1 C_2 L}{C_1 + C_2} = -1$$

Substituting $s = j\omega$:

$$\frac{j^2 \omega^2 C_1 C_2 L}{C_1 + C_2} = -1$$

But $j^2 = -1$, so:

$$\omega^2 = \frac{C_1 + C_2}{C_1 C_2 L}$$

$$\Rightarrow \qquad \omega = \sqrt{\frac{C_1 + C_2}{C_1 C_2 L}}$$

$$\Rightarrow \qquad f = \frac{1}{2\pi}\sqrt{\frac{C_1 + C_2}{C_1 C_2 L}}$$

Inductors with high inductances (for low frequencies) are usually impracticable, so this circuit is mainly used for high frequencies (e.g. RF). The Hartley oscillator (5.3.4) is similar, but the resonant network has a single capacitor and feedback is obtained from a tap on the inductor coil.

A Wien Bridge (Fig. 2.20) consists of an amplifier (here an op amp) with a series *RC* and a parallel *RC* network in its positive feedback path. There is also a negative feedback path to control gain, in which R_A is a filament lamp, a thermistor, or some other device for automatically varying the resistance.

The complex impedance of the series network is:

$$z_S = 1/sC + R = \frac{1 + sRC}{sC}$$

The complex impedance of the parallel network is

$$z_P = 1/sC \parallel R = \frac{R}{1 + sRC}$$

The positive feedback loop is a potential divider, with output:

$$v_+ = v_{OUT} \frac{Z_p}{Z_s + Z_P} = v_{OUT} \times \frac{sRC}{1 + 3sRC + s^2 R^2 C^2}$$

Put $s = j\omega$:

$$v_+ = v_{OUT} \times \frac{j\omega RC}{1 + 3j\omega RC - \omega^2 R^2 C^2}$$

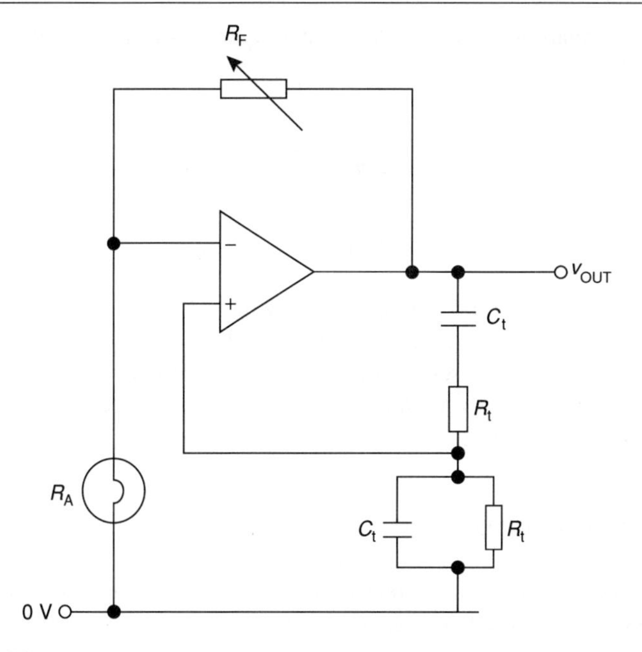

Figure 2.20

Multiply top and bottom by j:

$$v_+ = v_{OUT} \times \frac{-\omega RC}{j - 3\omega RC - j\omega^2 R^2 C^2} = v_{OUT} \times \frac{-\omega RC}{j(1 - \omega^2 R^2 C^2) - 3\omega RC}$$

In the negative feedback loop:

$$v_- = v_{OUT} \times \frac{1}{1 + R_F/R_A}$$

But the op amp acts to make $v_+ = v_-$:

$$\frac{-\omega RC}{j(1 - \omega^2 R^2 C^2) - 3\omega RC} = \frac{1}{1 + R_F/R_A}$$

Putting the imaginary parts equal on both sides of the equation:

$$1 - \omega^2 R^2 C^2 = 0$$

$\Rightarrow$ $$\omega = 1/RC$$

$\Rightarrow$ $$f = 1/2\pi RC$$

This is the equation for calculating the frequency of v_{OUT}. Putting the real parts equal:

$$\frac{-\omega RC}{-3\omega RC} = \frac{1}{1 + R_F/R_A}$$

$$\Rightarrow \quad \frac{1}{1 + R_F/R_A} = \frac{1}{3}$$

From D.2.2, the gain is exactly 3, for which $R_F = 2R_A$. Thus R_A stabilizes gain. If gain $< 3 \rightarrow$ oscillations die out. If gain $> 3 \rightarrow$ amplifier output saturates. Adjust R_F to set oscillator working correctly. If v_{OUT} rises $\rightarrow$ more current through $R_A \rightarrow R_A$ warmer $\rightarrow$ R_A resistance increases $\rightarrow$ gain reduced $\rightarrow$ v_{OUT} reduced.

2.4 Crystal oscillators

A quartz crystal cut to the correct dimensions, with its surfaces aligned with the planes of its crystal lattice and electrodes plated on opposite surfaces, is the electrical equivalent of the network drawn in Fig. 2.21. The resonant frequency of the crystal depends on its physical dimensions. If the crystal is included in a resonant circuit, which has a frequency close to that of the crystal, the crystal forces the circuit to resonate at the crystal frequency. Figure 2.22 shows another version of the Colpitt's oscillator (compare Fig. 2.19) using a JFET to pulse the supply of energy to the resonant network, which includes a crystal. The advantage of using a crystal is that it can be manufactured with much higher precision than inductors and capacitors. Typically crystals are available for set frequencies of 1 MHz and above, with a tolerance of $\pm 0.003\%$. They are used in high-precision oscillators, e.g. for generating RF, for TV circuits and for timing applications.

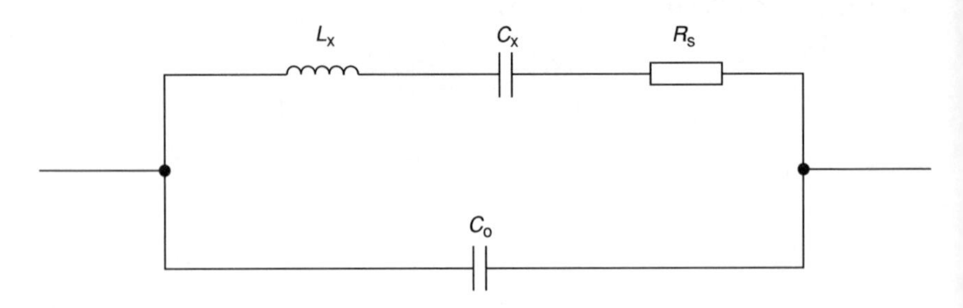

Figure 2.21

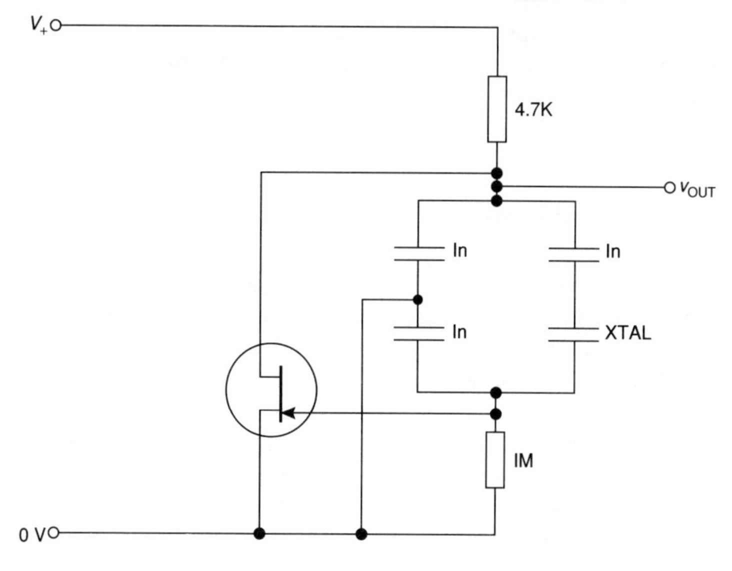

Figure 2.22

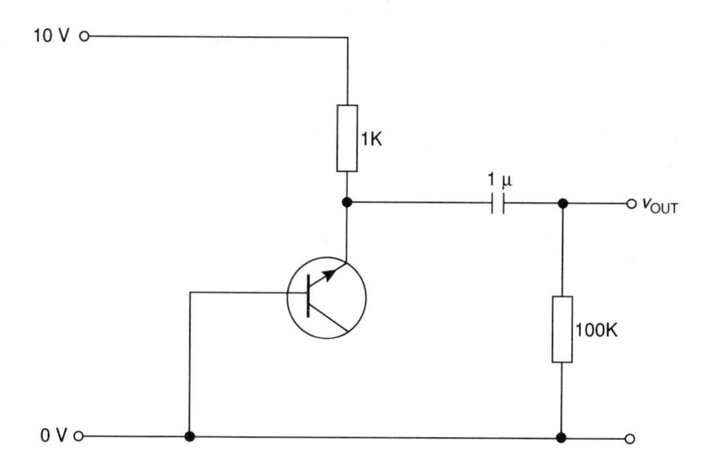

Figure 2.23

2.5 White noise generators

White noise is the analog of many randomly determined events in the real world, from the erratic motion of machine parts due to friction, to irregularities of illumination caused by dappled sunshine. Most electronic components have white noise inherent in their operation (7.2), so there are many devices that may be used as sources. A commonly used source is a reverse-biased base-emitter junction (Fig. 2.23). Alternatively, substitute a reverse-biased avalanche (Zener) diode for the transistor, with the pd across the diode well in excess of the breakdown voltage. Special 'noisy' diodes are available.

2.6 Analysis of signals

Consider a signal that is a pure sine wave, frequency f. Refer to this as the fundamental. The harmonics of this signal are sine waves that have frequencies $2f, 3f, 4f, \ldots$, all integral multiples of the fundamental, and are referred to as the 1st, 2nd, 3rd, ... harmonics.

Any periodic signal of whatever shape (even a square wave or saw-tooth wave) with frequency f can be considered to be the sum of a sinusoidal fundamental (frequency f) and one or more of its harmonics. The fundamental and the harmonics usually have different amplitudes; most often the fundamental has the largest amplitude, and the amplitudes of the harmonics decrease with increasing order.

The composition of a periodic signal is described by a frequency spectrum, which represents the frequencies of the component signals by an array of peaks or vertical lines, their heights representing amplitudes. A frequency spectrum is obtained by:

- Feeding a periodic electrical signal into a spectrum analyser and observing the peaks on the screen.
- Recording or calculating signal data, then plotting the results of a pen-and-paper Fourier analysis.

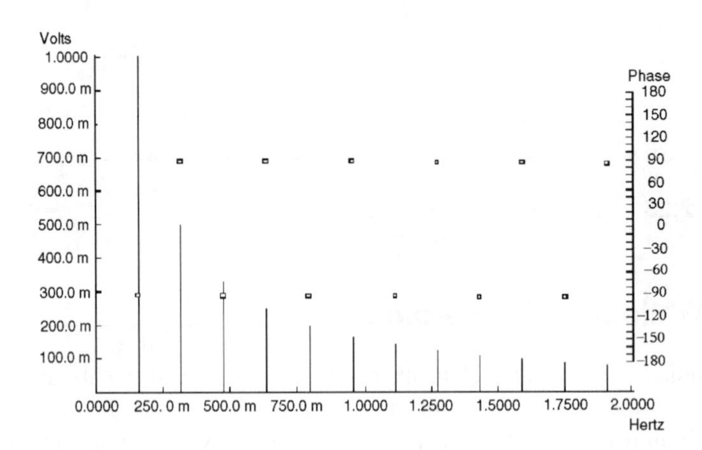

Figure 2.24

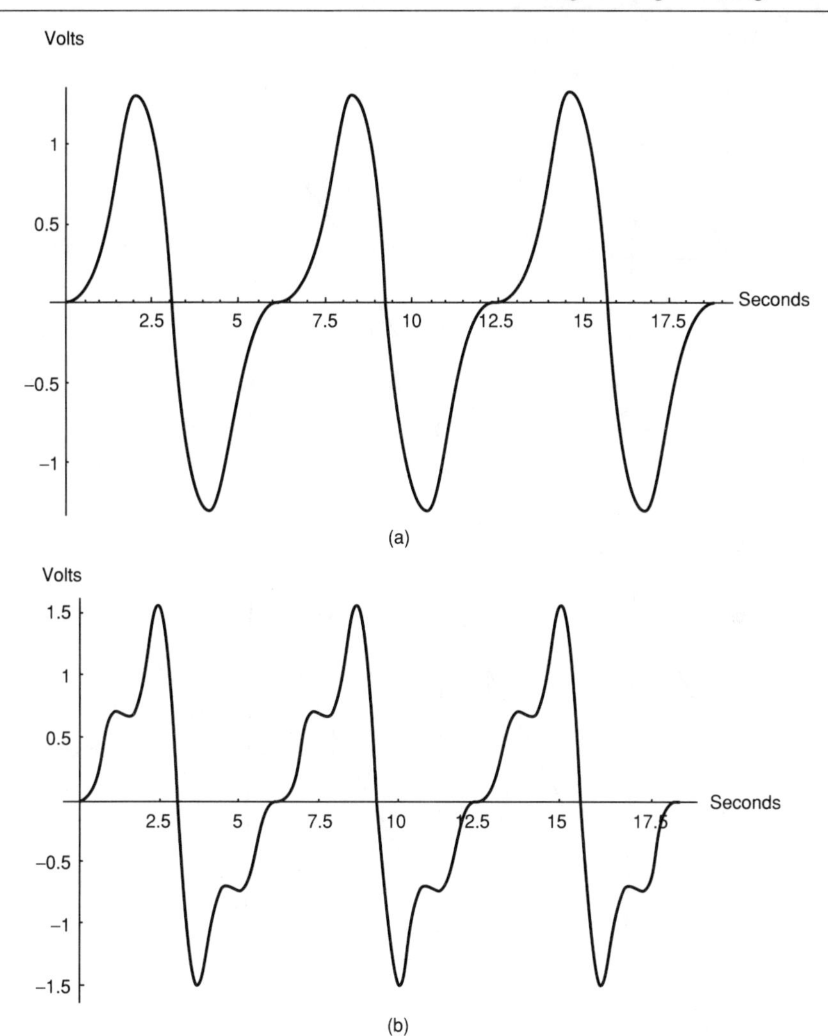

Figure 2.25

- Using a computer program to perform a Fourier analysis on signal data calculated by a circuit simulator.

Example: A Fourier analysis of a saw-tooth signal, amplitude 1.6 V, gives the spectrum shown in Fig. 2.24. The fundamental has a frequency 159 mHz and amplitude 1 V. The harmonics have amplitudes 0.5 V, 0.33 V, 0.25 V, 0.2 V, … 0.0833 V. Analysis shows that the signal is represented by this

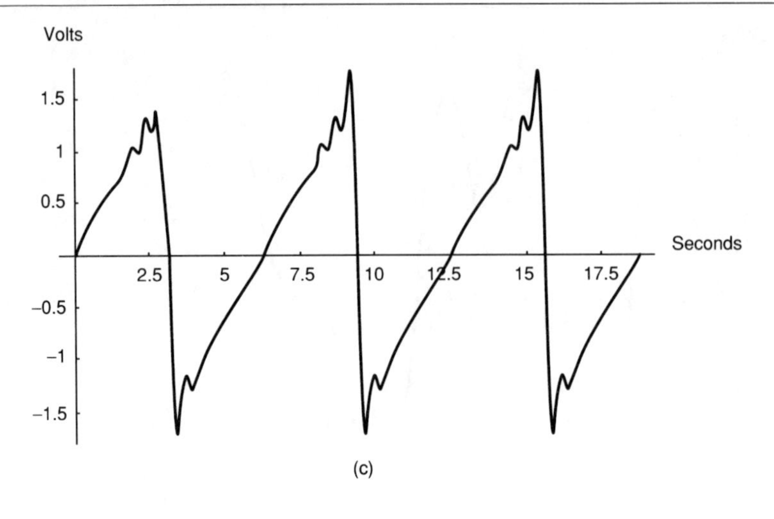

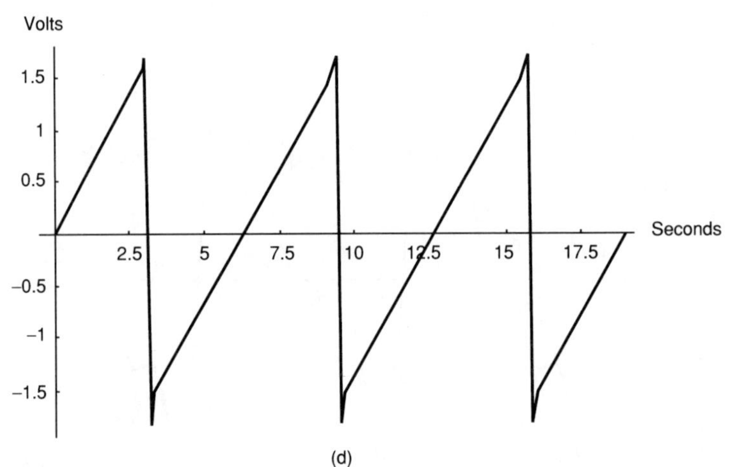

Figure 2.25 *continued*

Fourier series:

$$v = \sin \omega t - \frac{\sin 2\omega t}{2} + \frac{\sin 3\omega t}{3} - \frac{\sin 4\omega t}{4} + \frac{\sin 5\omega t}{5} - \cdots + (-1)^{n-1}\frac{\sin n\omega t}{n}$$

That a periodic linear ramp can truly be equivalent to a series of sine waves is demonstrated by Fig. 2.25 in which we synthesize a saw-tooth wave from sine waves. Graph (a) is a plot of the sum of the first two terms of the series: $v = \sin \omega t - (\sin 2\omega t)/2$. Figures 2.25b, c and d show sums of the first 4, 12 and 100 terms respectively. As the number of terms summed increases,

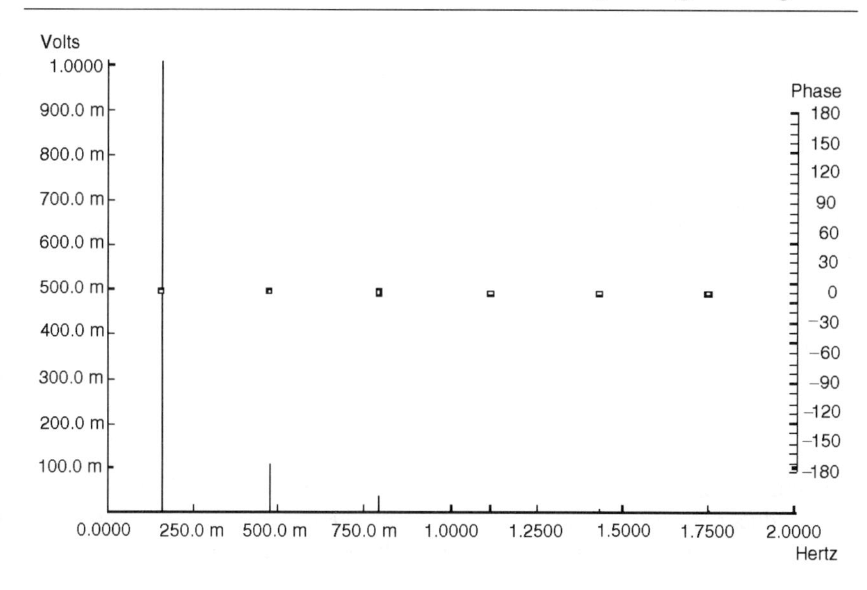

Figure 2.26

the plot approaches closer to a saw-tooth wave. The fact that the terms are alternately positive and negative is indicated by the phase plots (small squares) of the Fourier analysis (Fig. 2.24). These indicate the phase of each component relative to the fundamental which is taken in this analysis to be −90°. The even harmonics are all in phase with the fundamental (all positive) but the odd harmonics are all 180° out of phase, i.e. are negative.

For comparison, Fig. 2.26 is the spectrum of a triangular wave amplitude 1.2 V, frequency $1/2\pi$ Hz (159 mHz). Analysis shows its Fourier series to be:

$$v = \cos \omega t + \frac{\cos 3\omega t}{9} + \frac{\cos 5\omega t}{25} + \frac{\cos 7\omega t}{49} + \cdots + \frac{\cos n\omega t}{n^2}$$

where only odd values of n are included. Note the steep drop in harmonic amplitudes due to the divisor being n^2. The phase is shown as 0°; indicating that the series can alternatively be expressed in terms of sines, using the trigonometrical identity $\cos\theta \equiv \sin(\theta + 90°)$.

Knowledge of frequency spectra makes it easier to analyse circuit response to periodic waveforms. The principle of superposition (E.4) applies. The response of the circuit is analysed separately for the fundamental and each harmonic (see Chapter 6), and the results of the analyses combined.

3 Extracting analogs

This chapter examines ways of getting a usable analog signal from a sensor or transducer. In most cases this involves producing an electrical voltage or current signal analogous to one of five fundamental electrical quantities:

- emf
- current
- resistance
- capacitance
- inductance

Thus the circuits for interfacing sensors to electronic circuits fall into one of five categories.

3.1 Analogs of emf

Most sensors that produce a voltage output are transducers (1.2) such as thermocouples (1.3.3), pyroelectric sensors (1.3.5) and electret, electromagnetic (moving coil) and piezo-electric microphones (1.6.3–1.6.5). There are also some sensors which are not emf generators but in which the only accessible signal is a voltage analog of the measurand such as band-gap sensors (1.3.4), potentiometric position sensors (1.5.1), LVDTs (1.5.2) and Hall-effect devices (1.7.1). Occasionally the source of emf is non-electronic, as when we use probes to detect minute changes in body potentials due to muscle or brain activity. These may be interfaced in the same way as emf generators, as may signal generators of the kinds described in Chapter 2.

The main concerns in producing a usable analog signal from these devices are:

- Signal amplitude — to amplify it if it is too small (see Chapter 4), to attenuate it if it is too large.

- Signal DC level — to make this compatible with subsequent circuits.
- Impedance matching — to obtain sufficient transfer of power from source to subsequent circuits.
- Common mode rejection — when a small signal is superimposed on a larger and often varying DC level.
- Filtering — to remove unwanted signal components at the source — see Chapter 6.
- Noise — to reduce noise picked up from the environment or originating in the source — see Chapter 7.

3.1.1 Limiting amplitude

This is not likely to be a problem at the interface between a sensor and a measuring circuit, but it may be necessary as a safety measure at the input to an amplifier or similar circuit to prevent damage from an unexpectedly high input signal. Figure 3.1a is a diode signal *clipper* which limits the signal to $V_{REF} + 0.6\,\text{V}$. V_{REF} may be the positive supply line. Or substitute a potential divider to provide V_{REF} (Fig. 3.1b) when the supply voltage is higher. Clippers to set a minimum voltage or to limit the voltage between positive and negative levels are built on the same principles. Clipping results in signal distortion, if the signal amplitude is sufficiently great.

An *attenuator* reduces the amplitude of a signal, reducing its DC level proportionately. The simplest attenuator is a potential divider (Fig. 3.2), in which $v_{OUT} = v_{IN} \times R_B/(R_A + R_B)$. Its disadvantage is its high output impedance (drawing current lowers v_{OUT} significantly); to counter this, the current passing through the divider should be at least ten times that drawn from it. Preferably v_{OUT} should go to an FET, v_{OUT} varies if Z_{IN} of the following stage is affected by signal frequency. A resistive network such as Fig. 3.3a (O-type) is used for connecting a stage with $Z_{OUT} = R$ to a stage with $Z_{IN} = R$. Attenuation $a = v_{IN}/v_{OUT}$. Resistor values are $R_A = R(a^2 - 1)/4a$ and $R_B = R(a + 1)/(a - 1)$. Attenuators may be cascaded, the total attenuation being the product of the attenuations of each unit. This network is an example of various attenuators, used in test instruments to adjust signal level.

Take Fig. 3.3 as a case study in network analysis. Make the source 1 V to simplify calculations. Now write equations to describe everything known about the network and the conditions under which it is to operate.

In Fig. 3.3b, draw clockwise mesh currents i_1, i_2, i_3. Write mesh equations (E.3) using KVL (E.1):

$$(R + R_B)i_1 - R_B i_2 = 1 \tag{3.1}$$

$$-R_B i_1 + 2(R_A + R_B)i_2 - R_B i_3 = 0 \tag{3.2}$$

$$-R_B i_2 + (R_B + R)i_3 = 0 \tag{3.3}$$

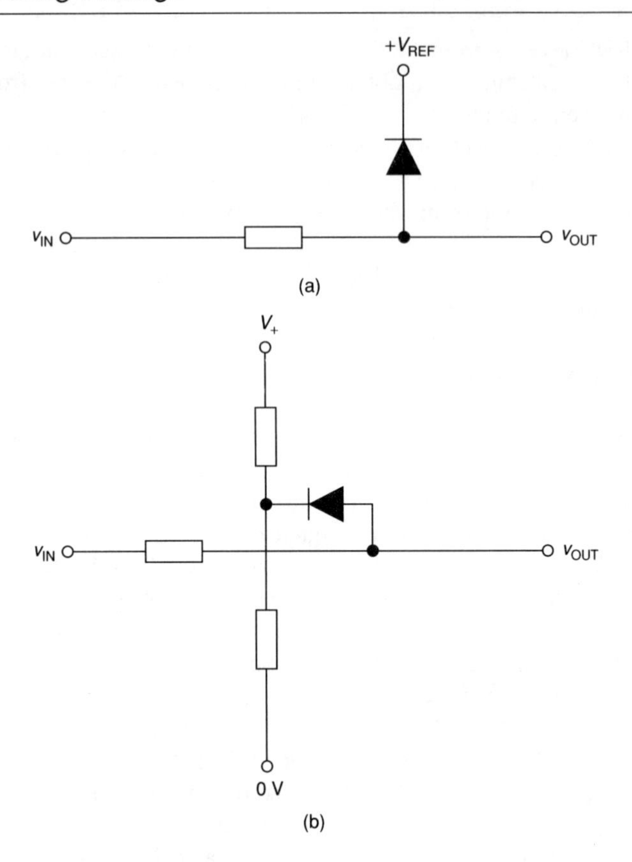

Figure 3.1

In mesh 1:

$$v_{\mathrm{IN}} = 1 - Ri_1$$

In mesh 3:

$$v_{\mathrm{OUT}} = Ri_3$$

The attenuation relationship gives:

$$a = \frac{v_{\mathrm{IN}}}{v_{\mathrm{OUT}}} = \frac{1 - Ri_1}{Ri_3}$$

$$\Rightarrow \qquad aRi_3 = 1 - Ri_1 \qquad (3.4)$$

This has eliminated variables v_{IN} and v_{OUT}, which are not wanted in the final result. The impedance looking into the network must equal the impedance of the source. Find the impedance of the network to the right of PQ by network reduction, using the standard formulae for resistances in parallel and

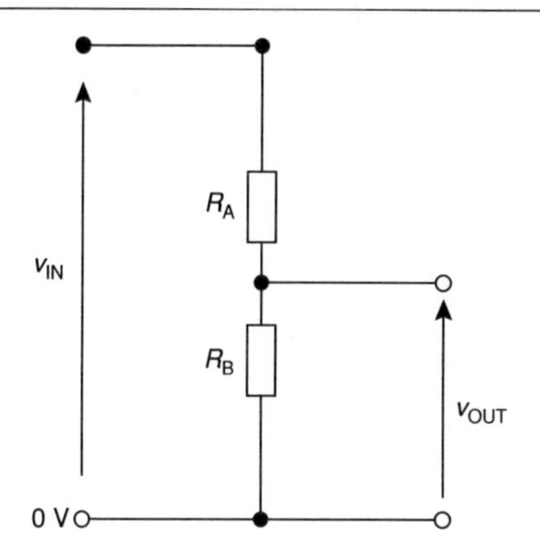

Figure 3.2

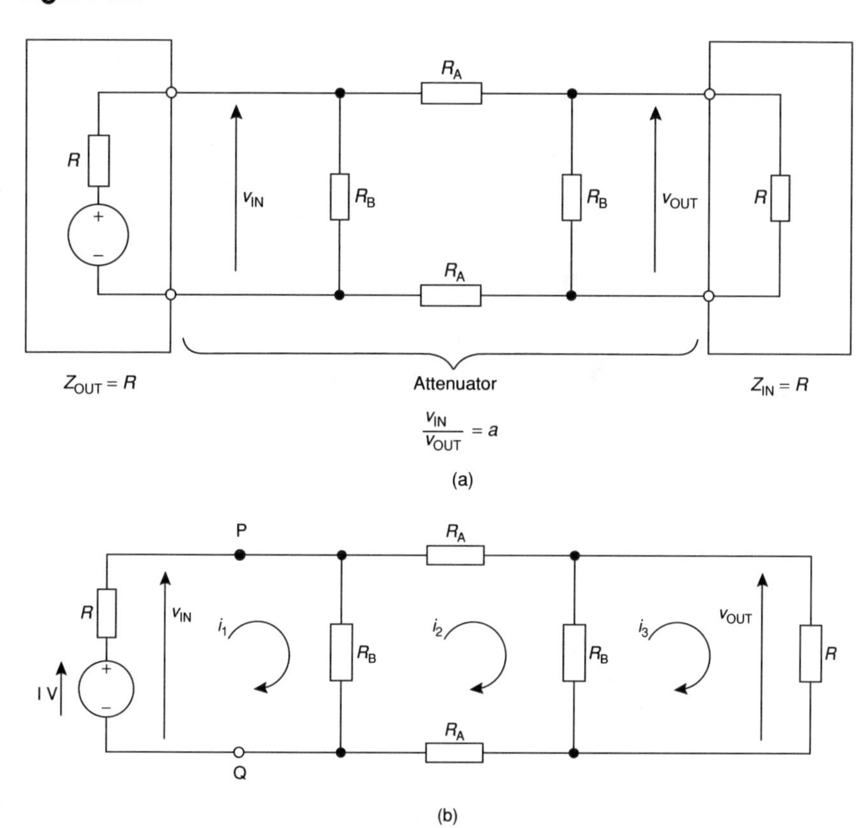

Figure 3.3

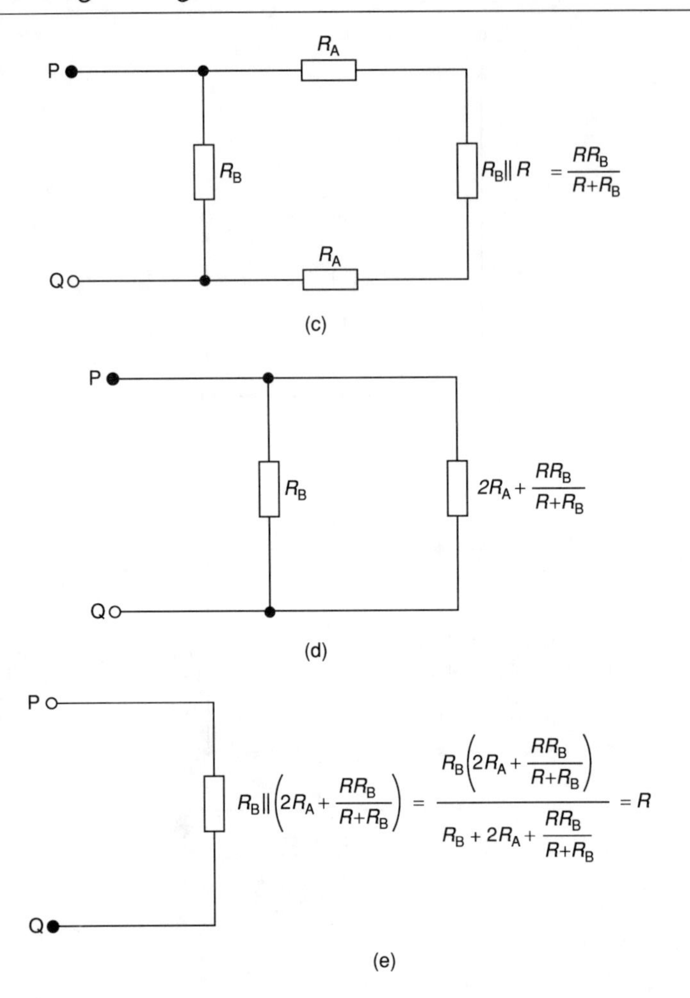

Figure 3.3 *continued*

series. First replace the right-hand R and R_B in parallel by their equivalent (Fig. 3.3c). Next replace this and the two R_A resistors with their equivalent in series (their sum, Fig. 3.3d). Finally, replace this and the remaining R_B with their equivalent in parallel. The impedance of this is R by definition (Fig. 3.3e). The equation on the right of Fig. 3.3e simplifies to:

$$R_A R_B^2 = R_A R^2 + R_B R^2 \tag{3.5}$$

We have five simultaneous equations, which can be solved to give R_A and R_B in terms of R, and a, eliminating the currents. Instead of several pages of error-prone pen-and-paper calculations, call on mathematical software, which produces the solutions in a few seconds:

$$R_A = \frac{R(a^2 - 1)}{4a}$$

$$R_B = \frac{R(a + 1)}{a - 1}$$

The same procedure can be used with other attenuators and with other networks.

Figure 3.4 is an attenuator which uses an FET as a variable resistor, shorting a variable proportion of the signal to ground. For linearity, this must operate in the resistor region of the i_D/v_{DS} curve, which limits amplitude of v_{IN} to about 1 V (Fig. C.14). Figure 3.5 shows v_{OUT} for a number of values of the control voltage, v_{GS}. As v_{GS} rises from -3.5 V to $+0.5$ V, Q1 is turned more fully on → resistance decreases → I_D increases → v_{OUT} amplitude decreases. In this JFET (2N3819) pinch-off voltage is -4 V (see Fig. C.17) and, as v_{GS} is reduced from -3 V to -4 V → Q1 gradually turned off → v_{OUT} amplitude increases but is clipped at about -800 mV with serious distortion.

Attenuator ics with voltage-controlled attentuation are available for use in audio equipment.

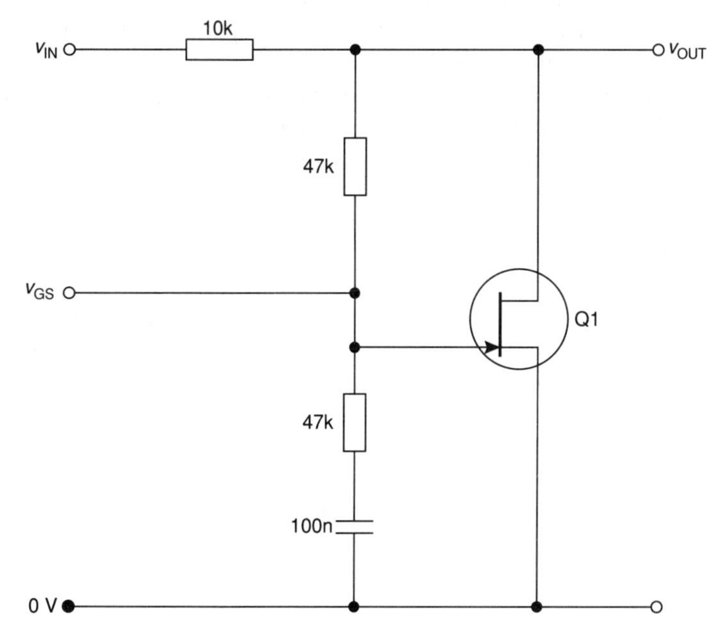

Figure 3.4

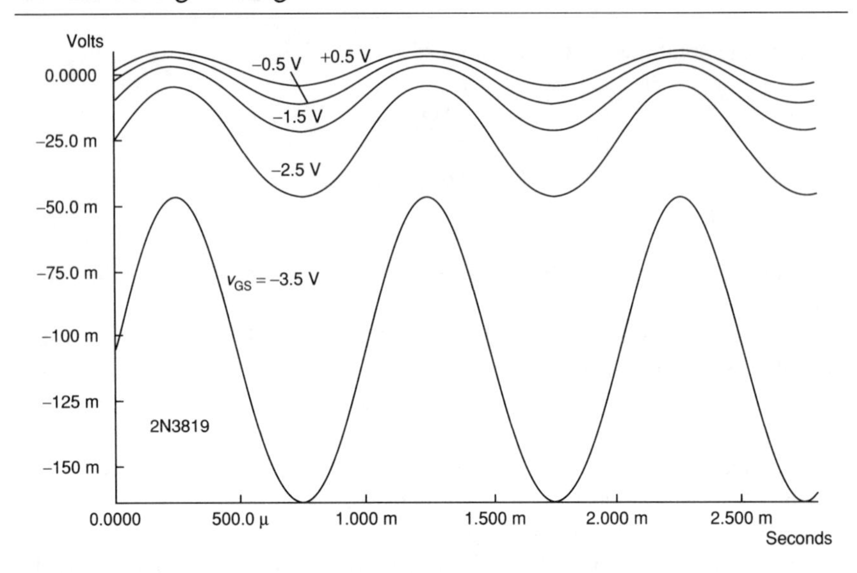

Figure 3.5

3.1.2 Changing DC level

A *clamp* centres a signal on a different DC level; it adds or subtracts a fixed amount to the signal at every instant, but the amplitude is unchanged. The capacitor (Fig. 3.6a) allows the DC levels of v_{IN} and v_{OUT} to differ, but AC signals pass across. In Fig. 3.6a, if v_{IN} has amplitude A and is centred on 0 V, and if the diode is ideal, the diode prevents charge flowing to ground → v_{OUT} raised until it never dips below 0 V → v_{OUT} centred on A. With a real diode, v_{OUT} is one diode drop below this → centred on $A - 0.6$. See Fig. 3.7, curve v_{OUTA} (capacitor starts uncharged to display initial charging).

In Fig. 3.6b the diode conducts when v_{IN} is positive → v_{OUT} pushed down until it never exceeds 0.6 V → centred on $0.6 - A$ (curve v_{OUTB}, Fig. 3.7). In Fig. 3.6c, where $V_{REF} = 10$ V, v_{OUT} is raised until it never falls below $10 - 0.6$ → centred on $10 - 0.6 + A$ (curve v_{OUTC}, Fig. 3.7). Note that clamping level depends on A; the capacitance must be large enough to smooth out the effects of amplitude changes (say, 100 nF for audio signals).

In Fig. 3.6a, if v_{IN} is centred on a level $>A$, diode never conducts and $v_{OUT} = v_{IN}$.

Figure 3.8 is an active clamp with settable clamp voltage, v_{CL}.

$$v_- = \frac{v_{IN} - v_{OUT}}{2}$$

$$v_+ = \frac{v_{CL}}{2}$$

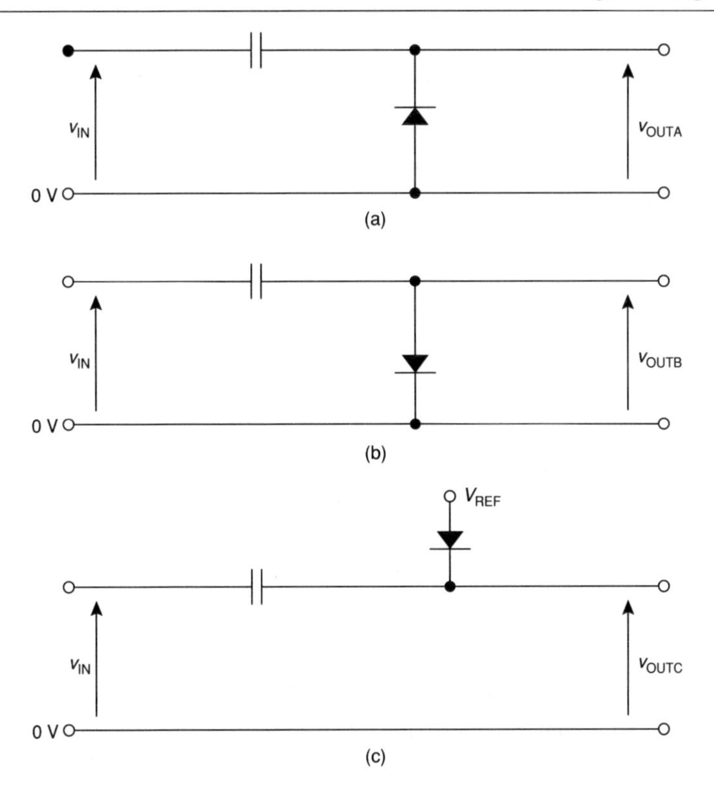

Figure 3.6

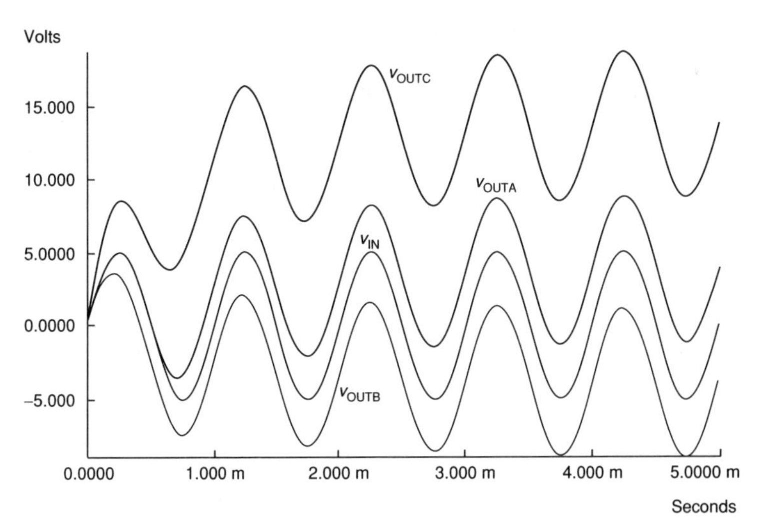

Figure 3.7

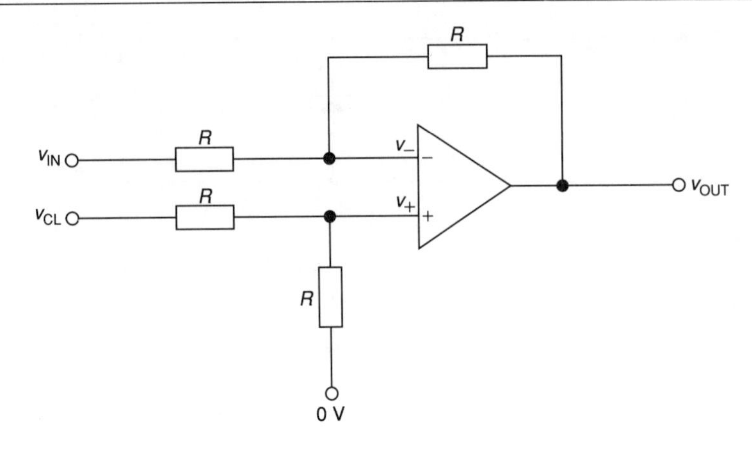

Figure 3.8

The op amp makes these equal:

$$\frac{v_{IN} - v_{OUT}}{2} = \frac{v_{CL}}{2}$$

$$\Rightarrow \qquad v_{OUT} = v_{IN} - v_{CL}$$

This avoids the effects of diode drop, but v_{OUT} rolls off at high frequencies (usually $> 10\,\text{kHz}$). $Z_{IN} = R/2$ at both inputs which minimizes effects of input offset.

3.1.3 Impedance matching

Voltage transfer from a source (e.g. sensor) to a receiver (e.g. amplifier) is illustrated in Fig. 3.9:

$$i = \frac{v - v_{IN}}{Z_{OUT}} = \frac{v_{IN}}{Z_{IN}}$$

$$\Rightarrow \qquad v_{IN} = v \times \frac{Z_{IN}}{Z_{IN} + Z_{OUT}}$$

To obtain optimum voltage transfer, v_{IN} must approach v. So $Z_{IN}/(Z_{IN} + Z_{OUT})$ must approach 1. This is achieved if Z_{IN} is much greater than Z_{OUT}. Feed the signal into a high-impedance network such as an emitter follower amplifier (4.2), otherwise voltage signal is lost.

Current transfer (Fig. 3.10):

$$v_{IN} = Z_{OUT}(i_{SOURCE} - i_{IN}) = Z_{IN} i_{IN}$$

$$\Rightarrow \qquad i_{IN} = i_{SOURCE} \times \frac{Z_{OUT}}{Z_{IN} + Z_{OUT}}$$

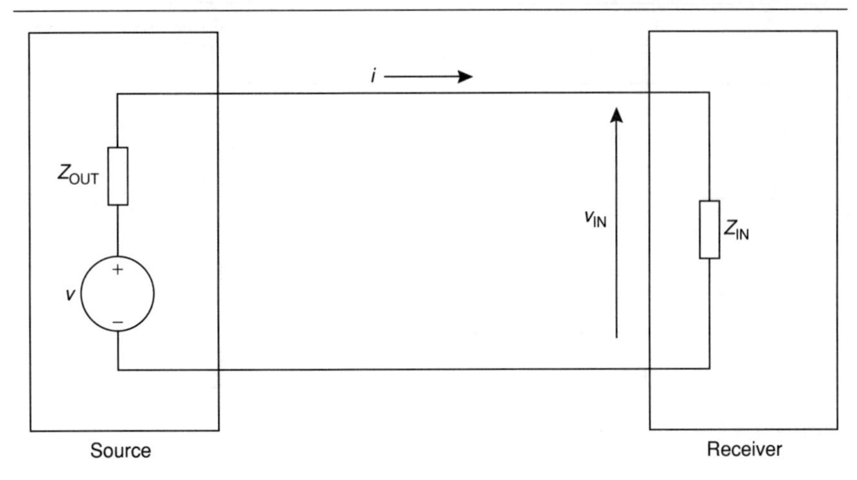

Figure 3.9

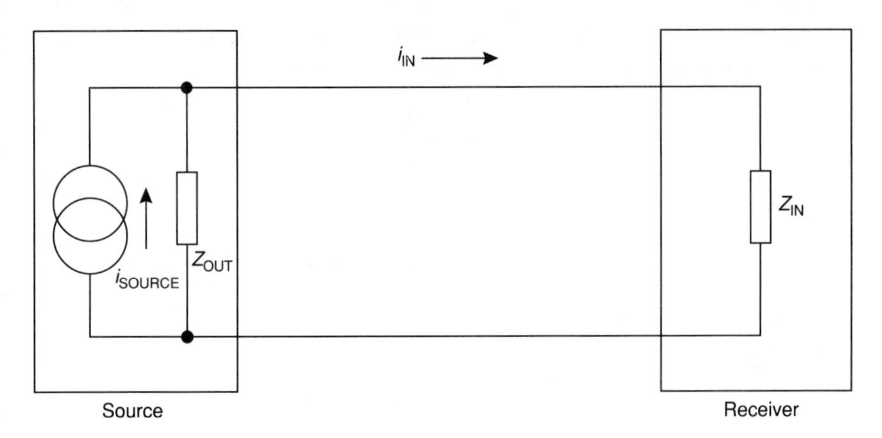

Figure 3.10

To obtain optimum current transfer, i_{IN} must approach i_{SOURCE}, which means that $Z_{OUT}/(Z_{IN} + Z_{OUT})$ must approach 1. Let Z_{OUT} be much greater than Z_{IN}. Feed signal into a low-impedance circuit. Example, see 3.2.

Power transfer (Fig. 3.11):

$$i = \frac{v}{Z_{IN} + Z_{OUT}} = \frac{v}{R_{OUT} + R_{IN} + j(X_{OUT} + X_{IN})}$$

$$\text{Power input} = p = |i|^2 R_{IN} = \left| \frac{v}{R_{OUT} + R_{IN} + j(X_{OUT} + X_{IN})} \right|^2 R_{IN}$$

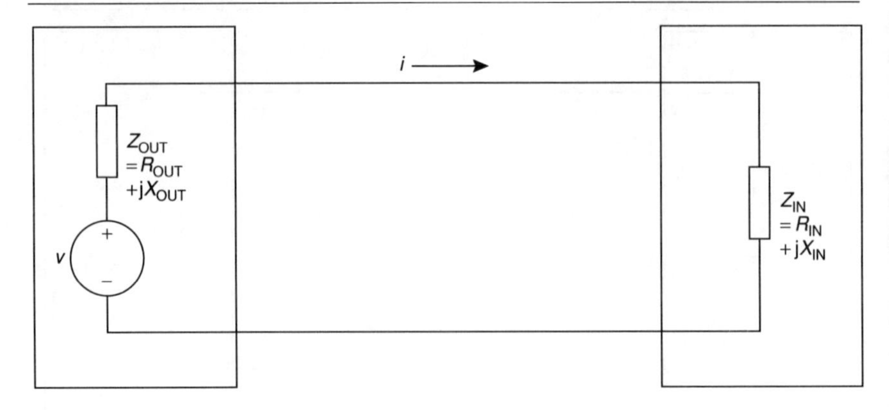

Figure 3.11

Maximum power transfer when $X_{OUT} + X_{IN} = 0$, i.e. when output and input are non-reactive, or when $X_{IN} = -X_{OUT}$, as when the output is capacitative and the input is inductive, or the other way about (in effect, an *LCR* circuit with maximum power transfer between *L* and *C*). Then:

$$p = \frac{v^2 R_{IN}}{(R_{IN} + R_{OUT})^2}$$

Maximum value when $\mathrm{d}p/\mathrm{d}R_{IN} = 0$:

$$\frac{\mathrm{d}p}{\mathrm{d}R_{IN}} = \frac{(R_{OUT} + R_{IN})^2 v^2 - 2v^2 R_{IN}(R_{OUT} + R_{IN})}{(R_{OUT} + R_{IN})^4} = 0$$

If numerator is 0:

$$\Rightarrow \qquad (R_{OUT} + R_{IN}) - 2R_{IN} = 0$$

$$\Rightarrow \qquad R_{OUT} = R_{IN}$$

For maximum power transfer, output and input impedances must be equal. But, substituting $R_{OUT} = R_{IN}$:

$$p = \frac{v^2}{4R_{OUT}}$$

which is only a quarter of the available power. In practice, matching for power transfer is used only when:

- Signals are large (as in radio transmitter output), to avoid waste of energy.
- Signals are small, to avoid increasing the signal-to-noise ratio (7.4).
- To avoid reflections at the ends of transmission lines (8.2.1).

A transformer is useful for impedance matching, because there is little power loss. Transformed voltage ratio = turns ratio, n:

$$\frac{v_{\text{OUT}}}{v_{\text{IN}}} = \frac{N_2}{N_1} = n \qquad (3.6)$$

In Fig. 3.12:

$$i_{\text{OUT}} = \frac{v_{\text{OUT}}}{R_{\text{L}}}$$

From power equation and (3.6):

$$i_{\text{IN}} = \frac{v_{\text{OUT}} i_{\text{OUT}}}{v_{\text{IN}}} = \frac{v_{\text{OUT}}^2}{v_{\text{IN}} R_{\text{L}}} \qquad (3.7)$$

When a signal is applied to the primary side it sees a resistance, R_{IN}, and, from (3.7):

$$R_{\text{IN}} = \frac{v_{\text{IN}}}{i_{\text{IN}}} = \frac{v_{\text{IN}}^2 R_{\text{L}}}{v_{\text{OUT}}^2}$$

$$\Rightarrow \qquad R_{\text{IN}} = \frac{R_{\text{L}}}{n^2}$$

R_{IN} is the reflected impedance of R. For maximum transfer, given R_{L}, match R_{IN} to the output impedance of the source by choosing suitable n.

Examples: To match a low-impedance (150 Ω) electromagnetic microphone to a high-impedance (10 kΩ) amplifier input. $R_{\text{IN}} = 150\,\Omega$, $R_{\text{L}} = 10\,\text{k}\Omega$, $n = \sqrt{(10\,000/150)} = \sqrt{66.67} = 8.16$. Use a step-up transformer ($n > 1$) with

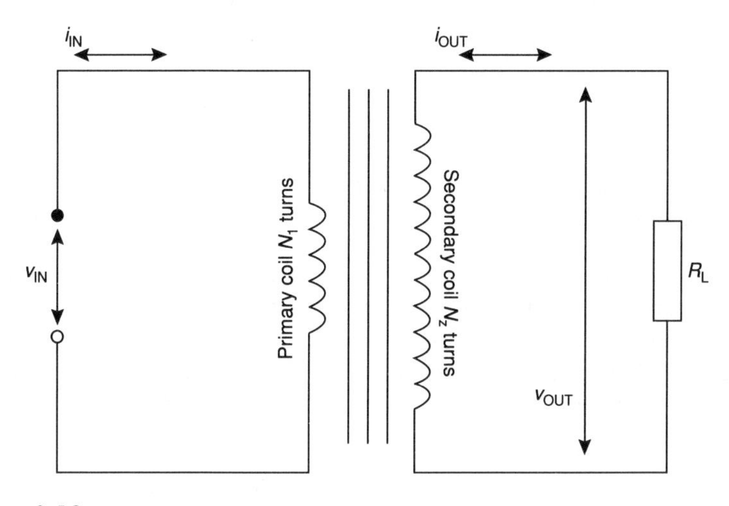

Figure 3.12

turns ratio 8:1. The amplifier sees the microphone as a $10\,\mathrm{k}\Omega$ impedance; conversely, the microphone sees the amplifier as a $150\,\Omega$ impedance. Similarly, use a step-down transformer for matching a loudspeaker ($8\,\Omega$) to the output ($75\,\Omega$) from an amplifier. $n = \sqrt{(8/75)} = \sqrt{0.1067} = 0.327$. Turns ratio is 1:3.

Practical transformers are not perfect, which may limit the efficiency of this technique, especially at high frequencies.

3.1.4 Common mode rejection

It is sometimes required to measure the (perhaps small) pds between two inputs to a circuit while equal (perhaps large) changes of potential are being applied to both points. The latter are *common mode* changes, which are to be eliminated as far as possible.

Example: Measuring the activity of a leg muscle by detecting the pds ($< 1\,\mathrm{mV}$) between two probes attached to the skin. The limb as a whole may be subject to overall change of potential, including mains-frequency signals induced by currents flowing in nearby equipment.

We use a *differential amplifier*, of which the simplest is the *long-tail pair* (Fig. 3.13). Transistors are identical, usually on same chip → same temperature → elimination of amplifier drift. They operate on a split supply with $V_{CC} = -V_{EE}$. With inputs at 0 V, quiescent current in each transistor is I_Q. Current in tail resistor R_t is $2I_Q$. This is a pair of emitter followers (4.2) → potential at the emitters (point A) is also close to 0 V. Pd across R_T is $V_{EE} = V_{CC}$.

For R_T:

$$2I_Q = \frac{V_{CC}}{R_T}$$

If $R_C = R_T$, pd across R_C is $I_Q R_C = I_Q R_T = V_{CC}/2$.

Quiescent output is $V_{CC}/2$.

(1) Common mode gain: Connect both inputs to same signal v_{in} (suppose this is positive, but could be negative without affecting the discussion) → potential across R_T increases by v_{in} (emitter followers) → current through R_T increases by v_{in}/R_T → collector current i_C through each transistor (and its R_C) increases by $v_{in}/2R_T$ → pd across R_C increases by $v_{in}R_C/2R_T$ → v_{out} decreases by $v_{in}R_C/2R_T$.

$$\text{Common mode voltage gain} = \frac{v_{out}}{v_{in}} = \frac{-R_C}{2R_T}$$

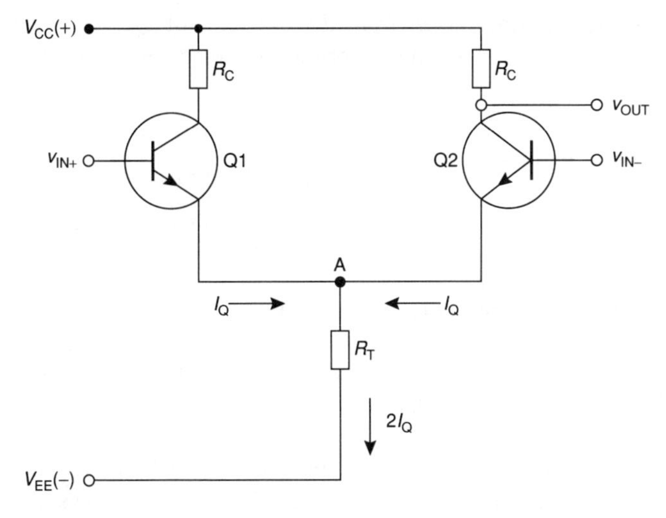

Figure 3.13

If $R_C = R_T$, common mode gain $= -0.5$. For other feasible values of R_C and R_T it is always small and negative. The higher the value of R_T, the lower the gain.

(2) Differential gain: Recall $r_e = 1/g_m$ and $g_m = 40i_E$ (C.2.4.2), and assume $R_T \gg r_e$. Inputs are connected to two initially equal signals v_{in+} and v_{in-}. Suppose v_{in+} rises by small amount v and v_{in-} falls by v to give differential input $v_{in+} - v_{in-} = 2v$. Total current through R_T remains unchanged (but more comes from Q1 and less from Q2) and potential at point A remains unchanged. Looking at Q2, v_{in-} decreased by $v \rightarrow$ voltage at base-emitter junction decreased by $v \rightarrow$ emitter current through r_e decreased by $v/r_e \rightarrow$ collector current decreased by $\approx v/r_e \rightarrow$ potential at collector of Q2 (v_{out}) increased by vR_c/r_e. Total change of input is $2v$, change of output is vR_c/r_e so:

$$\text{Differential voltage gain} = \frac{\text{change in output}}{\text{change in input}} = \frac{R_c}{2r_e} = \frac{g_m R_c}{2}$$

with g_m in mA/V and R_C in kΩ. Compared with common emitter amplifier (C.2.4.2), transconductance of differential amplifier is $g_m/2$(g_m is transconductance of each transistor):

$$g_m/2 \approx 40i_E/2 \approx 40i_T/4 \approx 10i_T$$

where i_T is the quiescent current through R_T, and is the sum of the two emitter currents. Substituting:

$$\text{Differential voltage gain} = 10i_T R_c$$

Gain is positive if output taken from the collector of Q2. $v_{\text{in}+}$ is the non-inverting input and $v_{\text{in}-}$ is the inverting input (compare with op amps (Fig. D.1)).

$$\text{Common mode rejection ratio} = \frac{|\text{differential voltage gain}|}{|\text{common mode voltage gain}|}$$

$$= 10 i_{\text{T}} R_{\text{C}} \times \frac{2R_{\text{T}}}{R_{\text{c}}} = 20 i_{\text{T}} R_{\text{T}}$$

Increase CMMR by increasing i_{T} and/or R_{T}. Involves large V_{EE}, but see below.

Example: Testing amplifier as Fig. 3.13, $V_{\text{CC}} = 15\,V$, $V_{\text{EE}} = -15\,V$, $R_{\text{C}} = R_{\text{T}} = 75\,\text{k}\Omega$, Q1 = Q2 = BC108BP.

Test 1, common mode: inputs are sine-wave signals amplitude 50 mV, frequency 1 kHz, in phase with each other. Figure 3.14 shows v_{out} having amplitude 2.5 mV, 180° out of phase with input. Common mode gain is –0.5, as is calculated from gain $= -R_{\text{C}}/2R_{\text{T}} = -75\,000/150\,000$.

Test 2, differential mode: inputs as before except $V_{\text{in}-}$ lags 90° behind $V_{\text{in}+}$ (Fig. 3.15). Inputs plotted on ×20 scale. Output amplitude is proportional to $v_{\text{in}+} - v_{\text{in}-}$ at each instant; measurements show differential gain is 131 and i_{T} steady at 192 μA. Calculation gives gain $\approx 10 \times 192 \times 10^{-6} \times 75\,000 = 144$, which is a reasonable approximation.

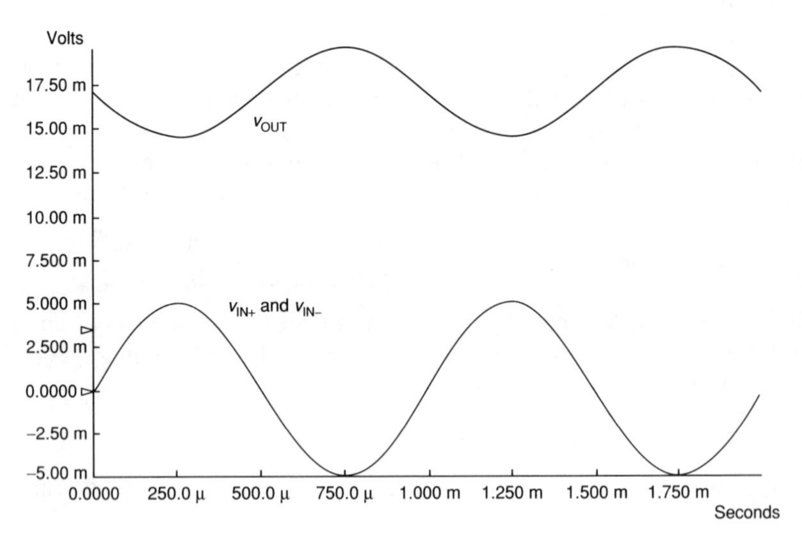

Figure 3.14

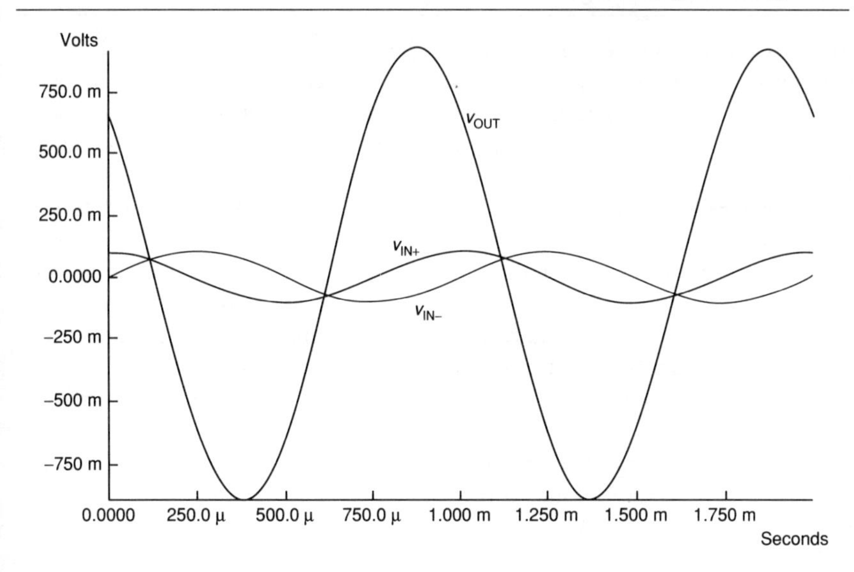

Figure 3.15

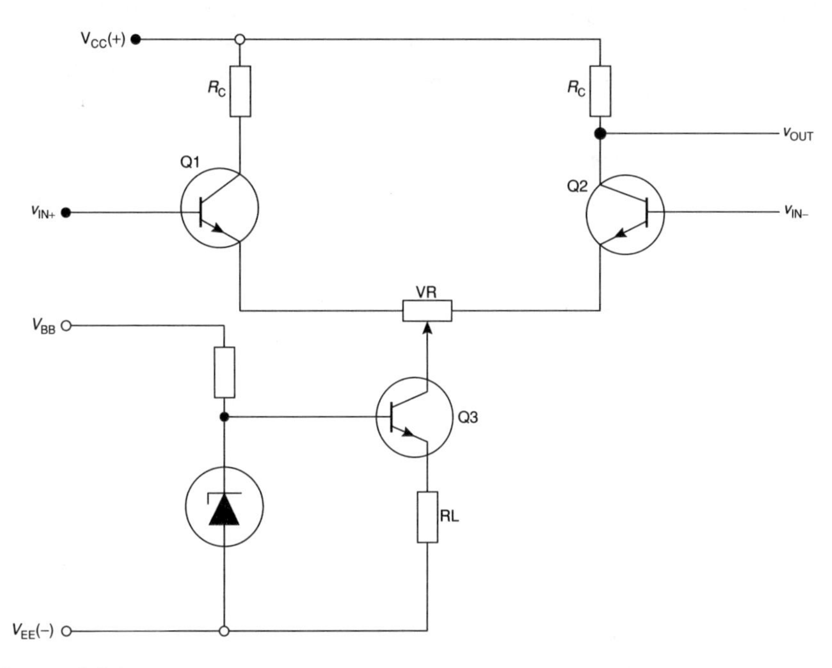

Figure 3.16

In a single common-emitter amplifier, voltage gain varies with signal level → varying i_E → varying g_m → distortion. In differential amplifiers, R_T is large compared with r_e → current through it is fairly constant → one i_E increases as the other decreases → increase in g_m of one transistor compensated for by decrease in the other g_m → approximately linear response → low distortion. Even better response is obtained by replacing R_T by a constant current generator (Fig. 3.16).

This also produces higher CMMR because the generator behaves as very high resistance, and relatively large current can be sunk, without requiring an unduly low V_{EE}. The variable resistor (optional) enables the two sides of the circuit to be balanced exactly, so further improving the CMMR.

To obtain greater input impedance, use FETs or Darlington transistors (4.2.6).

An operational amplifier can be wired as a differential amplifier (D.2.4); a typical op amp has a long-tailed pair at its inputs.

3.2 Analogs of current

If the analog is a varying current, the simplest technique is to use a resistor as a current-to-voltage converter, then measure the voltage, subject to the provisos of 3.1. In Fig. 3.17 the photodiode (1.8.2) is reverse biased. Incident

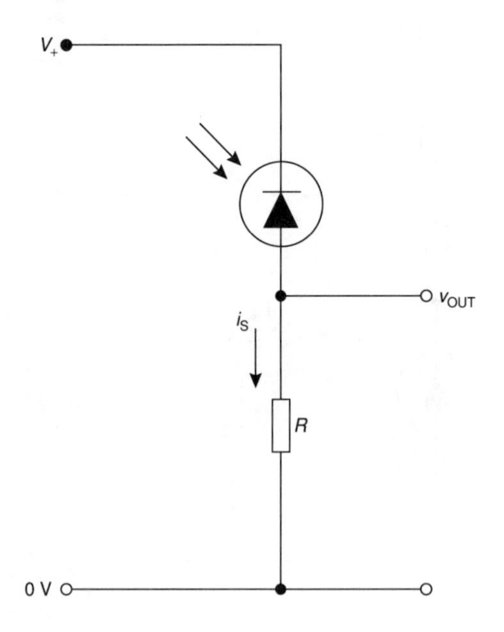

Figure 3.17

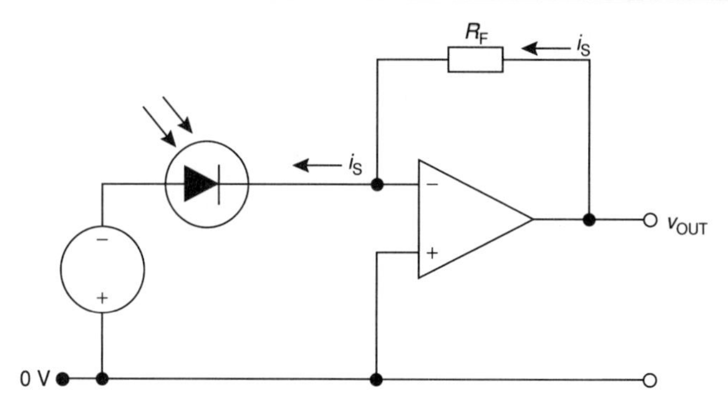

Figure 3.18

light energy liberates minority carriers (A.4, A.5), which carry a small reverse current i_S. This is linearly proportional to incident illumination over a wide illumination range. i_S generates a pd across R as it flows through it: $v_{OUT} = i_S R$. Figure 3.17 has high output impedance (R may need to be several hundred kilohms) so the voltage output must be fed to a high impedance stage (3.1.3).

Figure 3.18 uses an op amp as a current-to-voltage converter. The photodiode is reverse biased by a voltage source (which could be a Zener or other regulator) between the 0 V and V_- rails. As above, i_S is proportional to incident light intensity. v_- input is at 0 V. Current through $R_F = i_S$ and thus $v_{OUT} = -i_S R_F$.

Similar circuits are used with a phototransistor (1.8.3). Photodiodes are available with on-chip amplifiers (similar to Fig. 3.18), producing an output voltage proportional to light intensity.

3.3 Analogs of resistance

Devices which have a varying resistance as their analog output include platinum resistance thermometers (1.3.1), thermistors (1.3.2), strain gauges (1.4.1), carbon microphones (1.6.1), magnetoresistive devices (1.7.2), and light-dependent resistors (1.8.1). We may also wish to measure directly the resistance between a pair of probes (e.g. soil moisture, dampness in plaster walls, changes of resistance of skin as a means to lie detection).

Resistance as such cannot be processed further. It must be converted either into a current analog or a voltage analog. A device such as a carbon microphone, with appreciable but not large resistance and capable of producing relatively large changes in resistance, behaves as a variable current source.

Example: Connect the microphone in series with an electromagnetic earphone.

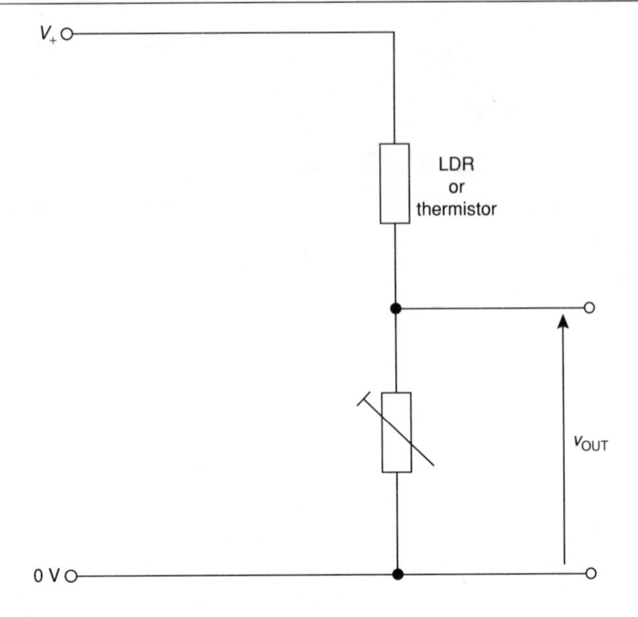

Figure 3.19

Thermistors or light-dependent resistors can be connected into a potential divider (Fig. 3.19). Increasing light or temperature → reduced resistance → v_{out} rises. Response level is adjusted by setting the variable resistor. A resistance cannot be measured without passing current through it. If the resistance is a temperature sensor, the self-heating raises temperature → error. It is essential to make current through thermistor as small as possible. Output is not linear but, if the sensor is operating in the region in which its resistance is a few tens of kilohms, the potential divider has an output impedance suitable for feeding to a BJT amplifier, or for tripping a BJT Schmitt trigger (10.1).

3.3.1 Bridge circuit

This sensitive circuit is used for measuring small resistances and small changes of resistance. It is applicable to platinum resistance thermometers and strain gauges, also for precision measurements with thermistors and light-dependent resistors. Derived from the Wheatstone bridge, the basic bridge (Fig. 3.20) has four arms, each carrying a resistor. v_{IN} may be constant or alternating.

The bridge consists of two potential dividers, ABC and ADB connected in parallel. Their outputs are:

$$v_{IN} \times \frac{R2}{R1 + R2} \quad \text{and} \quad v_{IN} \times \frac{R4}{R3 + R4}$$

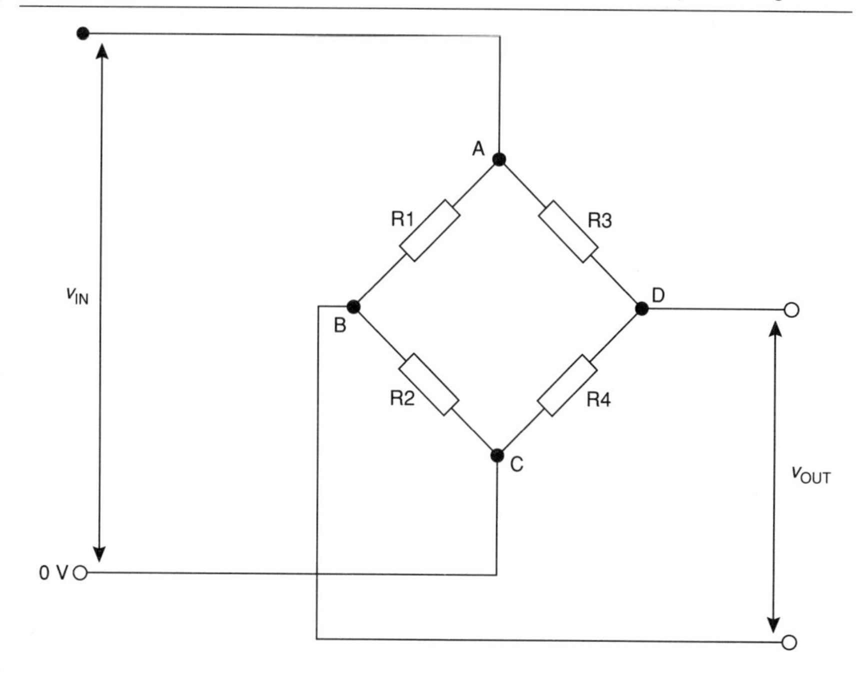

Figure 3.20

v_{OUT} is the difference between their outputs:

$$v_{OUT} = v_{IN}\left(\frac{R4}{R3 + R4} - \frac{R2}{R1 + R2}\right)$$

If the test resistor (the sensor) is R1 and the other resistors and v_{IN} are known, measure v_{OUT} and calculate R1. For greatest precision R2 = R3 = R4 ≈ R1. This technique differs from the traditional one in which the bridge is balanced by adjusting R3 until $v_{OUT} = 0$ (as measured by a sensitive meter). Or, if v_{IN} is an AF signal, adjust R3 until no sound is heard from a headset connected across the v_{OUT} terminals. Then R1 = R2 × R3/R4.

Figure 3.20 shows a *quarter-bridge* in which one arm holds a test resistance (strain gauge, platinum resistance, etc.). A half-bridge has a sensor in two arms (R1 and R4) → increased resistance lowers potential at B and raises potential at D → doubly increases v_{OUT} → increased sensitivity. If all four arms hold identical thermal sensors, R1 and R4 are used for sensing, R2 and R3 kept at room temperature → self-heating is compensated for. But currents are small in a bridge so self-heating is not a major problem. Alternative circuit (Fig. 3.21) has the sensor (R4) on a lead with a dummy lead made from the same cable inserted in series with R2. The dummy lead runs beside the actual lead and is

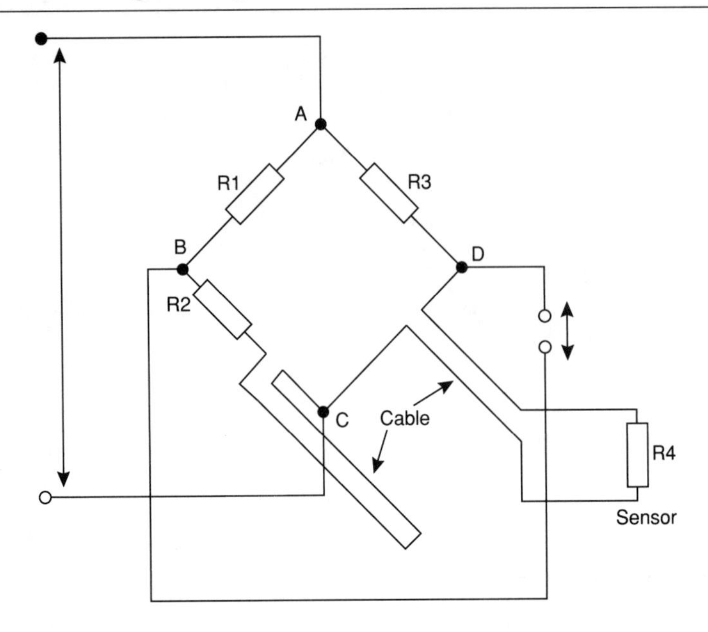

Figure 3.21

subjected to same temperatures. Changes in resistance of the actual lead are compensated by changes in resistance of dummy lead.

A *full bridge* has sensors in all four arms, as explained above for four thermal sensors. A full bridge can also be set up as two *half-bridges*, R1 and R4 operating in the opposite sense to R2 and R3.

Example: Four strain gauges positioned on the structure so that, when two are stretched, the other two are relaxed, and the other way about. This gives increased sensitivity but output is not linear. Pressure sensor ics have the full bridge and amplifier built-in on the same chip (1.4.2).

In Fig. 3.22 an op amp detects imbalance of the bridge. Given that R1 = R2; R3 is a standard resistor; R4 is the test resistor; $R4 = R3(1 + x)$. We find x.

$$v_+ = \frac{v_{IN}}{2}$$

Op amp settles when $v_- = v_+ = \dfrac{v_{IN}}{2}$

Current through R3 $= \dfrac{v_{IN} - v_{IN}/2}{R3} = \dfrac{v_{IN}}{2R3}$

Current through R4 $= \dfrac{v_{IN}/2 - v_{OUT}}{R4} = \dfrac{v_{IN}/2 - v_{OUT}}{R3(1 + x)}$

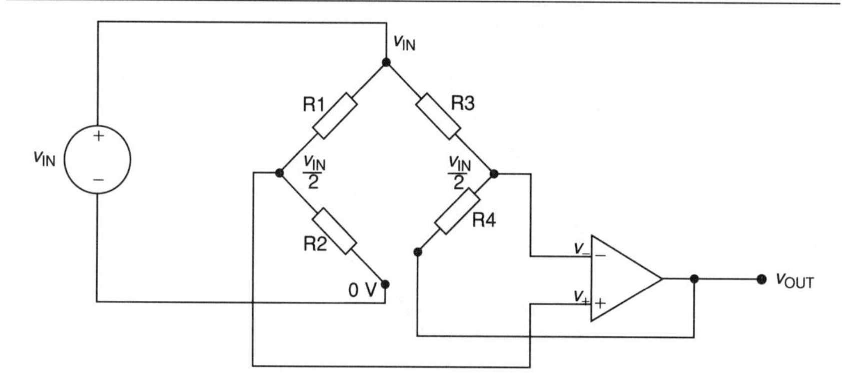

Figure 3.22

But the v_- input has high impedance so these currents are equal:

$$\frac{v_{IN}}{2R3} = \frac{v_{IN}/2 - v_{OUT}}{R3(1 + x)}$$

$$\Rightarrow \qquad \frac{v_{IN}R3(1 + x)}{2R3} = v_{IN}/2 - v_{OUT}$$

$$\Rightarrow \qquad x = \frac{-2v_{OUT}}{v_{IN}}$$

The relationship between v_{OUT} and x is linear.

Example: Given, $v_{IN} = 10\,V$, and $v_{OUT} = -1.7\,V$. Calculate $x = (2 \times 1.7)/10 = 0.34$. Then, if $R3 = 220\,\Omega$, $R4 = 220(1 + 0.34) = 294.8\,\Omega$.

3.4 Analogs of capacitance

A capacitor microphone (1.6.2) converts air pressure variations to varying capacitance, but usually this is immediately converted to a current or voltage analog as charge enters or leaves the capacitor plates. Capacitance is often the basis of other sensors, for example a displacement sensor. A plate of dielectric attached to an object slides between two fixed metal plates. As the object moves → area of dielectric between the plates changes → capacitance changes. Capacitance is measured by connecting the sensor into the circuit of an oscillator. The frequency then depends on the position of the object. The frequency is measured, perhaps by a digital counter and the display indicates the position.

Another technique for measuring frequency is to wire the capacitor into a bridge, with a matching capacitor in the opposite arm (Fig. 3.23). The bridge

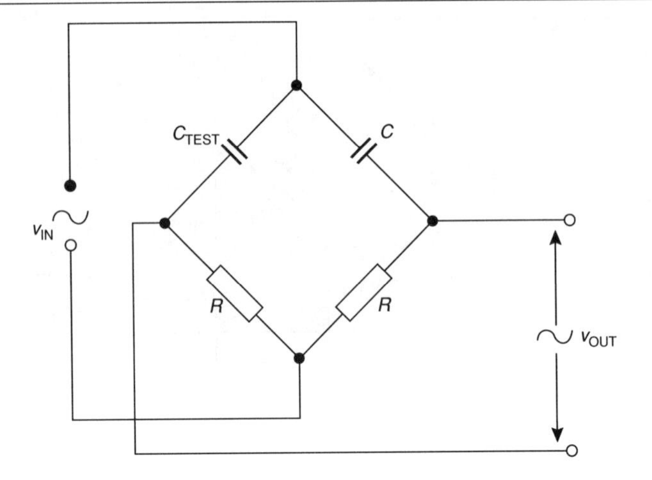

Figure 3.23

is energized with an alternating signal and the amplitude of the output signal varies with the changes in capacitance.

3.5 Analogs of inductance

Inductive analogs (such as produced by inductive displacement sensors) are converted to voltage or current analogs in ways similar to those used for capacitors, that is by wiring them into an oscillator, or by using a bridge with a matching inductor.

3.6 Analog switches

These are logically controlled CMOS devices, also known as *transmission gates*. The switch consists of two transistors (PMOS and NMOS) and an inverting logic gate, fabricated on the same chip (Fig. 3.24). When the control input is at logical high → Q1 and Q2 both switched on → analog signals pass from input to output. Actually signals can pass in either direction. Control input logical low → Q1 and Q2 both off → signals prevented from passing. Depending on the type, the ON resistance of the switch ranges from about 15 Ω to 200 Ω. The OFF resistance is up to 100 MΩ.

Figure 3.24 is a simple ON–OFF switch. Two or more such switches may be connected to produce other switching actions.

Example: Figure 3.25 is a *single-pole-double-throw* switch or *changeover* switch. When the control input is high, S1 is ON and S2 is off. The output

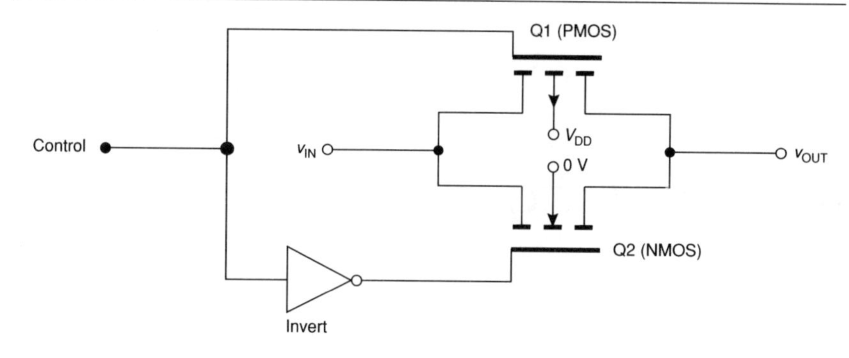

Figure 3.24

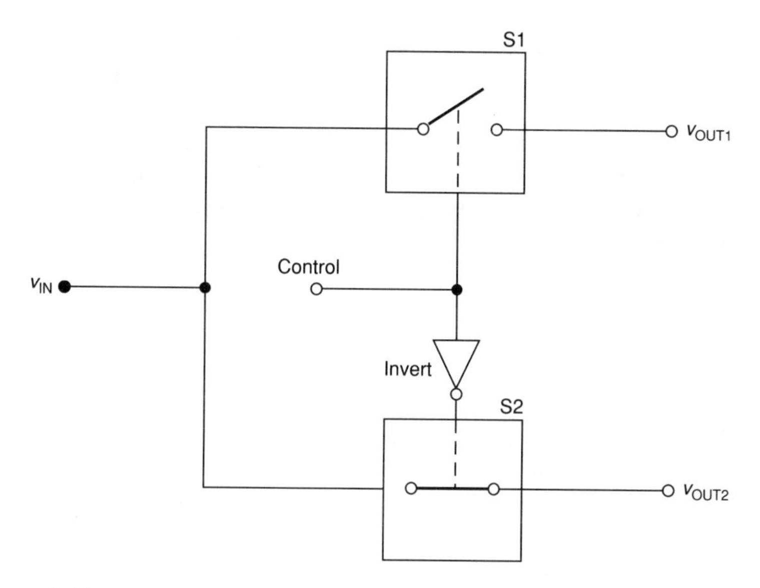

Figure 3.25

signal appears at $v_{OUT\,1}$. Conversely, signals appear at $v_{OUT\,2}$ when the control input is low. Usually several simple ON–OFF switches are constructed on the same chip making it easy to wire up combinations such as Fig. 3.25.

Analog switches are suitable for switching high-frequency signals without attenuation, some types being usable up to 120 MHz.

4 Amplifying analogs

This chapter deals with amplifiers based on discrete BJTs. We begin the chapter with an in-depth study of the common-emitter amplifier. Other BJT amplifiers are studied in less detail, attention being focused mainly on the ways in which they differ from the CE amplifier. Most of the analyses and tests described for the CE amplifier may also be applied to these other types. Amplifiers based on FETs and thermionic valves are described in Chapter 5, where we also describe applications of operational amplifiers.

4.1 Common-emitter amplifier

Use a CE amplifier for medium voltage and current gain, moderate Z_{in} and Z_{out}. These are usually *voltage amplifiers*. They produce phase change of 180° (they are inverting amplifiers). The representative CE amplifier in Fig. 4.1 has a resistor R_E between the emitter of Q1 and ground; we say it has *emitter degeneration*. If R_E is omitted → *grounded emitter* → greater gain but poorer stability (see later). Bias is provided by R_{B1} and R_{B2} acting as a potential divider. Input and output are capacitor coupled to preceding and following circuits, but direct connections can be used if DC levels are compatible.

4.1.1 Circuit operation

This is illustrated in Figs. 4.2 and 4.3, in which the curves have been shifted vertically to bring them all on the same plot. For the present, concentrate on signal shapes, not absolute values. v_{in} is a small signal (amplitude ≈ 10 mV) from a sinusoidal source. It passes across the input capacitor and the oscillations of the base voltage are identical. Taking v_{in}, v_b and other wholly lower-case symbols to mean 'fluctuations in ...':

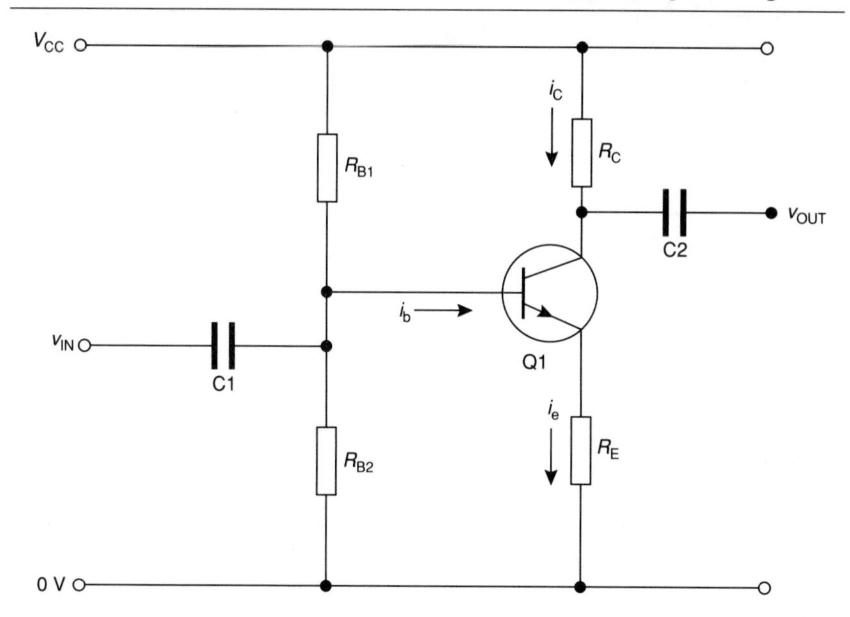

Figure 4.1

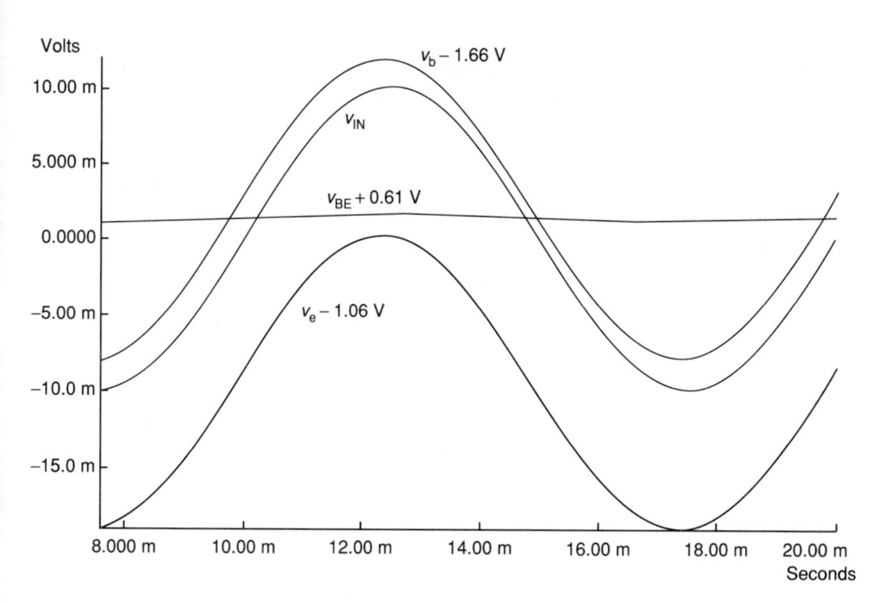

Figure 4.2

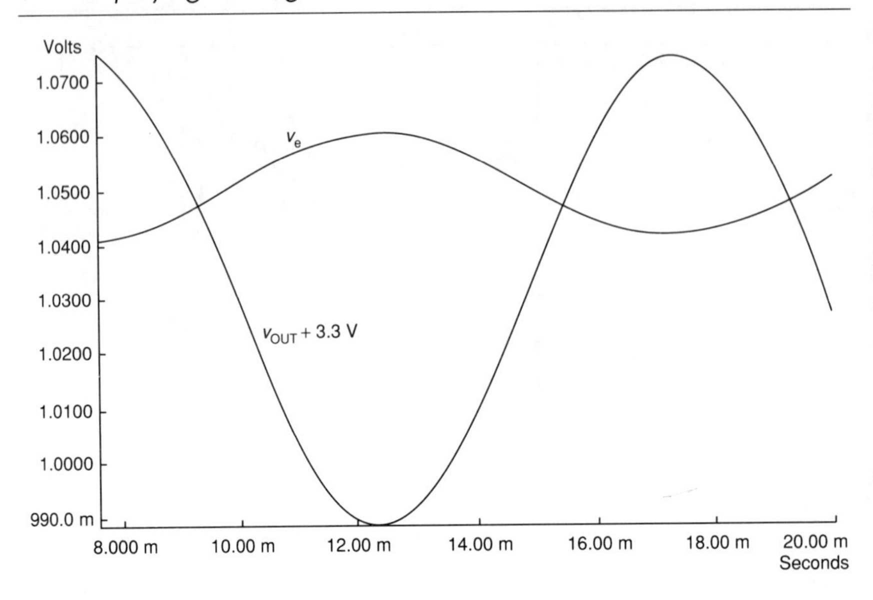

Figure 4.3

$$v_{in} = v_b$$

The base-emitter voltage V_{BE} remains more or less constant at about 0.6 V, so:

$$v_{in} = v_b = v_e$$

This is seen in Fig. 4.2 in which the v_{in}, v_b and v_e curves are more or less parallel. There is a slight phase shift; v_b and v_e lead v_{in} by a small angle. This increases if C1 has a lower value.

The fluctuating v_e produces a fluctuating current through R_E:

$$i_e = v_e/R_E$$

Since $i_e = i_c + i_b$, $i_c = h_{fe}i_b$, and since h_{fe} is large (100+), then $i_e \approx i_c$. For equal current changes, pd changes are proportional to resistance. Also, increase in current causes a fall in collector voltage (Fig. 4.3):

$$v_{out} = -i_c R_C$$

$$\Rightarrow \qquad v_{out} = -v_e R_C/R_E = -v_{in} R_C/R_E$$

$$\Rightarrow \qquad \text{voltage gain} = \frac{v_{out}}{v_{in}} = \frac{-R_C}{R_E}$$

Gain is negative, making the CE amplifier an *inverting amplifier*. The gain depends only on the ratio of R_C to R_E, not on the gain (h_{fe}) of the transistor

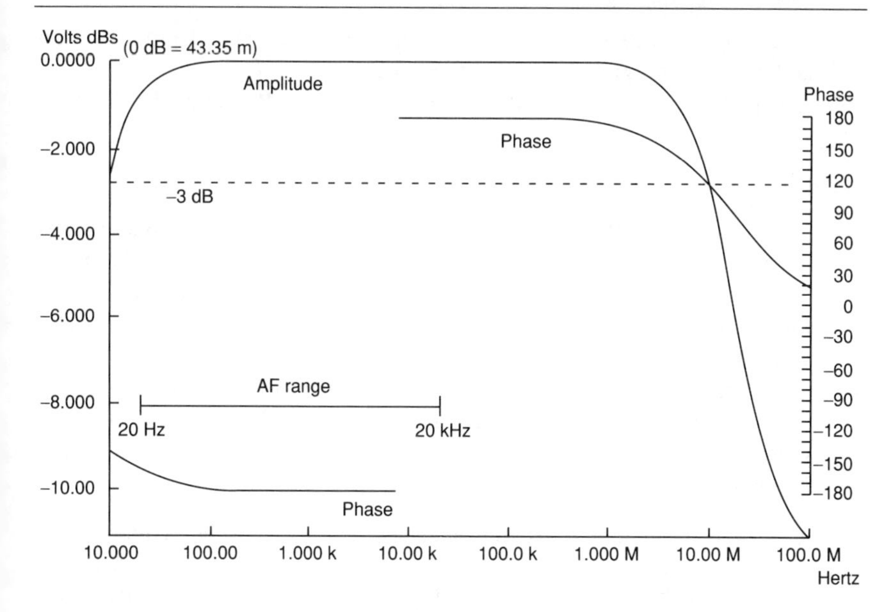

Figure 4.4

itself. Turning to Fig. 4.3, on which v_e is repeated, though with a smaller vertical scale than in Fig. 4.2, V_{out} is 180° out of phase with v_e (and hence with v_{in}, after allowing for the shift at C1), and its amplitude is 4.34 times that of v_e. In the simulation, R_C is 4.5 kΩ and R_E is 1 kΩ, so the calculated gain is 4.5. This small discrepancy is due to i_c actually being slightly less than i_e; also, i_e has r_e (about 25 Ω (C.2.4.2)) in series with it.

The CE amplifier is an example of stabilizing by negative feedback. Increased $v_B \rightarrow$ increased $V_{BE} \rightarrow$ increased $i_E \rightarrow$ increased $v_E \rightarrow$ reduced $V_{BE} \rightarrow$ reduced $i_E \rightarrow$ stability at the expense of reduced gain.

4.1.2 Design stages

As an example, design a CE amplifier (Fig. 4.1) given $V_{CC} = 9$ V, Q1 = BC548, to pass audio signals above 30 Hz.

(1) Decide on a quiescent value for I_C. 1 mA is a generally suitable value (7.2.3) unless there are reasons otherwise.

(2) Decide on a quiescent value for V_C. For most purposes a good value is $V_{CC}/2$ which allows V_{out} to swing freely in either direction without clipping or other distortion. In this example $V_{CC}/2 = 4.5$ V.

(3) Calculate R_C. $R_C = \dfrac{V_{CC}/2}{I_C} = \dfrac{4.5}{0.001} = 4.5\,\text{k}\Omega$

(4) Decide on a value for R_E to give required voltage gain. If required gain is 4.5, $R_E = R_C/\text{gain} = 4500/4.5 = 1\,\text{k}\Omega$.
(5) Assuming $I_E = I_C$, calculate quiescent $V_E = I_C \times R_E = 0.001 \times 1000 = 1\,\text{V}$.
(6) $V_{BE} = 0.6\,\text{V} \rightarrow V_B = V_E + 0.6 = 1 + 0.6 = 1.6\,\text{V}$.
(7) To provide V_B of 1.6 V, the ratio R_{B1}:R_{B2} is 7.4:1.6 $\rightarrow$ 4.625:1. If $I_C = 1\,\text{mA}$ and $h_{FE} = 100$, $I_B = 10\,\mu\text{A}$. The R_{B1}/R_{B2} potential divider must have at least 10 times I_B (= $100\,\mu\text{A}$) flowing through it. Maximum total resistance $= 9\,\text{V}/100\,\mu\text{A} = 90\,\text{k}\Omega$. A few trial calculations suggest $R_{B1} = 56\,\text{k}\Omega$ and $R_{B2} = 12\,\text{k}\Omega$ (preferred, as these are standard resistor values).
(8) On the input side, R_{B1} and R_{B2} in parallel (= $9.9\,\text{k}\Omega$), together with C1, form a high-pass filter (6.1.2). Calculate $C = 1/2\pi Rf = 536\,\text{nF}$. This gives $-3\,\text{dB}$ attenuation at 30 Hz. To obtain zero attenuation at 30 Hz, increase C about 10 times to $4.7\,\mu\text{F}$.

As a check on the calculations, run analyses to find quiescent voltages and currents (that is, no-signal values). Key voltages found are: $V_C = 5.132\,\text{V}$, $V_E = 870.8\,\text{mV}$. These are respectively higher and lower (slightly) than expected. Key currents are $134.3\,\mu\text{A}$ through R_{B1}, $123.1\,\mu\text{A}$ through R_{B2} and $859\,\mu\text{A}$ through R_C. These figures show that $I_B = 134.3 - 123.1 = 11.2\,\mu\text{A}$, which is verging on 10% of the main current through the potential divider. The divider has too high a Z_{out}, which is restricting I_B and making I_C too small. In turn, this adversely affects values of V_E and V_C.

Select another pair of values for R_{B1} and $R_{B2} \rightarrow R_{B1} = 20\,\text{k}\Omega$ and $R_{B2} = 4.7\,\text{k}\Omega$. Now $V_C = 4.330\,\text{V}$ and $V_E = 1.051\,\text{V}$, $I_C = 1.038\,\text{mA}$ and $I_E = 1.051\,\text{mA}$. These figures are much closer to specification. Summarizing: $R_{B1} = 20\,\text{k}\Omega$, $R_{B2} = 4.7\,\text{k}\Omega$, $R_C = 4.5\,\text{k}\Omega$, $R_E = 1\,\text{k}\Omega$, $\text{C1} = 4.7\,\mu\text{F}$, $\text{C2} = 1\,\mu\text{F}$.

Some results of analysing this circuit with the above component values have already appeared in Figs. 4.2 and 4.3. v_{in}, v_b, and v_e all have an amplitude of 10 mV, though their DC levels differ. v_{in} is centred on 0 V, since the generator has one terminal connected to this line. On the other side of C1, v_b is centred on +1.662 V. V_{BE} is an almost constant $0.62\,\text{V} \rightarrow v_e$ is centred on 1.05 V.

Summarizing the characteristics of this amplifier:

Voltage gain = theoretically 4.5 (approx.), in simulation 4.34.
Current gain: Measuring amplitudes of input and output signals shows current gain = 3.7.
Input impedance: This is the resistance of R_{B1}, R_{B2} and the impedance looking into the base of Q1. The base impedance is $h_{fe} \times R_E = 100\,\text{k}\Omega$.

$$1/Z_{in} = 1/20\,\text{k}\Omega + 1/4.7\,\text{k}\Omega + 1/100\,\text{k}\Omega$$

$$\Rightarrow \quad Z_{in} = 3.67\,\text{k}\Omega$$

Output impedance: This is the resistance of R_C in parallel with the impedance looking into the collector of Q1. But the latter is several megohms so ignore it. $Z_{out} = 4.5\,k\Omega$.

4.1.3 Frequency response

The result of a computer frequency analysis (Fig. 4.4) is a *Bode plot.* Given a fixed signal input, amplitude 10 mV, the plot shows the output amplitude and phase as frequency is swept over the range from 10 Hz to 100 MHz. It is plotted on logarithmic scales: frequency is marked in decades, amplitude is marked in decibels. This allows a wide range of frequencies and responses to be shown on a single diagram. In Fig. 4.4, the output amplitude is seen to be a steady maximum between 100 Hz and 10 MHz where the curve is horizontal. This is marked on the vertical scale as 0 dB, equivalent to an output of 43.35 mV, representing a voltage gain of 4.335. Gain is reduced at lower frequencies but not by much. The dashed line indicates the −3 dB or *'half-power'* level; response falls to this level at 10 Hz. For almost the whole audio range, 20 Hz to 20 kHz the amplifier delivers full power. Its response is suitable for audio applications and meets the design specification except for a small reduction (less than 1 dB) below 100 Hz. If it is essential for response to be completely flat from 30 Hz upward, some further increase of C1 (and possibly C2) is required. The fall in response at low frequencies is due to the filtering action of C1 and C2 and various resistors. Response falls again at high frequencies (above 1 MHz) due to capacitance at the junctions of Q1 (4.3.3).

Figure 4.4 also plots the phase response. Over most of the frequency range the output lags approximately 180° behind the input, which is characteristic of an inverting amplifier. There is a sudden apparent change to +180° around 10 kHz, due to the curve having reached the lowest scaled value. In fact, the lag continues to increase slightly, increasing beyond −180° with increasing frequency. Beyond 1 MHz, lag increases more rapidly. A horizontal or gently sloping phase curve is an advantage; signals of all audio frequencies show similar phase lag, so minimizing distortion. By contrast, a steep phase curve means that signals of different frequencies suffer different phase lags, and become out of step, leading to distortion.

The effect of increasing the value of C1 is quickly investigated by running a series of frequency response analyses. In Fig. 4.5 the analysis is carried out four times while the value of C1 is stepped over the range 5 μF to 35 μF. Stepping could have been linear (5, 15, 25, 35) but in this case a more useful spread of curves is obtained by stepping logarithmically, that is, the logarithms of the values are equally spaced:

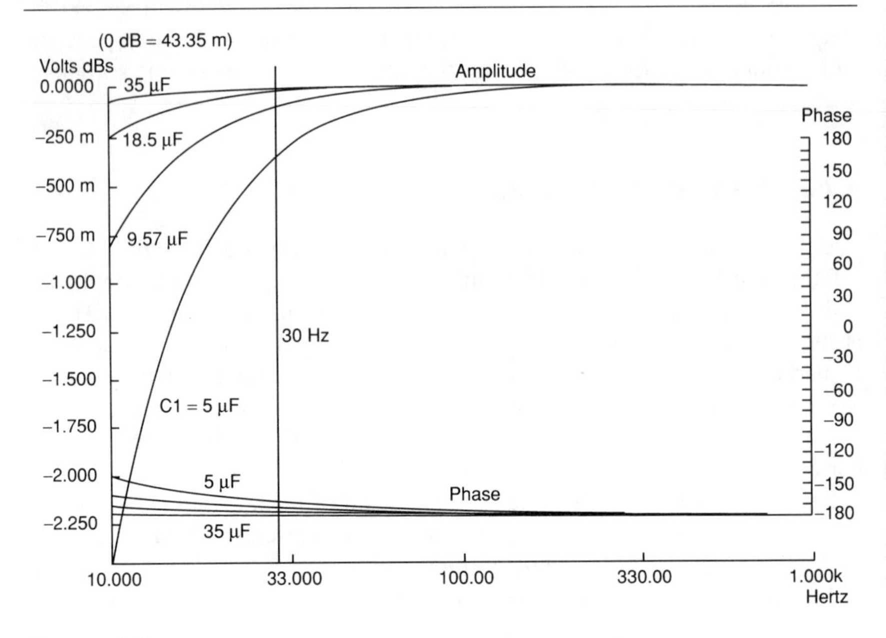

Figure 4.5

Step	C	$\log C$
1	5.00	0.6990
2	9.57	0.9807
3	18.30	1.2624
4	35.00	1.5441

From the Bode plot, an 18 μF capacitor results in an attenuation of only 26.1 mdB at 30 Hz, a power loss of only 0.3%, which is negligible.

4.1.4 Temperature

The tempco of V_{BE} is -2 mV/°C at room temperature and this causes the amplifier output to drift as room temperature changes or as the transistor becomes heated due to current passing through it. Figure 4.6 shows v_{out} plotted at temperatures ranging from -50°C to 100°C in steps of 25°C. Amplitude and phase are unaffected, but the DC level rises with increasing temperature, a total drift of 887 mV.

4.1.5 Distortion

With a pure sinusoidal input, the output signal seen on an oscilloscope or computer screen may appear to be sinusoidal, but the eye is not an infallible

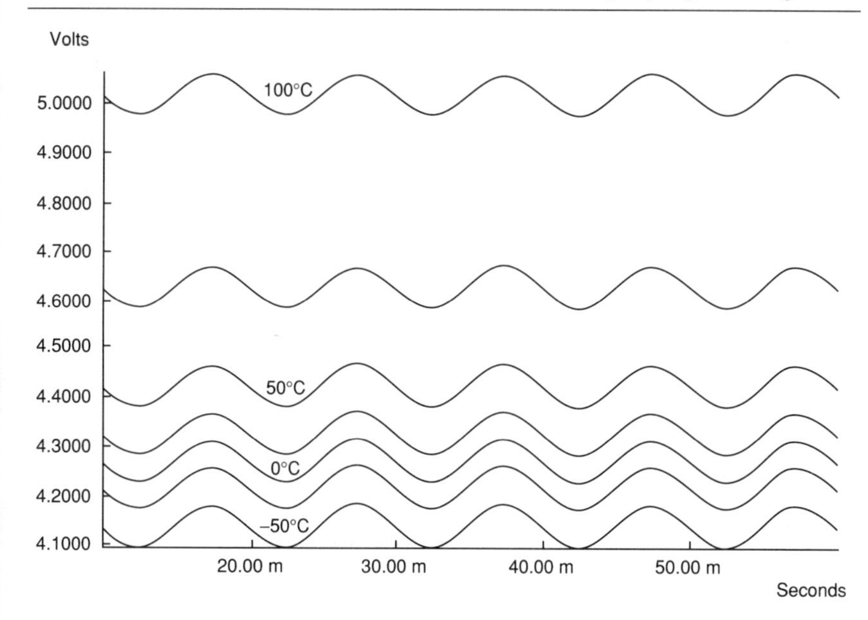

Figure 4.6

judge of its precise shape. The signal may be slightly distorted. A quick way to assess distortion is to use a spectrum analyser or to run a Fourier analysis on a simulator. In our test circuit (Fig. 4.1) the frequency spectrum of the input shows a single line at the fundamental frequency (2.6), proving that the voltage source produces a pure sine wave. If there is any distortion, the analysis of v_{out} (Fig. 4.7) shows harmonics, but none are seen here, so distortion is absent or at least insignificant. The analysis has a resolution of 10 Hz and shows a continuous spectrum from 10 Hz to 150 Hz, decreasing more or less steadily, with a prominent peak at 100 Hz, the signal frequency. The low-level continuous spectrum may be taken as flicker noise (7.2.3), but some of it could be mathematical 'noise' resulting from approximations and roundings in the calculations. The magnitude of the 100 Hz peak is 43.08 mV, which accounts for 99.4% of the amplitude of v_{out}, known to be 43.35 mV. The rest is noise.

For a more quantitative assessment of distortion, some simulators calculate the amplitudes of the lower harmonics with reference to the amplitude of the fundamental. If v_{in} consists of a mixture of two sinusoidal signals with frequencies f_1 and f_2, these result in *intermodulation distortion* at frequencies $(f_1 + f_2)$ and $(f_1 - f_2)$. The amplitudes of these are expressed in relation to the fundamental amplitude. Finally, there is intermodulation distortion between the fundamental and its harmonics. There are hundreds of ways in

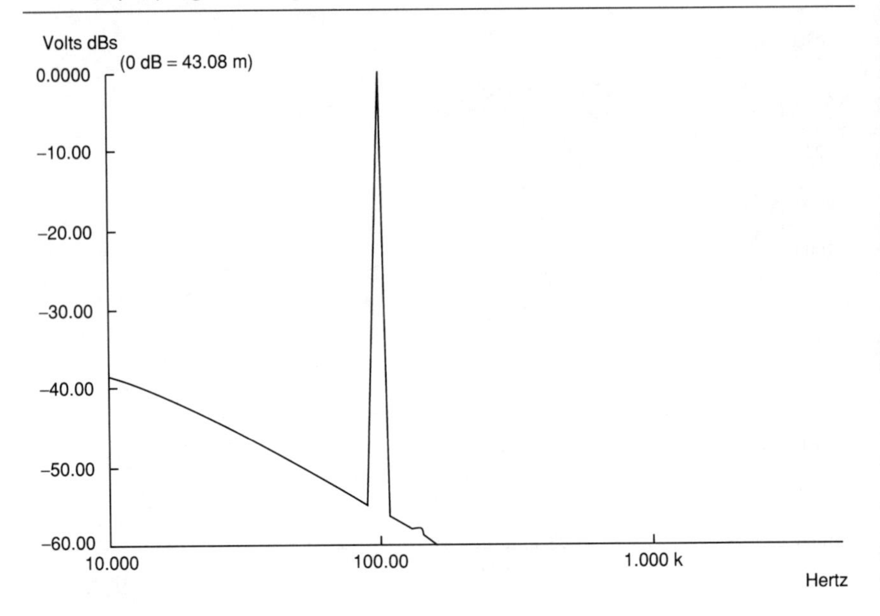

Figure 4.7

which harmonics and intermodulation frequencies may combine to produce distortion but most are insignificant and only those of the lowest frequencies are calculated.

Another measure of distortion is *total harmonic distortion*:

$$\mathrm{THD} = \frac{\sqrt{(a_2^2 + a_3^2 + \cdots + a_n^2)}}{a_1} \times 100$$

where a_1 is the amplitude of the fundamental and a_n is the amplitude of the nth harmonic.

4.1.6 Grounded emitter

If voltage gain $= -R_C/R_E$ (4.1.1), and R_C is determined by the need to set a suitable DC level for v_{out}, gain can be controlled by adjusting R_E. For maximum gain, omit R_E and ground the emitter directly. This leaves r_e between the base-emitter junction and 0 V. If $I_C = 1\,\mathrm{mA}$, then $r_e \approx 25\,\Omega$ (C.2.4.2) $\rightarrow$ gain $= -R_C/r_e = -4500/25 = -180$.

Figure 4.8 shows the output of a *grounded-emitter* C_E amplifier based on Fig. 4.1. R_E has been omitted entirely. R_{B1} and R_{B2} have been changed to $22.96\,\mathrm{k}\Omega$ and $1.74\,\mathrm{k}\Omega$, to maintain the values of V_{BE} and I_C as before. The input signal is unchanged. Gain is -161, of the same order as predicted. The output appears to be a sinusoid but the peaks are more rounded than the

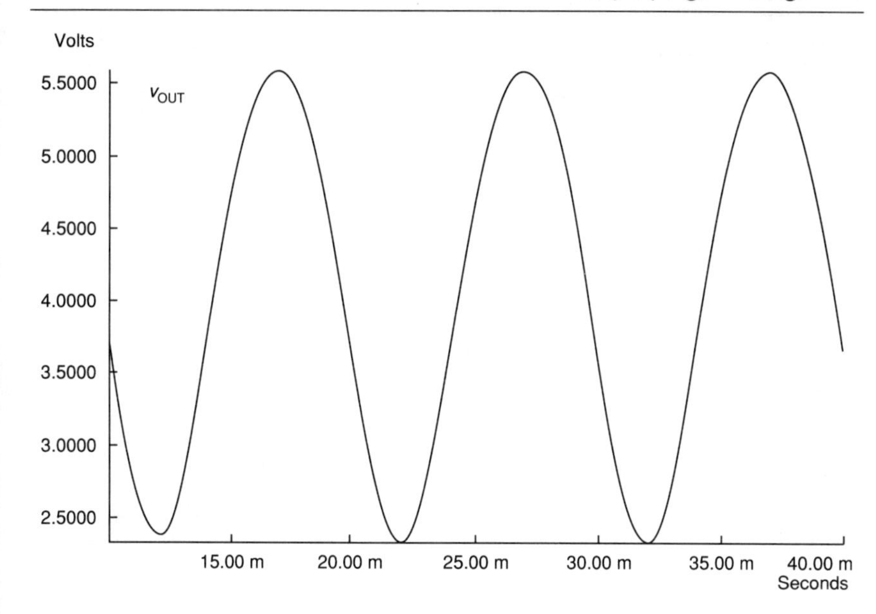

Figure 4.8

troughs. This is because the value of I_C, and hence the gain, varies with the signal; it is not constant at 1 mA as assumed above. Consequently on the peaks, when I_C is relatively low, gain is low while, in the troughs, I_C is relatively high, gain is high.The effect is to round the peaks and sharpen the troughs.

A Fourier analysis (Fig. 4.9) of the grounded-emitter amplifier output demonstrates that the output signal is distorted by the presence of harmonics at 200 Hz and 300 Hz (compare Fig. 4.7 in which no harmonic distortion is evident). Reading the amplitude of the second harmonic as −20 dB, the ratio of its amplitude and that of the fundamental is $\text{antilog}_{10}(-20/10) = 0.01$, or 1%. For the third harmonic, at −50 dB, the ratio is $\text{antilog}_{10}(-5) = 1 \times 10^{-5}$, or 0.001%. Thus high gain has been obtained at the cost of appreciable, though not excessive, harmonic distortion.

The input impedance is lower in a grounded-emitter amplifier because omitting R_E reduces the impedance looking into the base of Q1. For $I_C = 1$ mA, this impedance is 2.5 kΩ. With the bias resistor values quoted above:

$$1/Z_{in} = 1/22.96\text{k} + 1/1.74\text{k} + 1/2.5\text{k}$$

$$\Rightarrow \quad Z_{in} = 972\,\Omega$$

Compare with 4.2.1. Further, since r_e varies with I_C, the input impedance varies with the signal. If the signal source has low output impedance this leads to serious distortion.

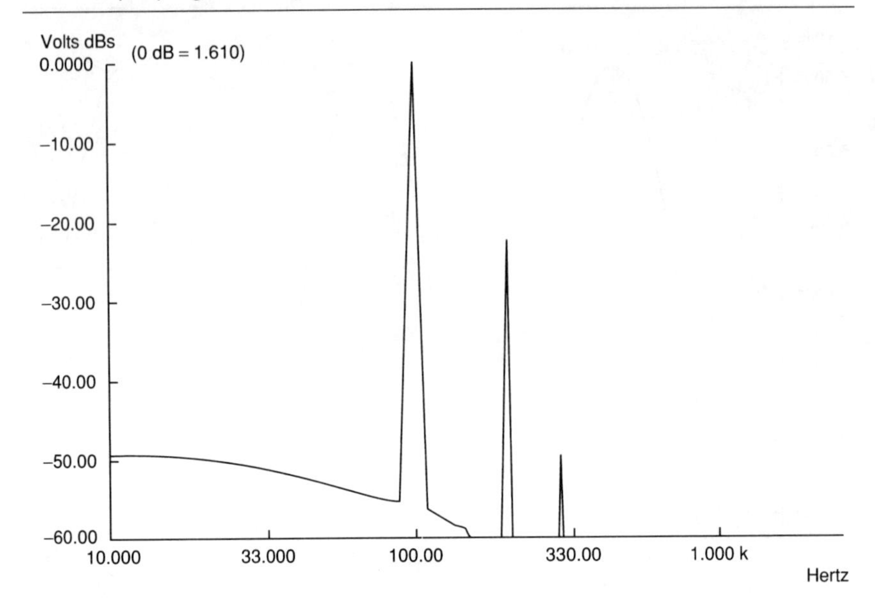

Figure 4.9

V_{BE} varies with temperature (C.2.6) which makes it difficult to bias the amplifier accurately. Without R_E to provide feedback, a small change in temperature has a relatively larger effect. A rise as small as 10°C can cause the transistor to saturate, leading to a clipped waveform.

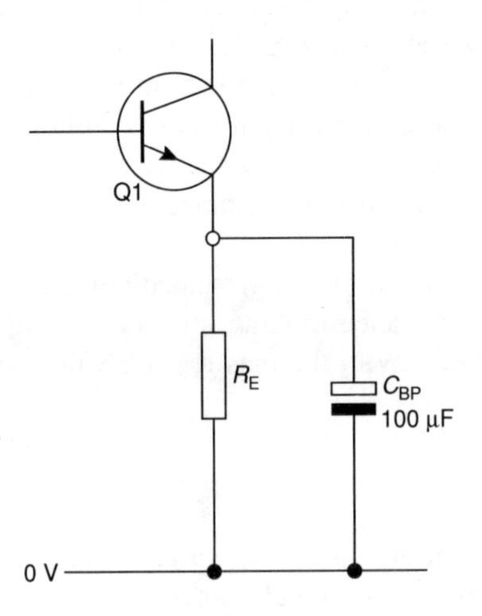

Figure 4.10

4.1.7 Bypass capacitor

In Fig. 4.1 the feedback action of R_E improves circuit stability, but results in reduced gain. This is because the signal fed back includes not only the DC component of V_E due to h_{FE}, and drifts in V_E due to temperature changes, but also the changes resulting from fluctuations in the magnitude of the signal. If a capacitor is connected across R_E (Fig. 4.10), it *bypasses* the signal to ground, so removing it partly or wholly from the feedback loop without affecting feedback of DC and drifts. In effect, the amplifier has a grounded emitter as far as the signal is concerned, but retains its independence of h_{fe}, and its temperature stability. Also, it is much easier, using resistors of standard values, to obtain a bias voltage within a few per cent of, say, 1.6 V or more, than it is to bias within the same percentage of 0.6 V. In the amplifier amended as in Fig. 4.10, the gain increases to 130 (compared with 4.33 when the capacitor is not there). This is of the order reported for the grounded emitter amplifier (4.1.6). However, bypassing does not remove the effect of the variations of I_C on r_e, so v_{out} shows harmonic distortion, almost equal to that in Fig. 4.9.

A compromise is reached by bypassing only part of the capacitance (Fig. 4.11). The gain is appreciably larger than obtained with a degenerated emitter but not as high as that attainable with a fully bypassed emitter.

Example: Gain is to be in the order of 80, using standard resistors. Given $R_C = 4.5\,k\Omega$ as before, $R_E = 4.5k/80 = 56.25\,\Omega$. Allowing for 25 Ω due to r_e, $R_{E1} = 31.25\,\Omega$. Make this 30 Ω. Add R_{E2}, typically this is enough to take the total emitter resistance to $R_C/10$. Let $R_{E2} = 450\,\Omega$. Total emitter resistance = 480 Ω. If quiescent $I_E = 1\,mA$, $V_E = 0.48\,V$. $V_B = 0.48 + 0.6 = 1.08\,V$.

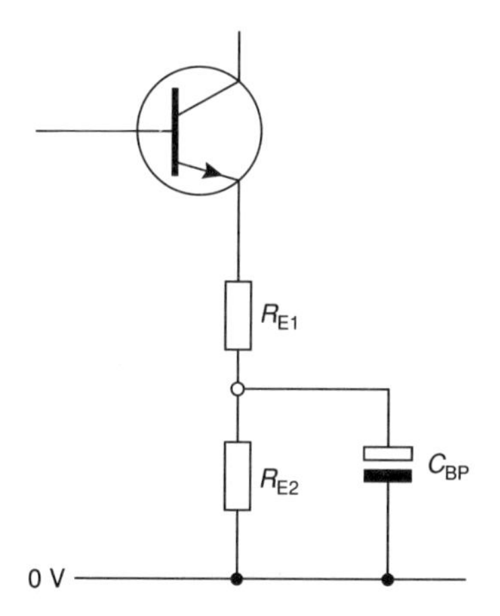

Figure 4.11

Biasing for 1.08 V, select $R_{B1} = 22\,k\Omega$ and $R_{B2} = 3\,k\Omega$ (when using a simulator, a quick check for DC quiescent conditions confirms that V_B is adjusted correctly). Make $C_{CB} = 100\,\mu F$ as before. Running the simulation with these component values gives a gain of 68. Further adjustments could set this to 80 if required. Running a Fourier analysis shows a peak at 100 Hz and a second peak at 200 Hz. But the amplitude of this is −35 dB, compared with −20 dB in Fig. 4.9. Also, there is no 300 Hz peak visible on the plot (made with −60 dB as minimum). The circuit has moderate gain of the required order and considerably less distortion than the circuit with the fully bypassed emitter.

When choosing a bypass capacitor, it should have a low impedance compared with the un-bypassed part of the emitter resistance, at the lowest signal frequency of interest. In these examples the lowest frequency has been taken to be 30 Hz ($\omega = 188$). If the un-bypassed resistance is 55 Ω, $C = 1/\omega X_C = 96\,\mu F$. The capacitor should be 96 μF or more, and the value selected is 100 μF.

Figure 4.12 describes a simple way of stabilizing V_B in a grounded-emitter amplifier. Instead of supplying I_B from V_{CC}, it is supplied from V_C. This introduces feedback to make the amplifier less dependent on h_{FE}. If h_{FE} is high → I_C larger → V_C lower → I_B reduced → I_C reduced. This allows h_{FE} to vary over an appreciable range without unduly affecting the quiescent operating conditions.

$$V_C = V_{CC} - I_C R_C$$

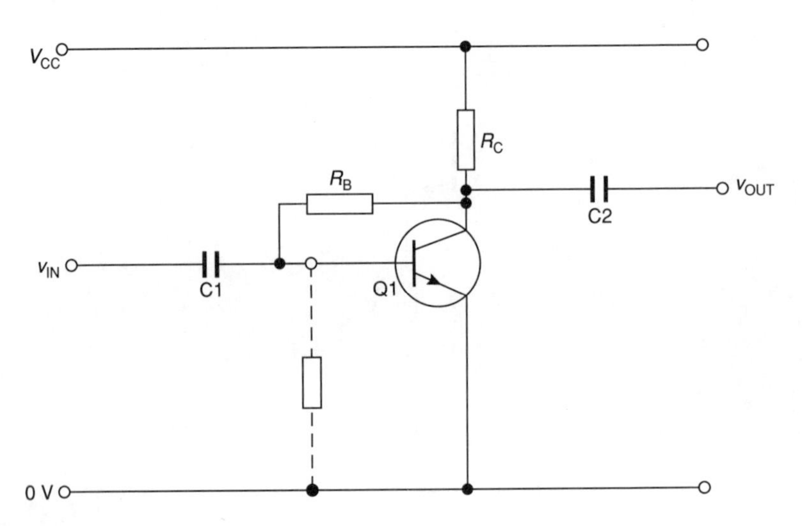

Figure 4.12

But $I_C = I_B h_{FE}$, so:

$$V_C = V_{CC} - I_B h_{FE} R_C$$

But $I_B = (V_C - 0.6)/R_B$, so:

$$V_C = V_{CC} - (V_C - 0.6)\, h_{FE} R_C / R_B$$

$$\Rightarrow \qquad V_C = \frac{V_{CC} + 0.6 h_{FE} R_C / R_B}{1 + h_{FE} R_C / R_B}$$

Although h_{FE} has an effect on the quiescent value of V_C, values of h_{FE} ranging from 100 to 400 or more still allow the amplifier to function without clipping, provided that input signals of moderate amplitude are applied. Harmonic distortion occurs, similar to that shown in Fig. 4.9.

4.2 Common-collector (emitter follower) amplifier

A CC amplifier has voltage gain just less than 1, high current gain, high Z_{IN} and low Z_{OUT}. CC amplifiers are used as buffers for impedance matching (3.1.3), and also as power amplifiers.

4.2.1 Circuit operation

A CC amplifier differs from the CE (Fig. 4.1) in having the collector directly connected to the V_{CC} rail, and always having an emitter resistor (Fig. 4.13). Output is taken from the emitter. Its action relies on the fixed voltage drop V_{BE} across the forward-biased base-emitter junction. Given that v_b and v_e are small fluctuations around the quiescent voltages V_B and V_E:

$$v_{in} = v_b = v_e$$

The 0.6 V drop between base and emitter means that V_E is 0.6 V lower than V_B. Often we bias the base so that $V_E = V_{CC}/2$, to allow maximum output swing. The internal emitter resistance r_e of the transistor, and the emitter resistor R_E form a potential divider (Fig. 4.14):

$$v_{out} = v_e \times \frac{R_E}{r_e + R_E} = v_{in} \times \frac{R_E}{r_e + R_E}$$

$$\Rightarrow \qquad \text{Voltage gain} = \frac{v_{out}}{v_{in}} = \frac{R_E}{r_e + R_E}$$

r_e depends on I_C (C.2.4.2).

Example: If $I_C = 1\,\text{mA} \rightarrow r_e = 25\,\Omega$. If $R_E = 4.5\,\text{k}\Omega$:

$$\text{Voltage gain} = \frac{4500}{4525} = 0.994$$

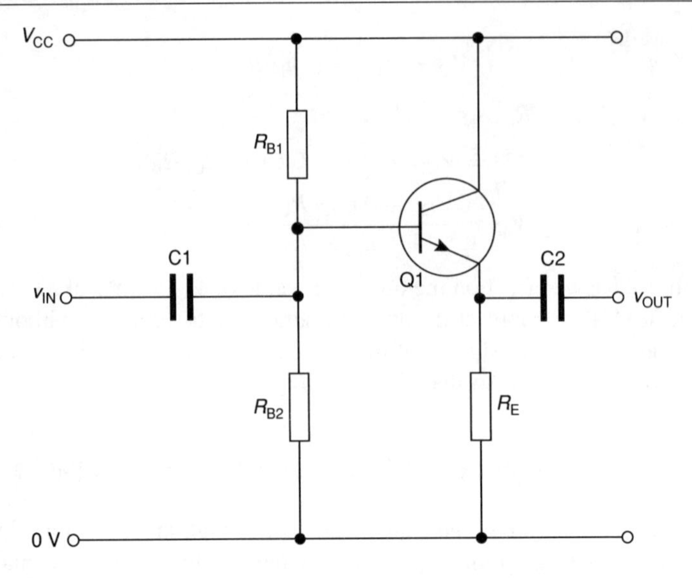

Figure 4.13

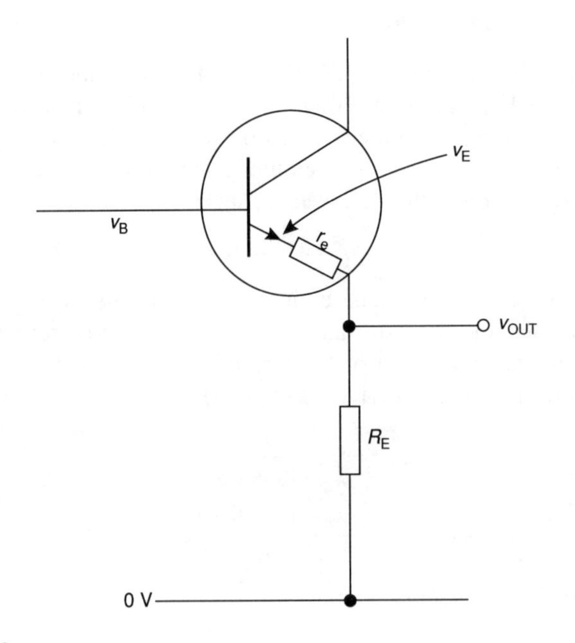

Figure 4.14

Voltage gain is almost 1, so an alternative name for the circuit is *emitter follower*.

4.2.2 Design stages

As an example, design a CC amplifier (Fig. 4.13) given $V_{CC} = 6\,V$, Q1 = BC548, to pass signals above 1 kHz.

(1) Decide on a quiescent value for I_C — 1 mA is a generally suitable value unless there are reasons otherwise.
(2) Decide on a quiescent value for V_E. For most purposes a good value is $V_{CC}/2$ which allows v_{out} to swing freely in either direction without clipping or other distortion. In this example $V_{CC}/2 = 3\,V$.
(3) Calculate $R_E \cdot R_E = (V_{CC}/2)/I_C = \dfrac{3}{0.001} = 3\,k\Omega$.
(4) $V_{BE} = 0.6\,V \rightarrow V_B = V_E + 0.6 = 3 + 0.6 = 3.6\,V$.
(5) To provide V_B of 3.6 V the ratio $R_{B1}{:}R_{B2}$ is 2.4:3.6 → 0.67:1. If $I_C = 1\,mA$ and $h_{FE} = 100$, $I_B = 10\,\mu A$. $10 \times 10\,\mu A = 100\,\mu A$ (4.1.1.2). Maximum total resistance $= 96\,V/100\,\mu A = 60\,k\Omega$. 24 kΩ and 36 kΩ would give the required V_B, but take half these values, 12 kΩ and 18 kΩ to give greater stability to V_B.
(6) On the input side, R_{B1} and R_{B2} in parallel (= 7.2 kΩ), together with C1, form a high-pass filter. Calculate $C = 1/2\pi Rf = 22.1\,nF$. This gives −3 dB attenuation at 1 kHz, so increase C about 10 times to 220 nF.
(7) On the output side, R_E and C2 form a high-pass filter. Calculate C = 53 nF. Make $C2 = 680\,nF$ or 1 μF.

A DC quiescent analysis shows that $V_B = 3.511\,V$, $V_E = 2.902\,V$ and $I_C = 0.967\,mA$, all of which conform to the design specification reasonably well. It might be desirable to bias the base to a slightly higher voltage, so increasing V_E and I_C slightly. With the values quoted and with v_{in} a 1 kHz signal, amplitude 2.5 V, a transient analysis of the amplifier gives Fig. 4.15. v_{out} swings freely between 0.47 V and 5.38 V, with amplitude 2.455 V, and a gain of 0.982. v_{out} is roughly in phase with v_{in}, but measurements on the graphs show v_{out} to lead v_{in} by 6.3°.

Input impedance is calculated as for the CE amplifier: the parallel combination of the two biasing resistors and the impedance looking into the base of Q1. Assuming $h_{fe} = 100$:

$$1/Z_{in} = 1/12k + 1/18k + 1/300k$$

$$\Rightarrow \qquad Z_{in} = 7.03\,k\Omega$$

This is for the amplifier alone, its output unconnected to a subsequent circuit. When connected, the input impedance of the connected circuit must be

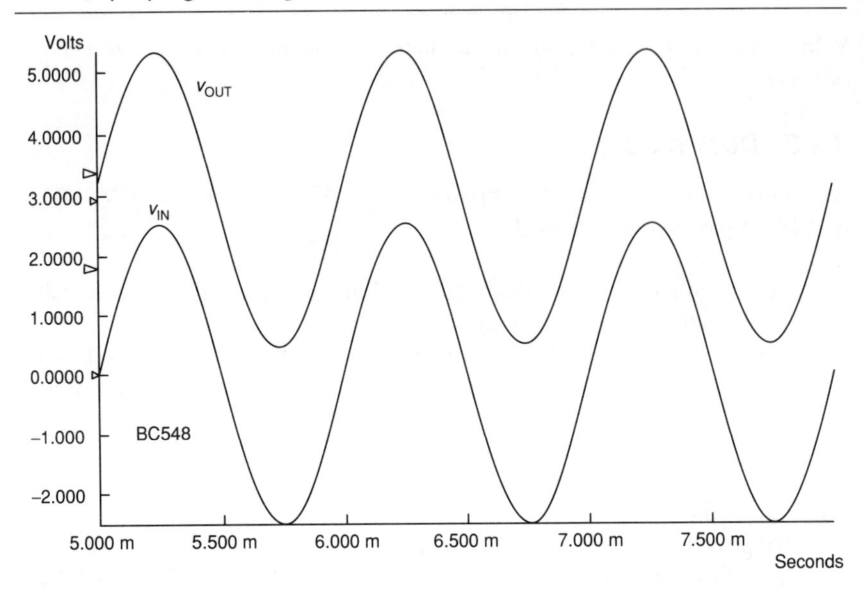

Figure 4.15

accounted for, in parallel with R_E. For example: if the subsequent circuit has input impedance $10\,k\Omega$:

$$10\,k\Omega \parallel 3\,k\Omega = 2.3\,k\Omega$$

$$\Rightarrow \quad 1/Z_{in} = 1/12\,k\Omega + 1/18\,k\Omega + 1/230\,k\Omega$$

$$\Rightarrow \quad Z_{in} = 6.98\,k\Omega$$

In both cases, Z_{in} is considerably reduced by the biasing resistors which, to provide ample base current, usually can be no larger than a few tens of kilohms. To increase Z_{in} (though at the expense of stability), omit R_{B2}. Now the value of R_{B1} becomes that which produces a suitable I_B. In this example, if $V_B = 2.4\,V$ and $I_B = I_C/h_{FE} = 10\,\mu A$, then $R_{B1} = 2.4/10\,\mu A = 240\,k\Omega$. $Z_{in} = 1/240\,k\Omega + 1/230\,k\Omega = 117\,k\Omega$. In general, the input impedance of a CC amplifier is high provided that only one biasing resistor is used. Bootstrapping (4.2.5) is used to combine high Z_{in} with superior stability.

Output impedance: ignore R_E because $R_E \gg r_e$ so $Z_{out} \approx r_e$. If $I_C = 1\,mA$, $Z_{out} = 25\,\Omega$, which is low. Taking into account the output impedance Z_{SOURCE} of a previous stage, the source of v_{in}, $Z_{out} = \dfrac{Z_{SOURCE} + r_e}{h_{FE} + 1}$. For example, if $Z_{SOURCE} = 500\,\Omega$ and $h_{FE} = 100$, $Z_{out} \approx 5 + 25 = 30\,\Omega$. In general Z_{out} of a CC amplifier is very low.

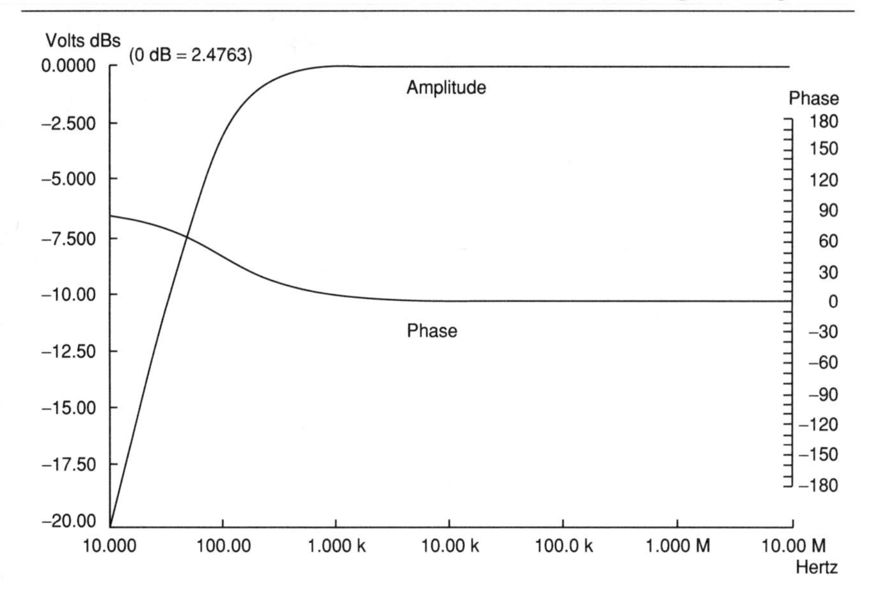

Figure 4.16

4.2.3 *Frequency response*

A Bode plot of the response of the CC amplifier of Fig. 4.12 reveals an interesting difference between CC and CE amplifiers. The response reaches its full amplitude (Fig. 4.16) at just below 1 kHz (as designed) then, in contrast to Fig. 4.4, remains at full amplitude up to at least 10 MHz. The good high-frequency response is due to the small junction capacitances in this connection. The base-emitter junction is forward-biased so capacitance is minimal. More important, the capacitance of the reverse-biased base-capacitor junction is ineffective since the collector is joined directly to the V_{CC} rail, which means there is no Miller effect (4.3.3).

Phase is little affected by frequency. v_{out} has a small phase lead below 1 kHz, but this is reduced to a few degrees when 1 kHz is reached and soon becomes zero.

4.2.4 *Distortion*

Because gain depends on r_e, which depends on I_C, gain varies with v_{in}. A Fourier analysis shows the frequency spectrum to be similar to that in Fig. 4.9, but the harmonic peaks are notably lower. The second harmonic (2 kHz) reaches only to -50 dB, while the third harmonic is only -60 dB. These results are for a signal swinging close to V_{CC} and 0 V, confirming that the CC amplifier is less subject to distortion than the CE amplifier.

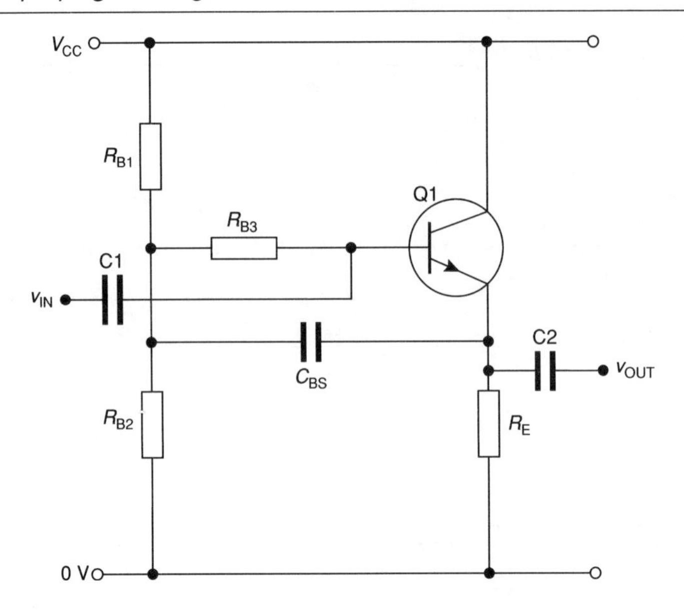

Figure 4.17

4.2.5 Bootstrapping

Bootstrapping makes use of negative feedback of v_{out} to the biasing chain. The circuit has an additional resistor R_{B3} in the biasing chain, and a bootstrap capacitor C_{BS} (Fig. 4.17). v_{in} first appears in the amplifier at the junction between R_{B3} and the base of Q1, having passed through C1. At the same time, given $v_{out} \approx v_{in}$, v_{in} is fed back through C_{BS} to the junction of R_{B1}/R_{B2}. In this way, R_{B3} is subjected to more-or-less identical signals at both ends and little current flows through it. In effect, but for signal frequencies only, it acts as a high impedance between the amplifier input terminal and R_{B1}/R_{B2}. In short, R_{B3} dominates the impedance of the biasing resistors. When calculating input impedance to signals, R_{B1} and R_{B2} may be ignored. Z_{in} is the parallel sum of R_{B3} (large) and the impedance of Q1 looking into the base (also large). Input impedance for signals is high. By contrast, DC levels do not pass through C_{BS}, so there is no DC feedback. Quiescent levels are stabilized, and quiescent base current flows through R_{B3} in the usual way.

4.2.6 Darlington pair

The Darlington pair is a two-stage CC amplifier, ideal for detecting very small currents. The emitter current from the first transistor becomes the base current of the second transistor. If each transistor has current gain h_{FE}, the current

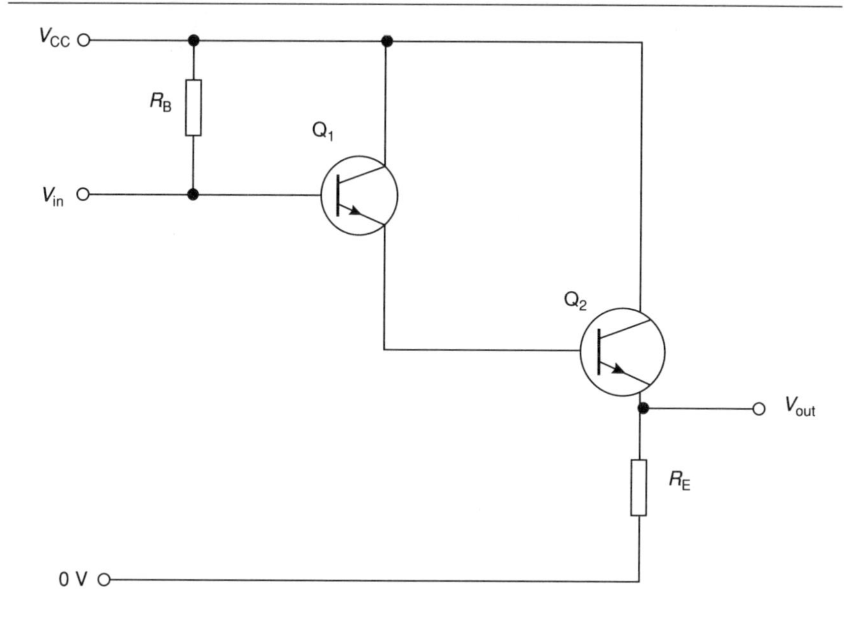

Figure 4.18

gain of the pair is h_{FE}^2. For example, if $h_{FE} = 100$, the pair have a gain of 10^4. If the amplifier is biased by a single resistor (Fig. 4.17), the base current required is so small that a resistor of many megohms may be used. Further, the impedance looking into the base of Q1 is h_{FE} times the impedance looking into the base of Q2, which is in turn h_{FE} times the impedance of R_E (and any load connected in parallel with it).

Example: If in Fig. 4.18, V_{CC} is 9 V and R_E is 4.5 kΩ, to give quiescent current 1 mA with $v_{out} = 4.5$ V, and if $h_{FE} = 100$, I_B of Q1 needs to be $(1 \times 10^{-3})/10^4 = 100$ nA. With two base-emitter voltage drops, V_B of Q1 is $4.5 + 1.2 = 5.7$ V. The voltage drop across R_B is 3.3 V so $R_B = 3.3/(100 \times 10^{-9}) = 33$ MΩ. The impedance looking into the base of Q1 is $4500 \times 10^4 = 45$ MΩ. Total $Z_{in} = 19$ MΩ. Compare with 4.2.1.

4.2.7 Power amplification

The second important use for CC amplifiers is as power amplifiers. Voltage amplification is only 1 but, with high Z_{in} and low Z_{out}, CC amplifiers are able to supply far more current than they take in and high power amplification is attainable. A typical application of CC power amplifiers is for driving loudspeakers. In its simplest form, a *Class A* amplifier is a CC amplifier in which a

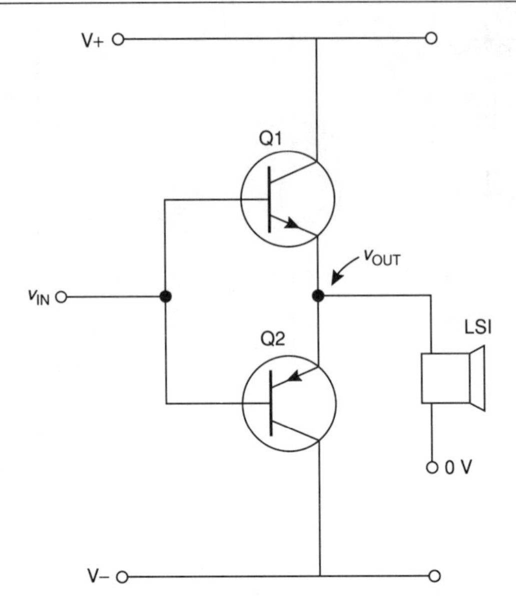

Figure 4.19

loudspeaker replaces R_E. It can be designed to pass high peak currents, using the procedure described in 4.2.2. We might design it with a quiescent current of, say, 50 mA but a high quiescent current is a continuous waste of power. With continuous large currents there is a risk of thermal runaway (C.2.6). Instead use a *Class B* amplifier (Fig. 4.19), with an npn and a pnp transistor.

When there is no signal Q1 and Q2 are off; no power is expended. As the signal goes positive, Q1 turns on and supplies power to the speaker; Q2 is off. The reverse occurs for a negative signal. But the transistors do not turn on until the signal exceeds 0.6 V in either direction. There is no output for a signal of amplitude less than 0.6 V. For signals of amplitude greater than 0.6 V there is zero output each time the signal changes from positive to negative or from negative to positive → *crossover distortion* (Fig. 4.20). Amplitude of v_{out} is less than that of v_{in} because, in the simulation, the amplifier is loaded with an 8 Ω speaker. The current supplied by the source has an amplitude of 0.5 mA; the current through the load has amplitude 50 mA, so current gain is 100(= h_{FE}). In terms of power, the source operates at 2 mW when voltage amplitude is 4 V. Power dissipation in the speaker is 1.5 V × 50 mA = 75 mW, → power gain is $75/2 = 37.5$.

A Fourier analysis of the output signal (Fig. 4.21) shows that several harmonics are present in strengths exceeding −60 dB, but these are solely the odd harmonics (3 kHz, 5 kHz ... 15 kHz). They are 180° out of phase with the fundamental.

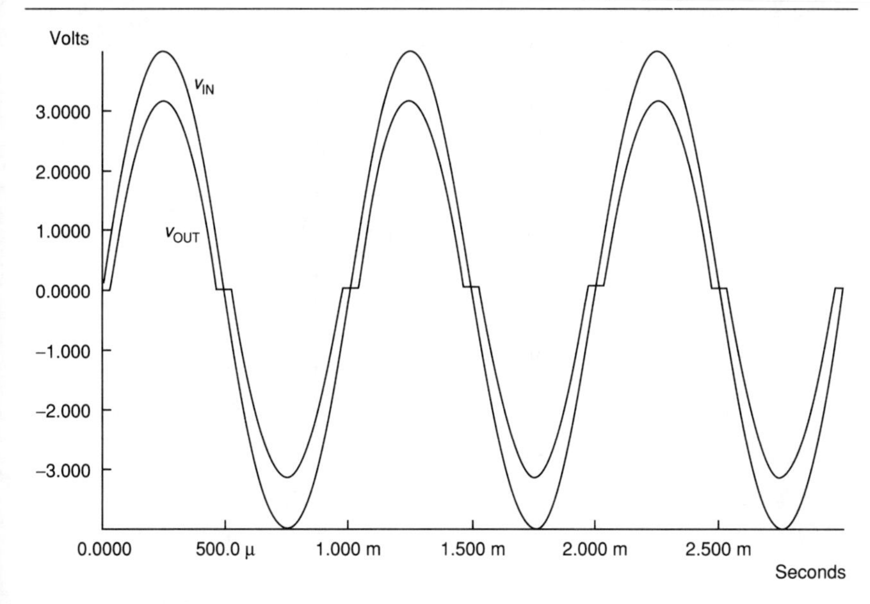

Figure 4.20

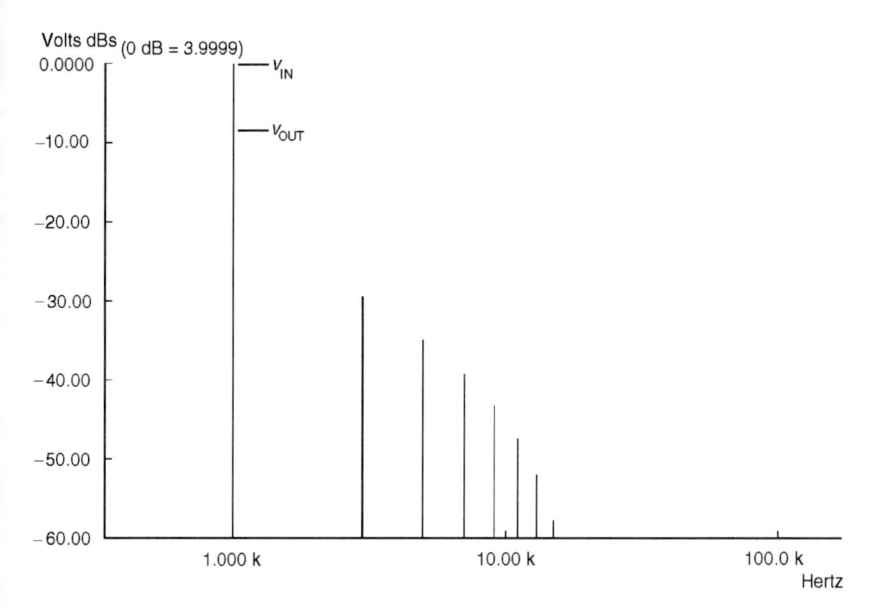

Figure 4.21

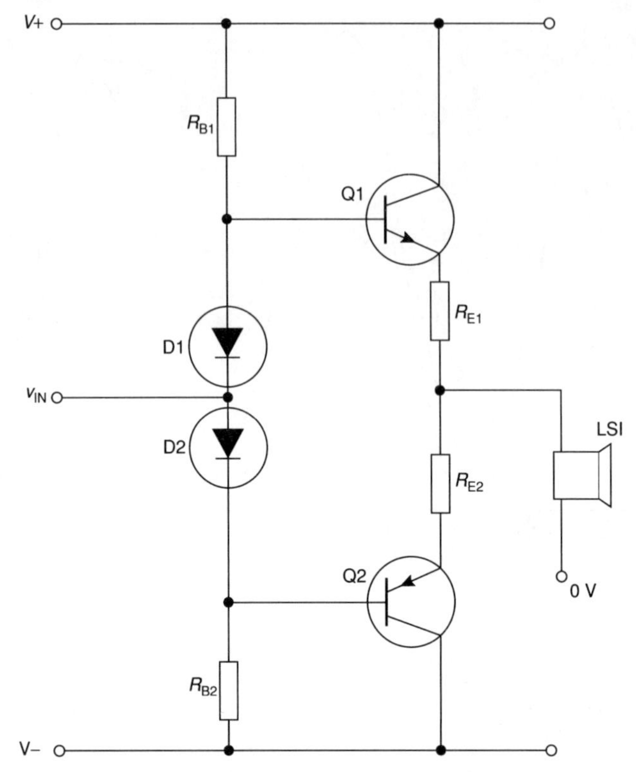

Figure 4.22

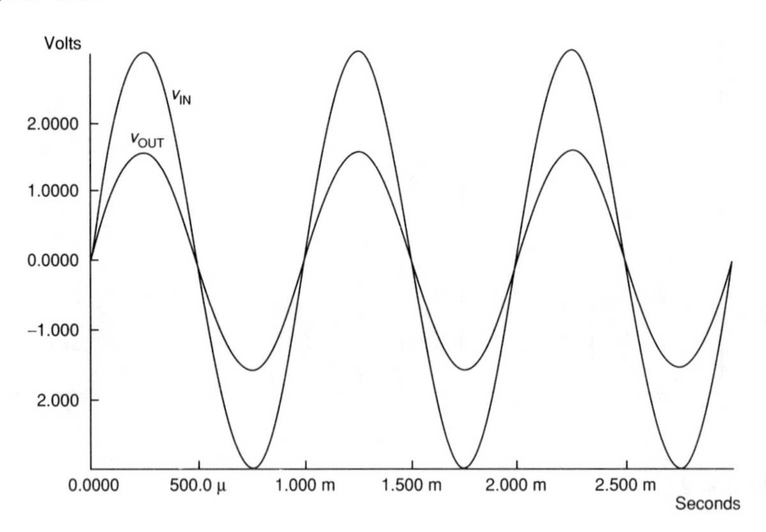

Figure 4.23

Crossover distortion is removed by biasing the transistors so that they are on the verge of conducting. This is done in Fig. 4.22 by using diodes. The voltage drop across a diode (approximately 0.6 V) matches the base-emitter voltage drop, so the transistors are on the point of starting to conduct when no signal is present. Biasing resistors are required and we introduce emitter resistors (of a few ohms) to improve stability. These changes lead to an immediate improvement in output (Fig. 4.23). The Fourier analysis of v_{out} has no harmonics down to -60 dB; the strongest harmonic is the 5th, at -61.1 dB. The action of the diodes compensates for temperature, since variations of V_{BE} of each transistor are balanced by increases of the pd across the corresponding diode.

An amplifier similar to Fig. 4.22 may also be built, using complementary Darlington pairs, to give increased current and power gain.

4.3 Common-base amplifier

This has low Z_{in}, fairly high Z_{out}, and high voltage gain but current gain < 1. Makes a good voltage amplifier provided source has low Z_{out} and load has high Z_{in}. The main application is as a high-frequency amplifier.

4.3.1 *Circuit operation*

In schematic diagrams of a CB amplifier, the transistor is usually drawn with its base connection vertically downward, as in Fig. C.9, but in Fig. 4.24 it is orientated as in the other amplifier schematics to help emphasize the similarities. As in the CE and CC amplifiers, I_B is provided by a potential divider $R_{B1} - R_{B2}$. As in the CE amplifier, I_C flows through R_C, generating the output potential $V_{CC} - I_C R_C$. I_C flows into Q1 and I_E flows out; $I_C \approx I_E$. The differences are that v_{in} is applied at the emitter, and that a bypass capacitor stabilizes V_B, the opposite situation to the bypassed CE amplifier (4.1.7). Alternating signals reaching the base are bypassed to ground. In the CE amplifier an upward swing in $v_{in} \rightarrow$ increase in $i_b \rightarrow$ increase in $i_c \rightarrow$ decrease in $v_{out} \rightarrow$ inverting amplifier. In the CB amplifier an upward swing in $v_{in} \rightarrow$ upward swing in $v_e \rightarrow$ decrease in $i_b \rightarrow$ decrease in $i_c \rightarrow$ increase in $v_{out} \rightarrow$ non-inverting amplifier.

Input impedance is r_e, the intrinsic emitter resistance of Q1. For an i_E of 1 mA, this is only 25 Ω, which ranks as a very small Z_{in}. Output impedance is R_C, which is usually of the order of a few kilohms.

4.3.2 *Design stages*

The design procedure is the same as for the CE amplifier: select R_C and R_E to give quiescent I_C (often 1 mA) and suitable gain. Select R_{B1} and R_{B2}

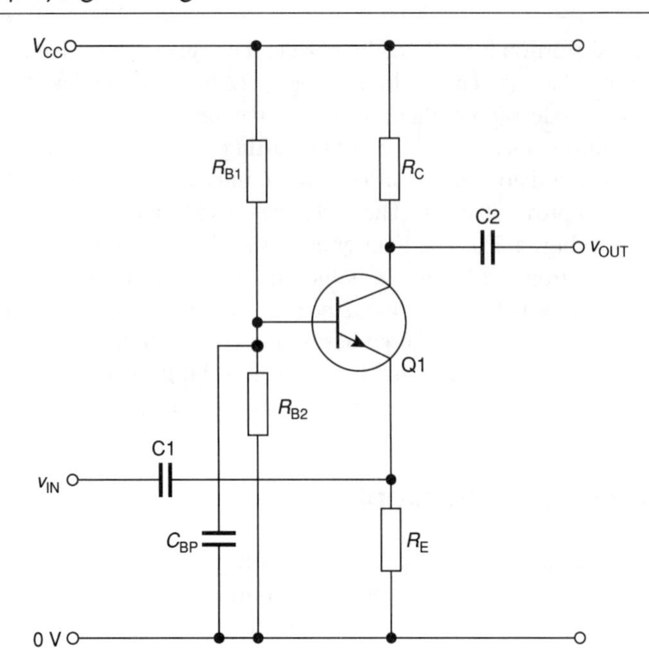

Figure 4.24

for correct bias to make $V_{BE} = 0.6\,V$, select C1 and C2 to pass frequencies required.

4.3.3 Frequency response

The CB amplifier has many applications as a high-frequency amplifier, for use in VHF and UHF circuits. This is because the grounded base isolates the input and output sides from each other, reducing capacitative feedback to the input side. In particular, it eliminates the *Miller effect* which is dominant in the CE amplifier at high frequencies, and arises from junction capacitance at the base-collector junction. In a CE amplifier the input signal is present on the base side of the junction and the output signal is present on the collector side. If G is the voltage gain of the amplifier (not of the transistor), as v_{in} fluctuates, v_{out} fluctuates G times as much and in the opposite direction. The pd across the junction capacitance varies by $(G + 1)$ times as much as v_{in}, and so does the current through the capacitance. The effective junction capacitance is $(G + 1)$ times the actual junction capacitance. Actual C_{be} may be only 2 or 3 picofarads, but effective C_{be} may be several hundred picofarads. This amplification of C_{be} is known as the *Miller effect*. Where it occurs, it severely attenuates high-frequency signals.

The Miller effect is significant in CE amplifiers, but is absent in CC amplifiers (4.2.3). In CB amplifiers, the input signal is applied to the emitter, so there is no capacitance amplification. There is only C_{be} (a few picofarads) between v_{in} and v_{out} and attenuation at high frequencies is minimal.

4.4 Single-input differential amplifier

A differential amplifier (3.1.4) may be used with advantage to amplify a single input signal. The signal is fed to the non-inverting input and the inverting input is grounded (Fig. 4.25, compare with Fig. 3.13). Omit the collector resistor of Q1 which makes no difference to the operation of the circuit. A collector resistor is required only for Q2, to produce the output voltage.

Removing the collector resistor on the input side means that the collector of Q1 is firmly held at V_{CC}. This completely eliminates the Miller effect, as in the CC amplifier (4.2.3). The other advantage of this amplifier is that temperature effects are much reduced. If the two transistors are manufactured on the same chip, they have identical characteristics and operate at equal temperatures. Temperature effects are largely rejected as being common mode.

4.5 Cascode amplifier

This uses another technique for eliminating the Miller effect. In Fig. 4.26, a common-emitter amplifier based on Q1 has an additional transistor Q2 in its

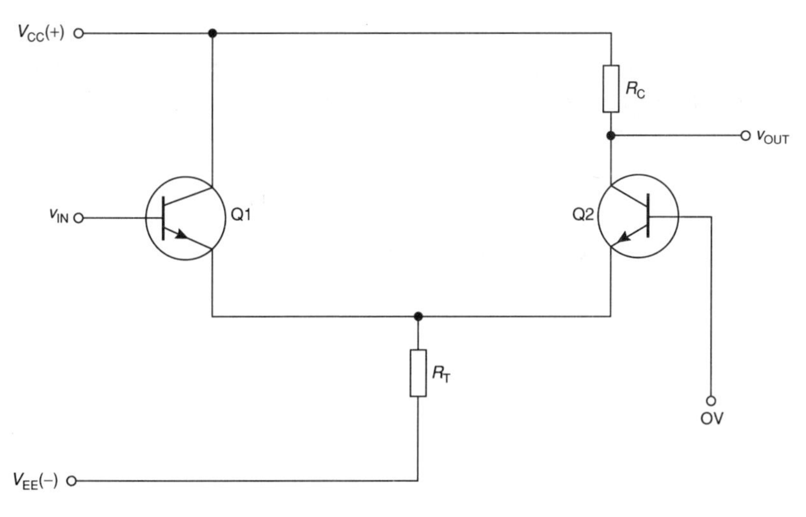

Figure 4.25

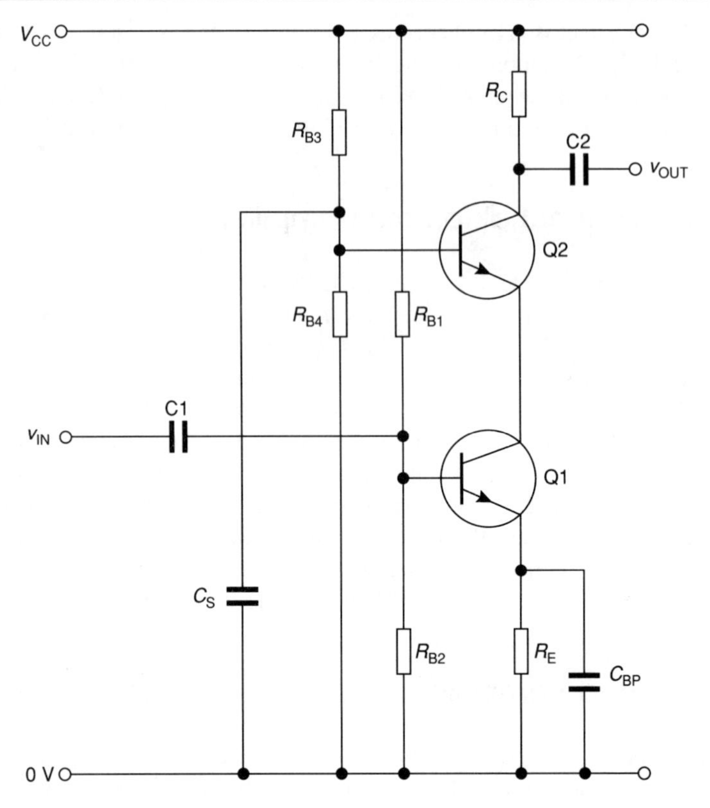

Figure 4.26

collector circuit. Q2 is biased by R_{B3} and R_{B4} with a capacitor C_S to stabilize it so that the collector of Q1 is held at a steady voltage. The voltage is set to keep Q1 in its operating region yet to allow the pd across R_C to swing widely in response to the signal. Because the collector of Q1 is held constant, there is no feedback effect of the base-collector capacitance and therefore no Miller effect (4.1.1.3).

The cascode connection is also employed with FETs.

5 More analog amplifiers

Common-emitter amplifiers based on discrete BJTs are described in Chapter 4. Here we look at a number of other ways of amplifying analog signals.

5.1 Common source amplifier

Amplifiers based on FETs have the same general plan and modes of operation as the BJTs described in Chapter 4, so we shall concentrate on the differences between FET and BJT amplifiers. The most significant difference is the high impedance of the gate, so that gate current is virtually zero. This makes biasing easier. There is no 0.6 V V_{BE} to provide for; all that is needed is to supply a suitable quiescent voltage just to turn on the transistor. This is negative of source for n-channel JFETs (C.3.2) and n-channel depletion-type MOSFETs, and positive of source (as with BJTs) for enhancement-type MOSFETs. P-channel FETs have complementary biasing requirements. Since gate current is effectively zero, bias resistors are in the order of megohms $\rightarrow$ amplifier Z_{in} is high, even with bias resistors taken into account (4.1.2, summary).

Some other differences between BJTs and FETs (C.4) are significant when these devices are used in amplifiers.

5.1.1 Circuit operation

When an FET is being used as a variable resistor (3.1.1) we operate it in the linear region of its characteristic curve (C.13). When used as an amplifier (Fig. 5.1), it is operated in the saturation region. CS amplifiers exist in several variations, as do CE amplifiers. For example, in Fig. 5.1 R_S is bypassed by a capacitor (compare Fig. 4.10) and for the same reason.

In the saturation region, i_D is proportional to V_{GS}, but not linearly (C.3.4). For this reason, v_{in} should be small to avoid distortion, and it is implicit in

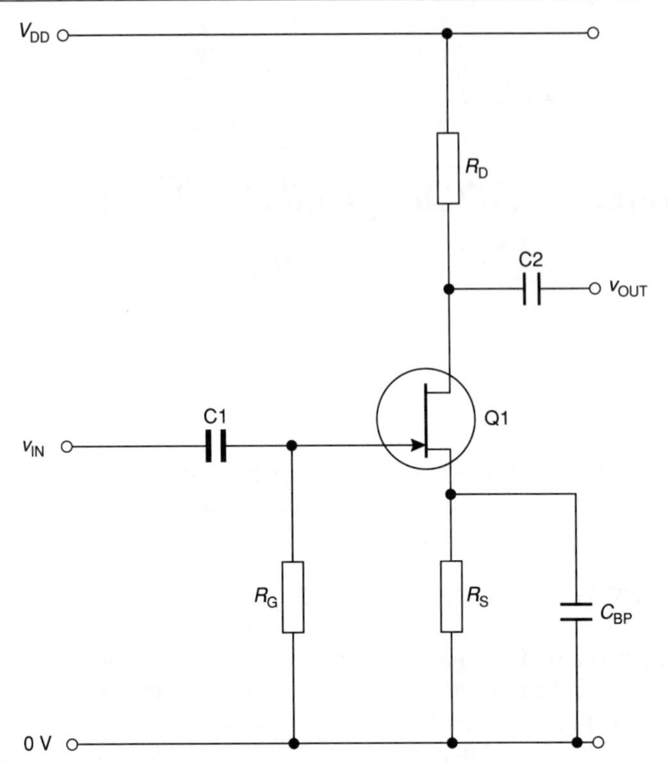

Figure 5.1

the discussion below that g_m is the transconductance at and around the chosen quiescent point.

If input is v_{in}, a fluctuating current i_d is flowing through R_S, and the quiescent voltage drop between gate and source is zero (there being no current flowing into the gate):

$$v_{gs} = v_{in} - R_S i_d = v_{in} - R_S g_m v_{gs}$$

$$\Rightarrow \qquad v_{gs} = \frac{v_{in}}{1 + R_S g_m} \tag{5.1}$$

On the output side:

$$v_{out} = -i_d(R_D \parallel R_L)$$

where R_L is a load resistance connected to C2. The negative sign indicates that v_{out} falls as i_d increases. Substituting $i_d = v_{gs} g_m$ and also from (5.1) above:

$$v_{out} = \frac{-(R_D \parallel R_L) g_m v_{in}}{1 + R_S g_m}$$

$$\text{Voltage gain} = \frac{v_{out}}{v_{in}} = \frac{-(R_D \parallel R_L)g_m}{1 + R_S g_m}$$
$$= \frac{-(R_D \parallel R_L)}{1/g_m + R_S}$$

This equation applies when there is no bypass capacitor. If the bypass capacitor is present then, for alternating signals, $R_S = 0$ and:

$$\text{Voltage gain} = -g_m(R_D \parallel R_L)$$

This is an inverting amplifier. In general, FET amplifiers have lower voltage gain than BJT amplifiers, but have the advantage of high Z_{in}. Since g_m depends on V_{GS}, only small signals may be amplified without distortion. Amplifiers are often designed with an FET input stage (with high Z_{in}, small signal for low distortion, low voltage gain) followed by a BJT stage or stages (for voltage gain and possibly for power gain in the last stage).

5.1.2 Design stages

5.1.2.1 JFET amplifier

As an example, design a CS amplifier (Fig. 5.1), given $V_{DD} = 15\,\text{V}$, $Q1 =$ 2N3819, to pass signals above 200 Hz.

(1) Decide on a quiescent value for I_D, to put quiescent point near to the centre of the saturation region (Fig. C.14 reproduced in Fig. 5.2). A good option is to set I_D to $I_{DSS}/2$, in this case $I_D = 7\,\text{mA}$. Interpolation between curves indicates $V_{GS} = -1.4\,\text{V}$. Replot the characteristic graph for $V_{GS} = -1.4\,\text{V} \rightarrow$ when $i_D = 7\,\text{mA}$ then $V_{DS} = 5.58\,\text{V}$.
(2) Decide on a quiescent value for V_D. For most purposes, a satisfactory value is $V_{DD}/2$ which allows v_{OUT} to swing freely in either direction without clipping. In this example $V_{DD}/2 = 7.5\,\text{V}$.
(3) Calculate $R_D = (V_{DD}/2)/I_D = 7.5/0.007 = 1071\,\Omega$. The nearest standard value is $1.1\,\text{k}\Omega$.
(4) Calculate $V_S = V_{DD} - V_D - V_{DS} = 15 - 7.5 - 5.58 = 1.92\,\text{V}$.
(5) Calculate $R_S = V_S/I_D = 1.92/0.007 = 274\,\Omega$. Nearest is $270\,\Omega$.
(6) A suitable value for R_G is $1\,\text{M}\Omega$.
(7) On the input side R_G and C1 form a high-pass filter. Calculate $C = 1/2\pi Rf = 723\,\text{pF}$. Take about 10 times this: 10 nF.
(8) On the output side R_D and C2 form a high-pass filter. Calculate $C = 723\,\text{nF}$. Use $10\,\mu\text{F}$.

The data sheets supply a value of g_m to enable calculations of gain to be made. But this is quoted for $I_D = I_{DSS}$. In practical circuits $I_D < I_{DSS}$ and g_m is less

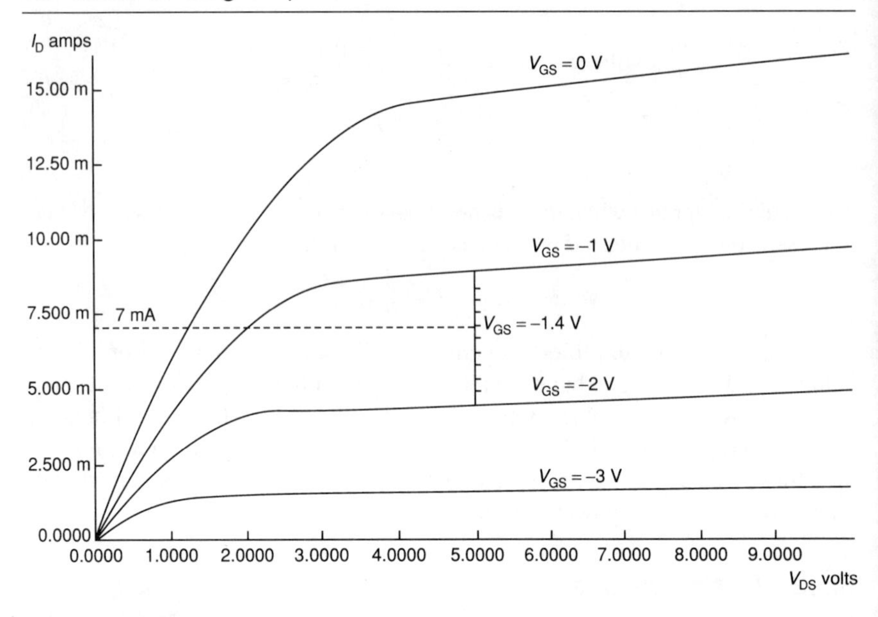

Figure 5.2

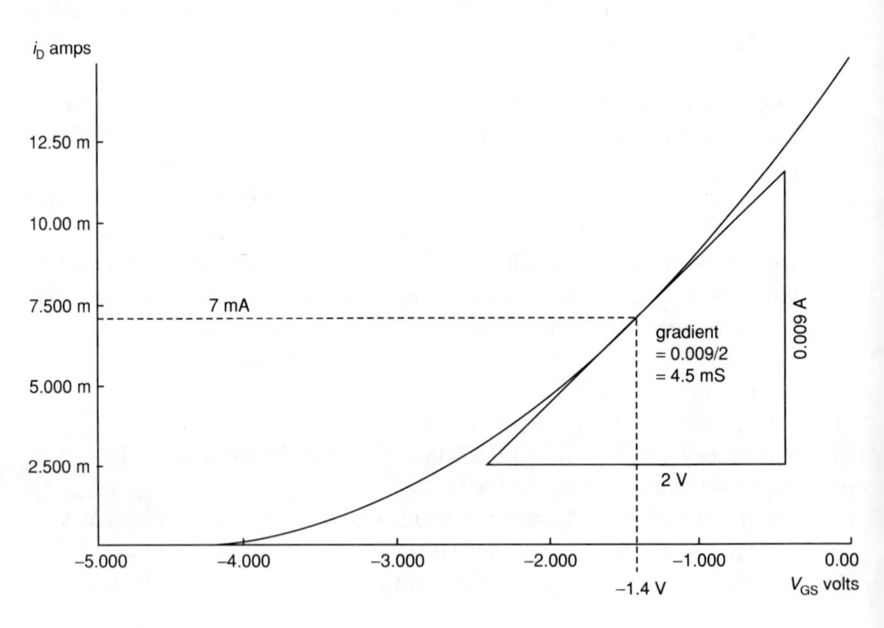

Figure 5.3

than quoted. Alternatively, with a simulator, using the same netlist as for Fig. C.17, change v_{DS} to 5.58 V and replot to obtain Fig. 5.3. Draw a tangent to the curve at $V_{GS} = -1.4$ V. Measure i_D and the gradient i_D/v_{GS} ($= g_m$). Results: $i_D = 7$ mA, $g_m = 4.5$ mS. If there is no R_L, then voltage gain $= 4.5 \times 10^{-3} \times 1100 = -4.95$. When the amplitude of v_{in} is 100 mV, a simulator plot of v_{out} shows a gain of −4.3.

Summarizing characteristics of this amplifier:

Voltage gain = theoretically – 4.95 (approx.), in simulation −4.3.
Input impedance: This is the resistance of $R_B = 1\,\text{M}\Omega$. Gate leakage current increases exponentially with temperature → Z_{in} decreases.
Output impedance: This is the resistance of $R_D = 1.1\,\text{k}\Omega$.

A Bode plot of the frequency response shows a flat response from 100 Hz to 30 MHz, with phase lag equal to −180° from 200 Hz to 7.5 MHz.

5.1.2.2 MOSFET amplifier

An n-channel enhancement-type MOSFET amplifier (Fig. 5.4) requires biasing for a positive gate voltage, usually about 2 V above the threshold voltage. With virtually no gate current, the biasing resistors are of high value. In Fig. 5.4, for example, with $V_{DD} = 15$ V, make $R_{G1} = 10\,\text{M}\Omega$ and $R_{G2} = 1.5\,\text{M}\Omega$ to obtain a gate voltage of 2 V, with $Z_{in} = 1.3\,\text{M}\Omega$. An additional resistor, $R_{G3} = 10\,\text{M}\Omega$, can be used to give even higher $Z_{in}(11.3\,\text{M}\Omega)$. Design procedure for the remainder of the circuit is as for the JFET amplifier. In general, MOSFETs have higher g_m than JFETs, sometimes expressed in hundreds of millisiemens or even siemens (JFETs in a few millisiemens). An amplifier in which Q1 is an IRF710, biased to 2 V, with other resistors and capacitors as in the JFET amplifier, has a voltage gain of 109, corresponding to $g_m = 109/1.1\text{k} = 100$ mS.

Analysis of Fig. 5.4 shows a quiescent gate current of 33 pA, a reminder that although gate current can usually be ignored, it is not actually zero. Gate current increases exponentially with increase of temperature, approximately doubling for a temperature increase of 10°C. A rough calculation shows that if (e.g.) $I_G = 33$ pF at 27°C (the simulation temperature), it is 32 times greater at 77°C, reaching 1 nA. This produces a voltage drop across the biasing network of $1\,\text{nA} \times 11.3\,\text{M}\Omega = 11$ mV, which is of the same order as the small signals typically measured with instruments having JFET inputs.

Dynamic gate current is the result of gate capacitance, which is relatively large in FETs when compared with BJTs. It applies to AC signals, acting at high frequencies as a capacitance that conveys a substantial part of the signal to ground, thereby seriously attenuating it.

A common-drain or source-follower amplifier is used as a buffer, especially at the input stage to testmeters and other measuring circuits. Its high Z_{in}

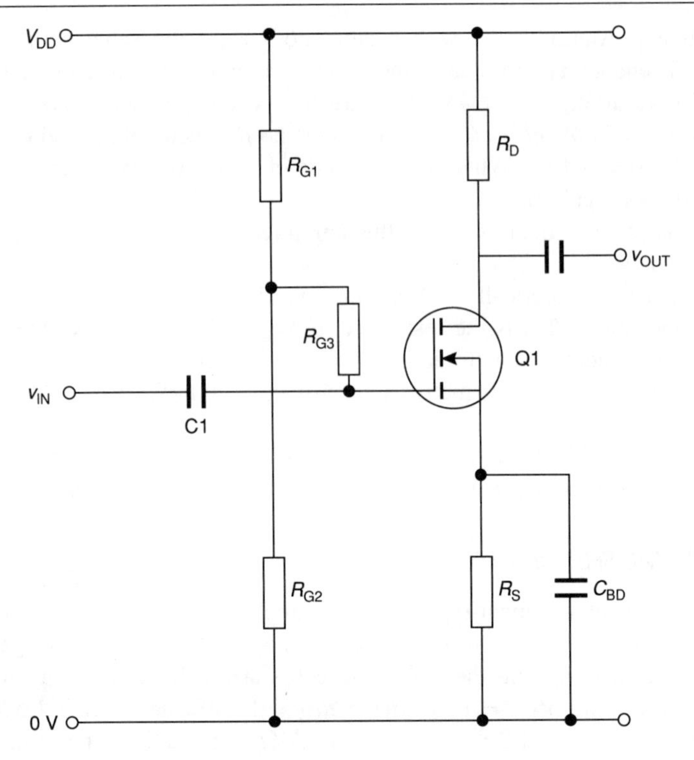

Figure 5.4

is of benefit when measuring the output of high-impedance devices such as capacitor microphones.

5.2 Operational amplifiers

Basic op amp circuits are explained in Appendix D. Here we describe more advanced applications.

5.2.1 Integrator

The output is proportional to the sum of the voltage over a period of time. Current flows through R (Fig. 5.5) to plate x of the capacitor. Since plate x is at 0 V, current at any instant is $i_R = v_{IN}/R$. As charge accumulates on plate x, v_{OUT} falls → reduces potential of plate y → keeps plate x at 0 V.

Looking more closely at this, we note that the (−) terminal remains at 0 V and no current flows into it. Current flowing to plate x is i_C which equals i_R.

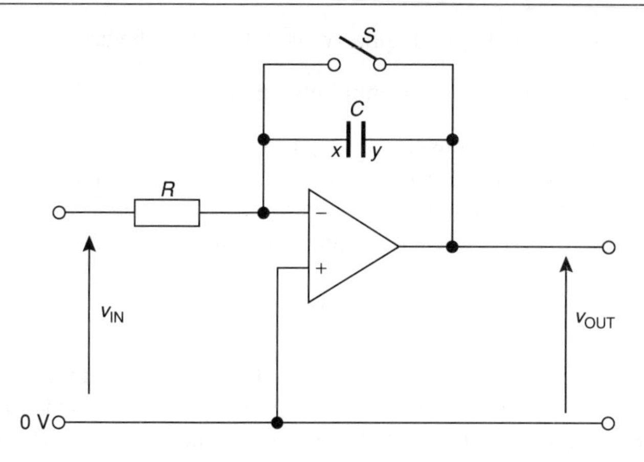

Figure 5.5

Current flowing from plate y to amplifier output is $-i_C$, leaving y negatively charged.

$$-i_C = \frac{-dq_C}{dt}$$

where q_C is the charge on the capacitor. For a capacitor, $q = Cv$, and so:

$$-i_C = C \times \frac{-dv_C}{dt}$$

Plate x is held at 0 V, so plate y falls to $-v_C$ and therefore v_{OUT} must have fallen to the same level:

$$v_C = v_{OUT}$$

$$\Rightarrow \quad C \times \frac{-dv_{OUT}}{dt} = -i_C = i_R = \frac{v_{IN}}{R}$$

$$\Rightarrow \quad dv_{OUT} = \frac{-1}{RC} \times v_{IN}\, dt$$

Integrating both sides:

$$v_{OUT} = \frac{-1}{RC} \int_0^t v_{IN}\, dt + k$$

where k is the initial charge, if any.

If v_{IN} is constant and positive, v_{OUT} ramps linearly down. Ramp is linear because the potential of plate x is constant at 0 V → pd across R is constant → constant current flow → constant increase of negative charge on plate y → constant fall of v_{OUT}. Circuit is reset by briefly closing S to discharge the capacitor. S may also be an FET switch.

If v_{IN} is varying, perhaps irregularly, as in Fig. 5.6, the area of each strip is:

the present value of v_{IN} × an infinitely small increment of time $= v_{IN}\,dt$

The total area under the curve is $\int v_{IN}\,dt \propto v_{OUT}$. In words, v_{OUT} is proportional to the area under the curve of v_{IN} against time. The dashed line in Fig. 5.7 encloses an equal area. Its area is $v_{AV} \times t$, where v_{AV} is the average or mean value of v_{in} between zero time and time t:

$$v_{AV} = \frac{1}{t}\int_0^t v_{IN}\,dt = -v_{OUT} \times \frac{RC}{t}$$

The integrator circuit is a way of finding the average voltage over a given period. Because the integrator is an averaging circuit, it responds to slow fluctuations in input, but not to rapid changes. It has the action of a low-pass filter.

In a practical circuit, v_{OUT} is in error because of the input offset voltage V_{os} and the input bias current I_b. Such errors accumulate with time. Even with zero input signal, the output ramps steadily up or down until the amplifier is saturated. This limits the time for which a signal can be integrated with precision. To reduce or correct these errors, use an op amp with low input offset voltage, and low input bias current (e.g. an op amp with FET input). Limit the length of the integration time. Use R_D (Fig. 5.8) to reduce the effect of voltage offset V_{os} to $V_{os} \times R_D/R$, compared with V_{os} × open-loop gain in the absence of R_D. This works well at high frequencies but at low frequencies,

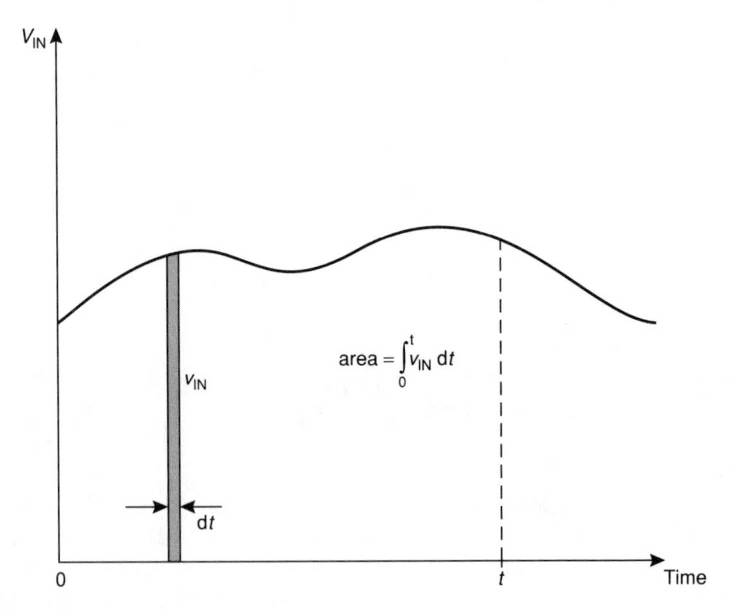

Figure 5.6

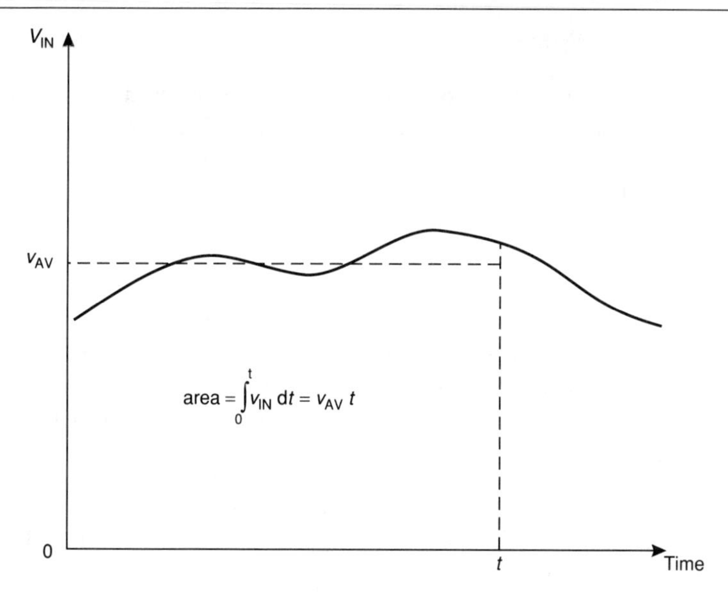

Figure 5.7

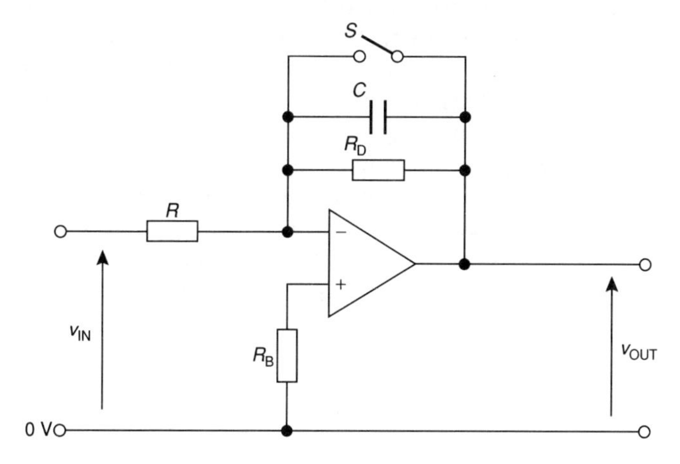

Figure 5.8

when the impedance of the capacitor is high, current flows mainly through R_D, and the integrating effect of C is outweighed. Thus, the integrator loses precision at low frequencies. Use R_B, equal to R and R_D in parallel, to correct the error due to I_b.

Integrating and summing functions may be combined by substituting a capacitor for R_F in Fig. D.6.

5.2.2 Differentiator

If the rate of change of the pd across a capacitor is dv/dt (V/s), the rate of change of charge is $C \cdot dv/dt$ (C/s). But 1 C/s equals 1 A, so the current flowing into or out of the capacitor is $i = C(dv/dt)$ amps. In Fig. 5.9 this current flows through R and:

$$v_{out} = -iR = -RC\frac{dv_{in}}{dt}$$

The circuit is a differentiator, with output proportional to the rate of change of v_{in}. Its frequency response (according to the equation above) is that of a high-pass filter. However, the roll-off of gain with frequency that is characteristic of an operational amplifier limits response at the highest frequencies, so that its response is that of a band-pass filter.

Because noise includes sharp spikes which have a high rate of change, the circuit is particularly subject to noise. Including capacitor C_R (Fig. 5.10) starts high-frequency roll-off at a lower frequency, $1/2\pi RC_R$, helping to reduce the effect of noise. The series resistor R_S sets the frequency $1/2\pi R_S C$ at which the frequency response levels off and differentiation ceases. Figure 5.11 shows the frequency response of the amplifier when the resistors and capacitors are such as to give increasing gain (6 dB per octave (6.1)) from about 1 kHz to about 10 kHz, roughly constant gain from 10 kHz and 100 Hz (above the −3 dB level) and with a sharp roll-off (−6 dB per octave) above 100 kHz. The circuit acts as a differentiator in the 1 kHz to 10 kHz range, as shown by its output, given a 5 kHz triangular-wave input (Fig. 5.12). At the first and subsequent peaks, where the rate of change of v_{IN} is most rapid in the negative direction, v_{OUT} begins a steep upward climb (remember this is an inverting differentiator). At the troughs of v_{IN} a steep downward fall in v_{OUT} occurs. The same effect is seen when V_{IN} is a square wave (Fig. 5.13) with very sharp upward and downward spikes whenever v_{IN} instantly changes level.

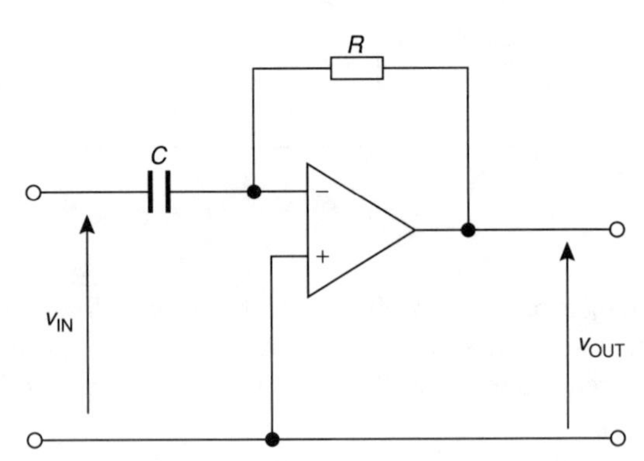

Figure 5.9

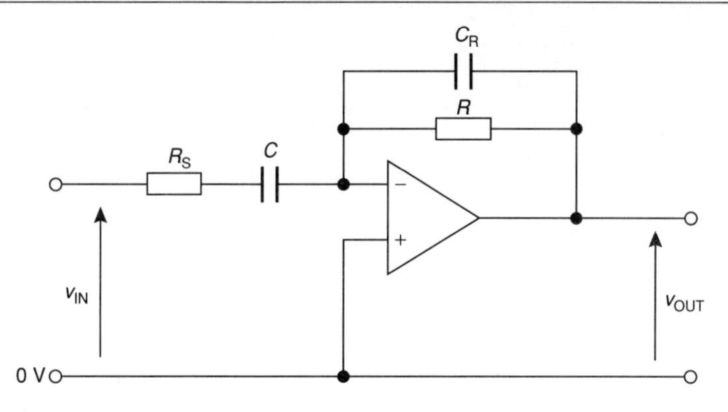

Figure 5.10

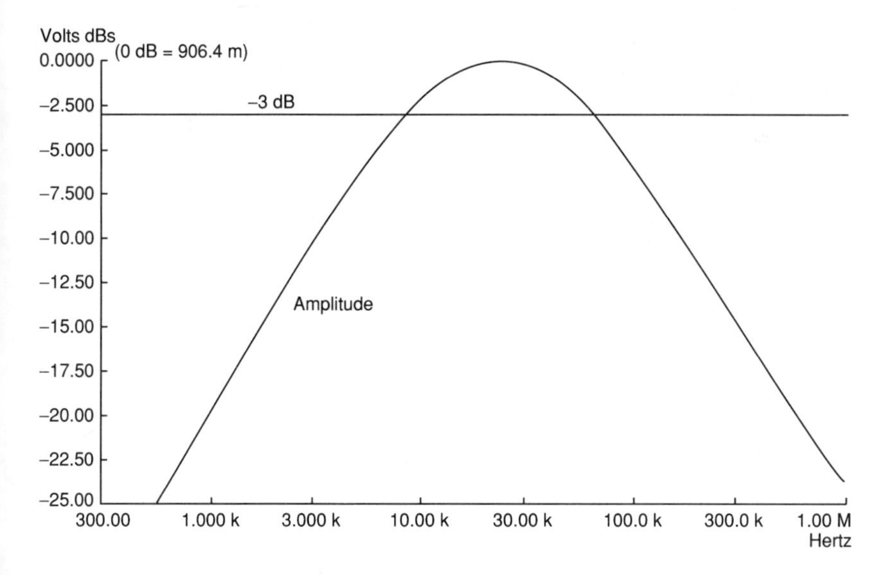

Figure 5.11

Between 10 kHz and 100 kHz, the circuit acts as a band-pass filter, with a centre frequency of about 30 kHz. Above 100 kHz, C takes on the role of a coupling capacitor, and the circuit acts as an integrator (compare Figs. 5.8 and 5.10). Its effect on a 500 kHz square-wave signal illustrates this (contrast Figs. 5.13 and 5.14). High-frequency noise is mainly eliminated by the integrating action.

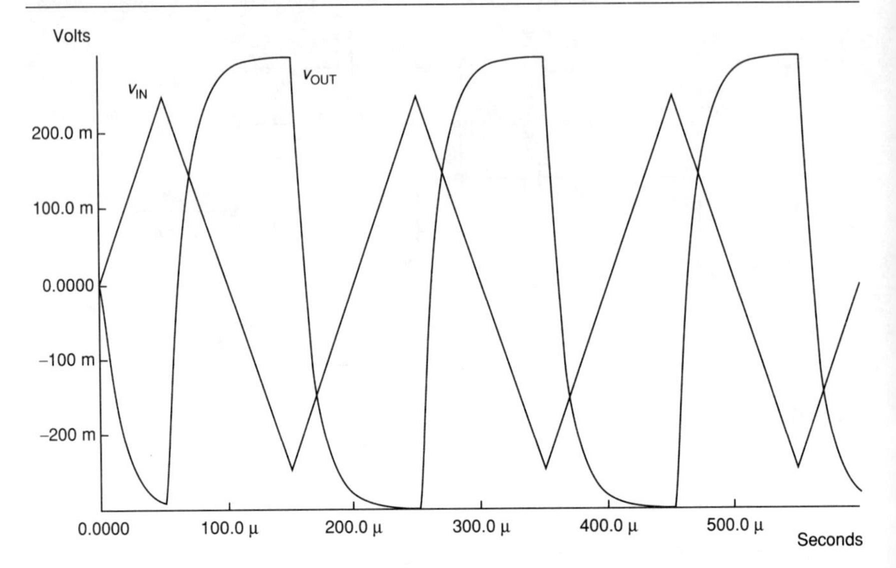

Figure 5.12

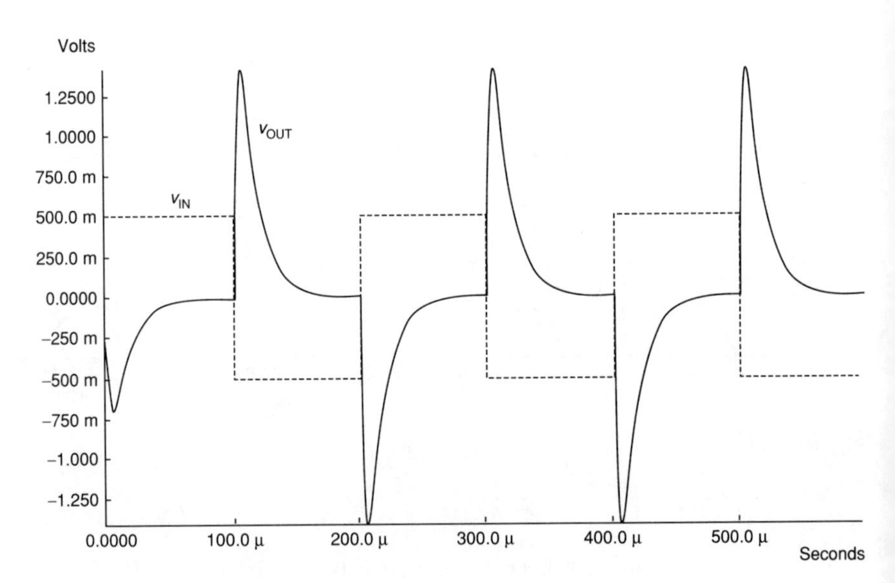

Figure 5.13

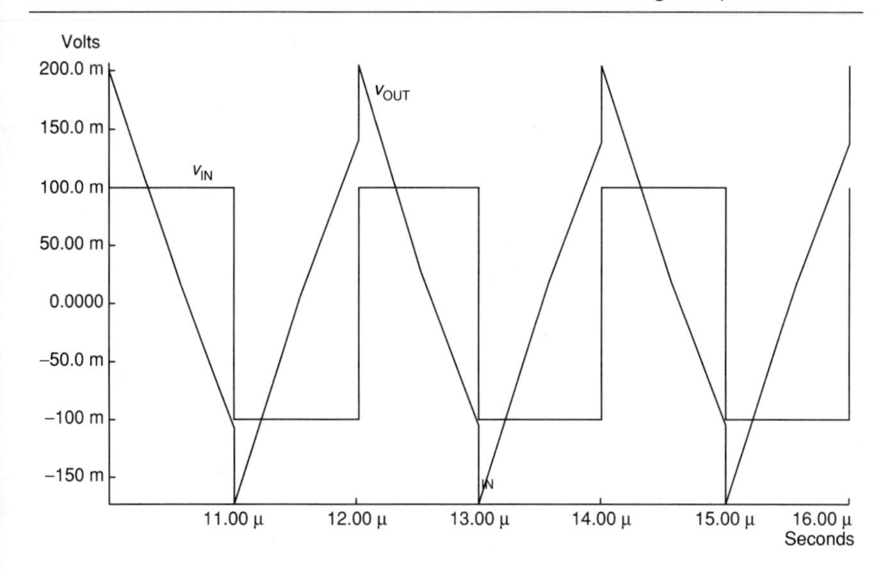

Figure 5.14

5.2.3 Instrumentation op amps

Precision op amps have an assortment of desirable features such as low offsets and high CMMR to give them good performance. Instrumentation amplifiers (in-amps) consist of two or more op amps on the same chip, as in Fig. 5.15. All resistors R are equal in value and, since these and the three op amps are on the same chip, it is easier to obtain balance. Both inverting and non-inverting inputs have high impedance (compare Fig. D.5, where they have not). The CMMR of the input stage is 1. The gain of this stage is $1 + 2R/R_G$, where R_G is the gain-setting resistor. This is an external resistor, used to fix the gain. Or an assortment of R_Gs of different values may be fabricated on the chip, and one of them brought into the circuit by connecting appropriate pairs of pins. If R_G is open circuit, the gain is unity. The differential stage usually has a gain of unity.

5.2.4 Transconductance amplifiers

The output current is proportional to the difference between v_+ and v_-. Typical transconductance is 10 mS.

5.2.5 Comparators

An op amp may be used as a comparator (D.2.2.3), but purpose-designed comparators have better characteristics, especially high slew rate. Outputs are

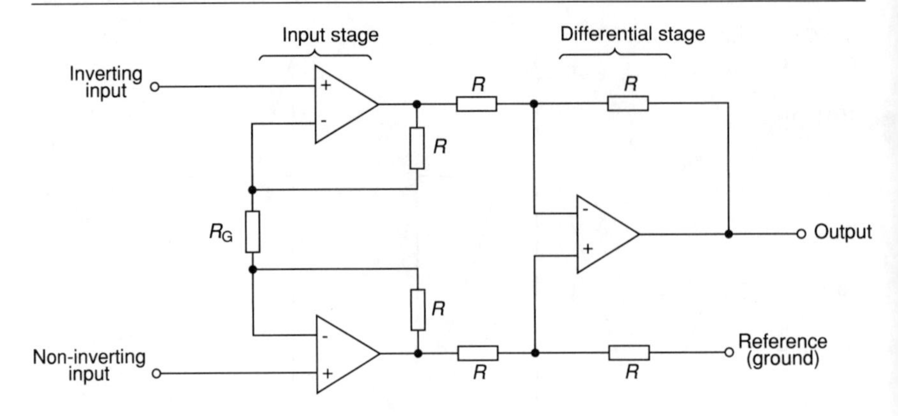

Figure 5.15

often open-collector, requiring an external pull-up resistor. If $v_+ > v_-$, output swings to V_+. If $v_+ < v_-$ output swings to 0 V.

5.2.6 Chopper amplifiers

Chopper amplifiers are designed to reduce input offset and drift so as to produce a high-precision amplifier suitable for instrumentation. Since offset and drift are temperature dependent, corrections must be applied continually as the ambient temperature and the temperature of the amplifier change. The adjective 'chopper' is applied to a number of different designs, all intended to produce the same outcome, but in slightly different ways. Figure 5.16

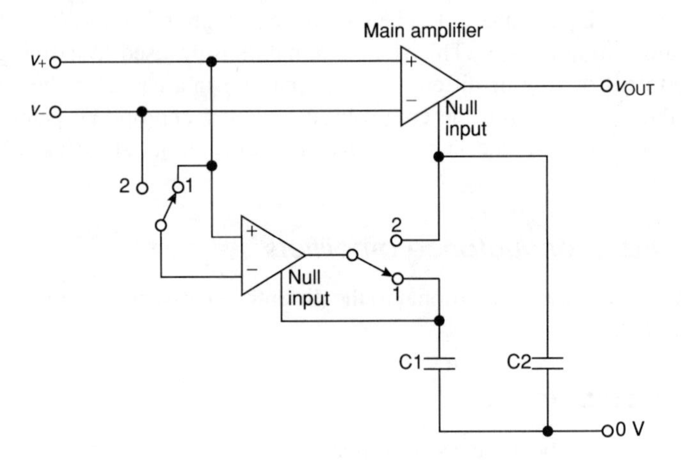

Figure 5.16

illustrates the principle of one of the more recent designs, available as an integrated circuit. It is also known as an *auto-zeroing* amplifier, or *chopper-stabilized* amplifier.

The amplifier consists of two amplifiers on the same chip, plus a square-wave oscillator running at a few hundred hertz. The main amplifier functions in the normal way and its output is subject to errors resulting from input offset voltage (D.1) and drift. The purpose of the second amplifier, the null amplifier, is to detect these effects and correct them, as well as correcting for its own errors. In short, the purpose of the null amplifier is to bring the output of the main amplifier to zero when the inputs of that amplifier are at equal voltages. Considered as a module, the circuit of Fig. 5.16 is an amplifier with two inputs, (+) and (−), and a single output; it may be connected in any of the configurations of an ordinary op amp, such as inverting amplifier, differentiator, or summer, so there is usually feedback from output to one of the input terminals.

The correction cycle is controlled by the oscillator which operates two changeover analog switches (3.6). There are two phases to the operation. With the switches set as in Fig. 5.16, both to position (1), the inputs to the *null amplifier* are shorted together. This puts both inputs at the same voltage. Unless the amplifier is already corrected, its output is at a slightly positive or negative voltage. The other switch connects the output to capacitor C1, which stores this positive or negative voltage. The voltage across C1 is also applied to the offset null terminal, and acts to correct for input offset voltage and for drift. The switches are set to position (2) in the second phase of each cycle. The null amplifier has been set to zero output when its input terminals are at equal voltage, in other words, its output has been corrected for offset voltage and drift. If the main amplifier is operating in one of the usual op amp configurations, its terminals too should be at equal voltages, now shared with both terminals of the null amplifier. But, as long as the main amplifier remains uncorrected for offset voltage or drift, the voltages are unequal. The null amplifier detects these unequal voltages and there is a corresponding output from the null amplifier. In this phase the output of the null amplifier is switched to C2 and to the null terminal of the main amplifier. Thus the null amplifier applies a correction to the main amplifier to compensate for the null amplifier's errors.

Correction for offset voltage and temperature changes is far better with a chopper amplifier than with any high-precision bipolar op amp, though there are disadvantages such as spikes at clock frequency on the output.

5.2.7 Logarithmic amplifier

A diode is included in the feedback loop (Fig. 5.17). Because of the exponential relationship between the current I_D through the diode and the pd v_D

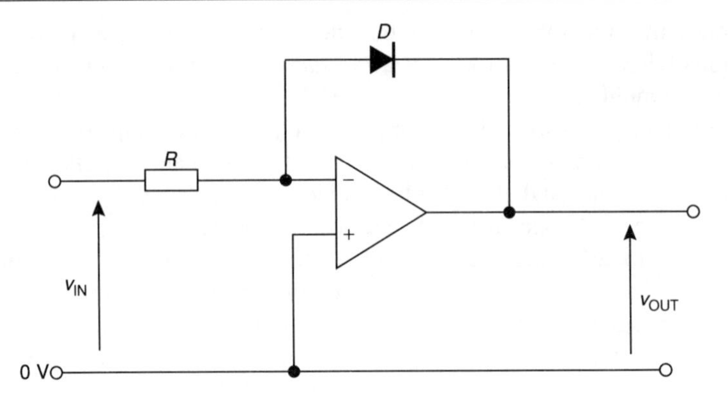

Figure 5.17

across it (B.1):

$$i_D = I_S(e^{v_D/V_T} - 1) \approx I_S e^{v_D/V_T}$$

Assuming $n = 1$. Taking logarithms:

$$\ln i_D = \ln I_S + v_D/V_T$$

$$\Rightarrow \quad v_D = V_T(\ln i_D - \ln I_S) = V_T \ln(i_D/I_S)$$

The (−) terminal is at 0 V so $v_{OUT} = v_D$. The current through the diode equals the current through R, so $i_D = v_{IN}/R$:

$$\Rightarrow \quad v_{OUT} = V_T \ln(v_{IN}/RI_S)$$

At 300 K (27°C), $V_T = 25.875$ mV. I_S is of the order of 1 μA. The output is proportional to the natural logarithm of the input.

Logarithmic amplifiers have applications in instrumentation and measurement. When a physical quantity varies over a range of several decades, its electronic analog taken from the output of a logarithmic amplifier has a relatively small range that can be accommodated within the working range of a testmeter or oscilloscope. If the quantity is rising or falling exponentially, the output is linear, a feature which makes it easy to detect exponentially changing quantities.

The base-emitter junction of an npn BJT may be used in Fig. 5.17 in place of the diode. V_T is temperature dependent so temperature correction is required for precision.

The reverse operation is performed by an *antilog amplifier* in which the resistor and diode of Fig. 5.17 are interchanged:

$$v_{OUT} = -RI_S \text{ antiln } (v_{IN}/V_T)$$

A version of a logarithmic amplifier is used for compressing the dynamic range of an audio signal. In Fig. 5.18, D1 and D2 allow both positive-going and negative-going signals to be compressed. R_F passes small signals (less than 0.6 V in either polarity) without compression, which is necessary because the logarithm of zero cannot be evaluated. It may be expanded to its original form by a circuit in which D1 and D2 are connected in parallel with R.

5.2.8 Precision voltage reference

A voltage reference may be used as a standard for comparison with a fluctuating analog voltage. A simple Zener reference consists of a reverse-biased Zener diode and a series resistor (Fig. 5.19). R_S drops the supply voltage to V_Z. Regulation is affected by the slope resistance (B.4.2) of the Zener.

Example: If $\Delta V/\Delta I = 3\,\Omega$, increase in i_D of 1 mA → increase in V_Z of 3 mV. With a reasonably high load, a change of 100 mA in i_D → change of 0.3 V in V_Z. The reference is not suitable when i_L is liable to change significantly during operation.

Calculating R_S: decide on maximum i_L and minimum i_Z (5 mA or more), then:

$$R_S = \frac{V_+ - V_Z}{i_L + i_Z}\ \Omega$$

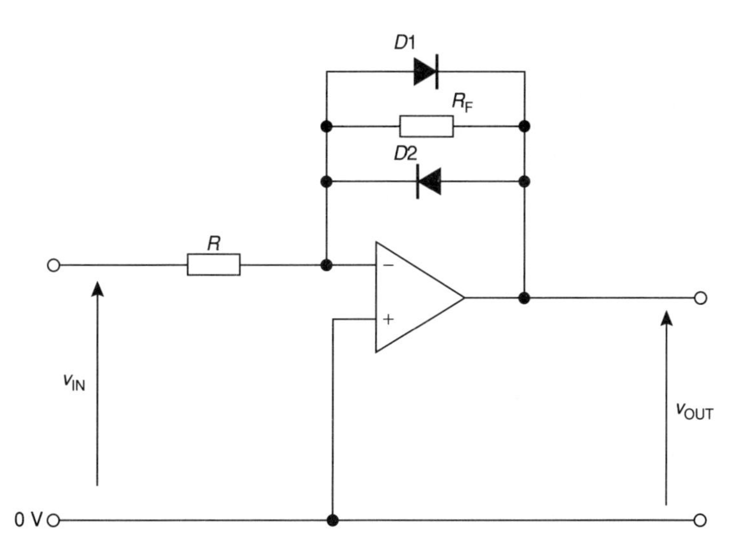

Figure 5.18

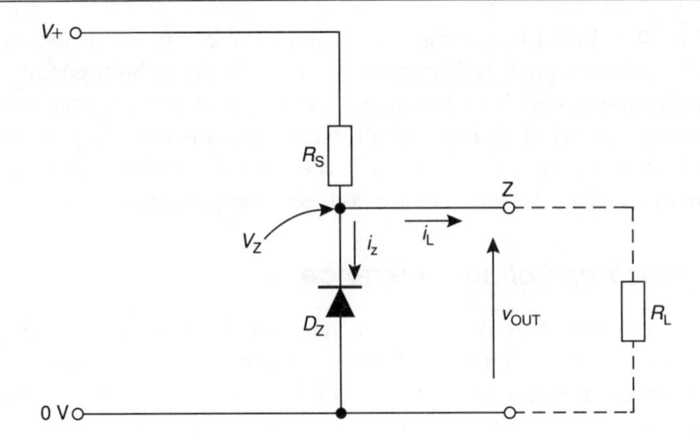

Figure 5.19

Zener must be rated to dissipate $P = V_Z(i_L + i_Z)$ W. Slope resistance introduces error, especially if the load current is not reasonably constant. Other disadvantages are that the circuit has low Z_{OUT}, and that power is wasted in the diode when the load is drawing less than its maximum current. The diode may need a heat-sink.

Figure 5.20 is an improved reference based on a non-inverting op amp (D.2.2):

$$v_{OUT} = v_Z \times \frac{R_A}{R_A + R_F}$$

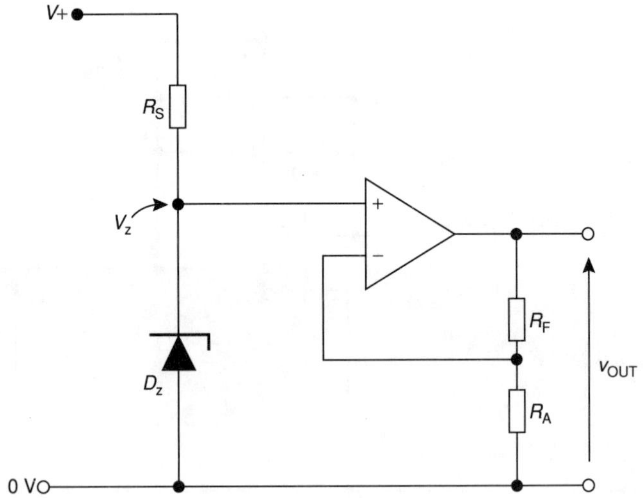

Figure 5.20

R_S is chosen to allow a current of slightly more than 5 mA to pass through the diode. Provided the source is regulated, the current is steady. Variations in load current have no effect on diode current → there is no variation in V_{OUT} resulting from changes in load current. Only a low-power diode is required. V_{OUT} depends on V_Z and also on the resistor values → V_{OUT} is not restricted to the standard range of values. It is possible to use a Zener with $V_Z = 5.6$ V or 6.2 V, giving the advantage of minimum slope resistance and tempco (B.4.2).

5.2.9 Precision rectifiers

To measure the amplitude of an alternating signal it is often best to rectify it, then measure the maximum voltage reached by the DC signal. Conventional half-wave and full-wave rectifiers (Fig. 5.21) reduce the pd across the load by 1 or 2 diode drops respectively. The size of the voltage drop depends on the diode current and also on temperature (B.1), a further source of error. Signals of amplitude less than 1 or 2 diode drops are not passed.

Figure 5.22 is a *half-wave precision rectifier* in which the effect of diode drop is eliminated. D1 is the rectifying diode, conducting during negative half-cycles. The output of an op amp stabilizes when the voltages at its two inputs are equal (D.2.1). Ignoring input bias currents for the moment, both inputs settle at 0 V. The current through the input resistor is v_{IN}/R and the current through the feedback resistor is $-v_{OUT}/R$. These currents are the same current:

$$v_{OUT} = -v_{IN}$$

The voltage drop of D1 has not entered into this calculation. Note that the diode drop occurs between the op amp output terminal and the circuit output terminal. The output of the op amp is actually one diode drop higher than v_{OUT}. As the drop varies in size with current or temperature, the op amp compensates automatically by adjusting its output voltage so as to keep its two input terminals always at equal voltage.

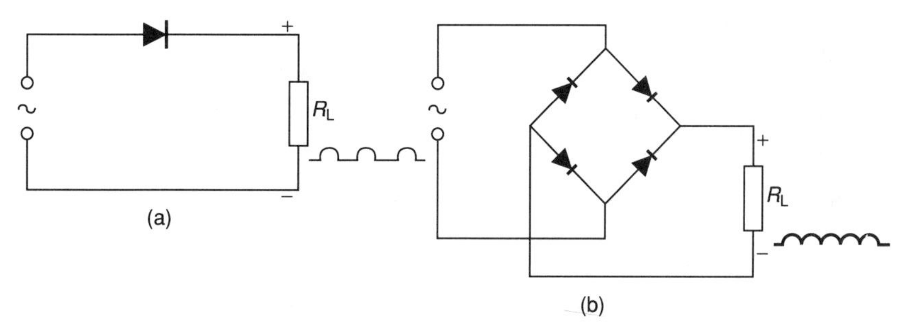

Figure 5.21

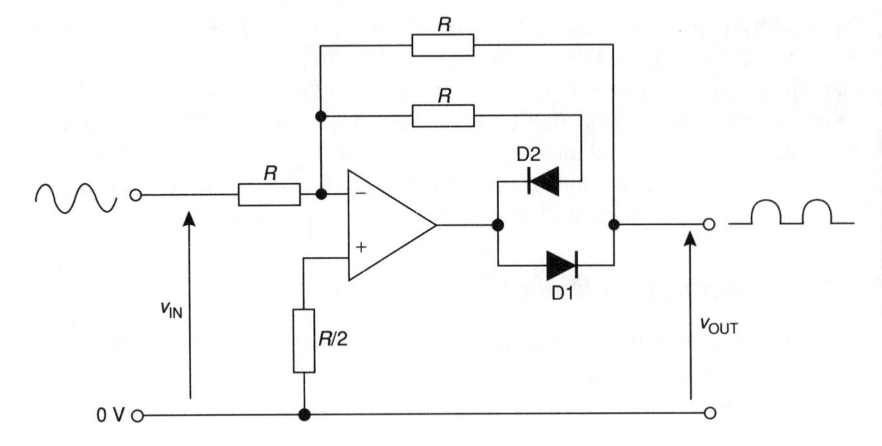

Figure 5.22

Input bias current is compensated for by the $R/2$ resistor at the (+) input. The other diode D2 does not take part in rectification. Its function is to prevent the amplifier from becoming overloaded during the positive half-cycle and therefore taking too long to recover at the beginning of the negative half-cycle.

A *full-wave rectifier* appears in Fig. 5.23. The effects of diode drops are eliminated for the same reasons as above. In the positive half-cycle, the output of OA1 goes negative → D1 conducts → input (+) of OA2 made negative → output of OA2 goes negative. In the negative half-cycle, the output of OA1 goes positive → D2 conducts → input (–) of OA2 made positive → output of OA2 goes negative. The result is a continuous series of negative-going half-waves → full-wave rectification with negative polarity.

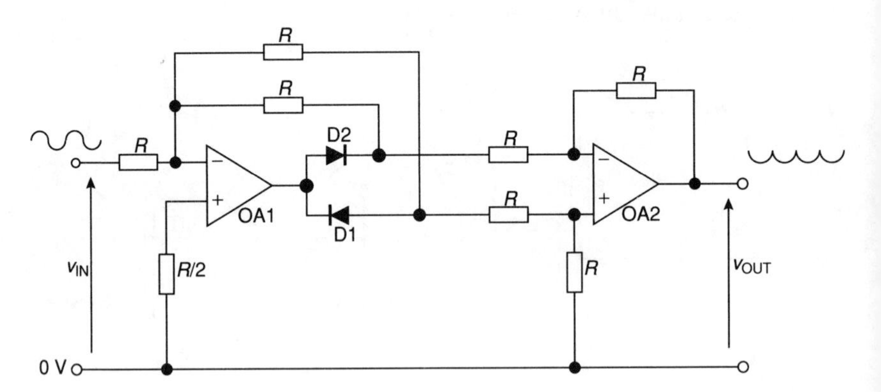

Figure 5.23

5.2.10 Voltage peak detector

This registers the highest voltage level reached recently. The circuit (Fig. 5.24) consists essentially of two unity-gain followers (D.2.2, Fig. D.3b). With increasing $v_{IN} \rightarrow$ output of OA1 increases $\rightarrow$ D1 conducts $\rightarrow$ C charges to $v_{IN} \rightarrow v_{IN}$ fed to OA2 $\rightarrow v_{OUT} = v_{IN}$. Ignore R_{LEAK} for the moment. Both inputs of OA2 are at v_{OUT}. If v_{IN} falls $\rightarrow$ D1 and high Z_{IN} of OA2 prevent C from discharging (OA2 acting as a buffer) $\rightarrow v_{OUT}$ remains constant. Current flows through R_F and D2 to prevent OA1 from saturating. Thus, each time v_{IN} rises above its previous highest level, v_{OUT} rises equally (Fig. 5.25).

R_{LEAK} is a high-value resistor (1 MΩ or more), allowing the charge to steadily leak away, so that the detector responds only to the most recent high

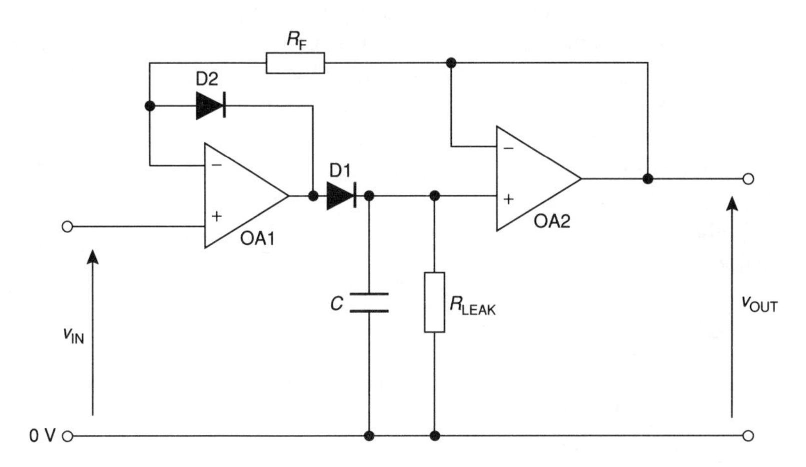

Figure 5.24

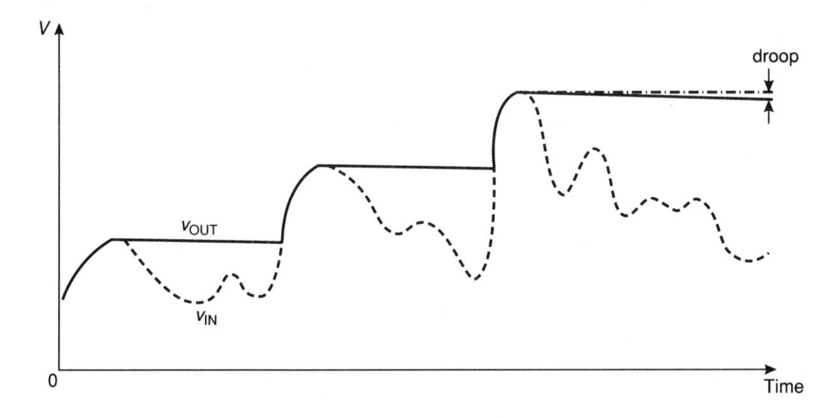

Figure 5.25

input levels. Alternatively, replace R_{LEAK} with a switch (mechanical or FET) by which the circuit may be reset periodically. With a switched detector, errors occur due to unwanted leakage: through D1, through C, and into the (+) input of OA2. Leakage causes a steady slow fall in v_{OUT}. This is known as *droop*. Reduce droop by using op amps with low bias currents, using a low-leakage diode, using a low-leakage capacitor, or resetting frequently.

Reverse the diodes to obtain a lowest-voltage peak detector.

5.3 Thermionic valve amplifiers

Thermionic valves (vacuum tubes) have been replaced as active devices by solid-state semiconductor devices such as diodes and transistors. They have several disadvantages, such as their need for high supply voltage and substantial heating current, and their significant heat generation, large size, fragility, high cost, and limited life-span. Yet thermionic values are still used in certain circuits, such as high-power oscillators in radio transmitters and in the output stages of power audio amplifiers. They are in favour with individual hi-fi connoisseurs who claim that the quality of sound produced from a valve amplifier is superior to that produced by a solid-state amplifier. There are also several other useful devices, such as oscilloscope tubes, which share many of the features of valves.

5.3.1 Diode

The most elementary thermionic valve is the *diode*, shown in diagrammatic form in Fig. 5.26. It consists of a wire filament (the *cathode*) and a metal plate (the *anode*) sealed in a vacuum in a glass or metal envelope. The cathode is heated to red heat by passing a current through it → thermally excited electrons escape from the cathode → form a negatively charged cloud (a space charge) around the cathode. They leave the cathode positively charged. Eventually the cathode begins to attract electrons back from the cloud and a state of equilibrium is reached.

If the anode is made positive of the cathode, as in Fig. 5.26, electrons are attracted to the anode, reducing the space charge so that more electrons can escape from the cathode. A continuous current of electrons flows through the diode from cathode to anode. In conventional terms, current i_D flows from anode to cathode. Conversely, if the cathode is made positive of the anode, electrons of the space charge are repelled back toward the cathode and no current flows. The diode is thus a device through which current flows in one direction but not in the other. This is the same action as in a semiconductor diode (Appendix B) to which the thermionic diode gives its name. The electrons from the cathode are majority carriers. There are no minority carriers, so there in no reverse leakage, as there is in a semiconductor diode.

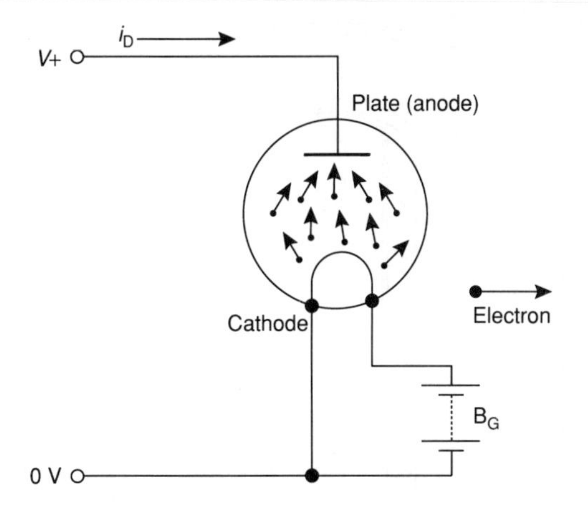

Figure 5.26

Physically, the diode consists of an evacuated cylindrical glass or metal envelope containing the two electrodes which are connected to metal pins set in a composition base, or into the moulded glass base of the envelope. Usually the filament lies along the axis of the envelope and is surrounded by the plate which is cylindrical. Typically the heater voltage is 6.3 V DC, supplied from a battery, and the current is a few hundred milliamps. The pd between cathode and anode is of the order of 100 V to 250 V. In early diodes the filament was made of pure tungsten but better emission of electrons at lower temperature (dull red heat) is obtained if the tungsten is sintered with a small quantity of thorium or is coated with oxides of strontium and barium.

Diodes are used as rectifiers in high-voltage, high-current power supplies.

5.3.2 *Triode*

The triode is the thermionic equivalent of the transistor. It contains three electrodes, *cathode, grid* and *anode*, as shown in half-section (omitting the envelope and base) in Fig. 5.27. The large heating current requires a substantial battery (e.g. a rechargeable lead-acid battery), particularly if the circuit includes several valves. In such cases it is preferable to take the heating current from a low-voltage winding of a mains transformer. AC can be used for heating but this introduces 50 Hz 'mains hum' into the circuit, which is then difficult to eliminate. Additionally, AC $\rightarrow$ fluctuations in filament temperature $\rightarrow$ fluctuations in electron emission $\rightarrow$ 100 Hz hum. The solution is indirect heating, in which the heater wire is electrically separate from the cathode. The heater wire is twisted as shown, or in some other way, to minimize inductive effects. It is surrounded by a cylindrical cathode. This in

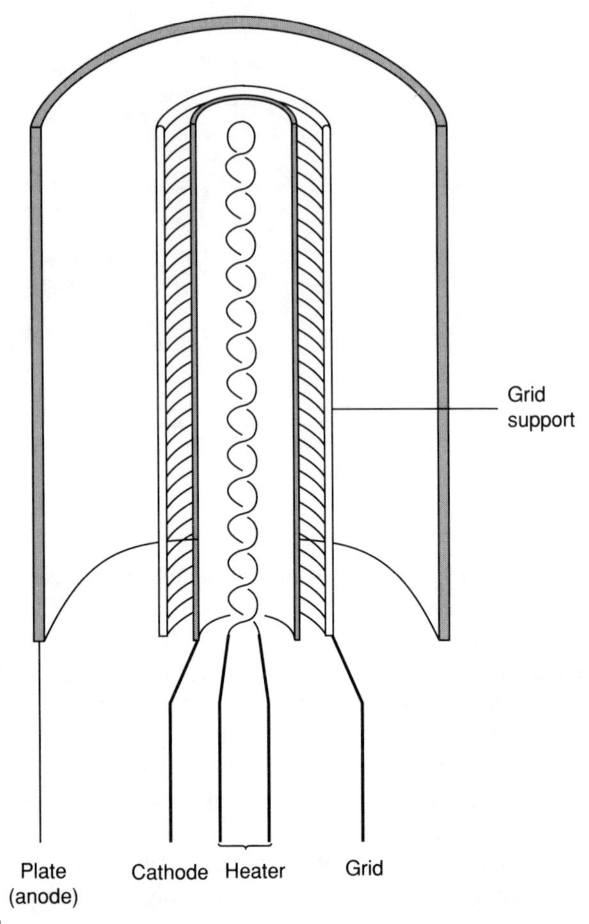

Figure 5.27

turn is surrounded by a grid of fine wire mounted on insulating supports, and the whole is surrounded by the cylindrical anode plate. This is the control grid. Some types of valve have electrodes that are rectangular in section instead of cylindrical.

The action of the triode is better explained by reference to its schematic symbol, Fig. 5.28. With current flowing through the heater (it takes several seconds to warm up → a delay when switching on valve equipment) and the anode positive of the cathode, electrons flow from cathode to anode, as explained for the diode. The grid is held a few volts negative of the cathode. Depending on the grid voltage, a smaller or greater proportion of the electrons is repelled back toward the cathode. If the grid voltage increases slightly → fewer electrons repelled → current increased.

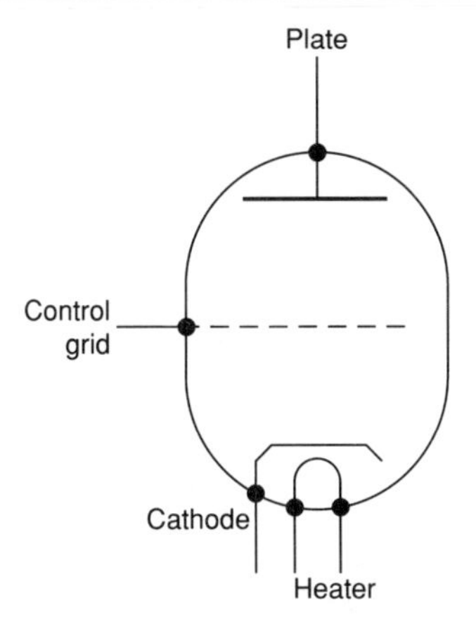

Figure 5.28

The action of a triode is very similar to that of an FET. The flow of current is controlled by the electric field produced by the charge on the grid (comparable to the gate of an FET). The grid does not emit electrons, neither does it attract them, so no current flows in or out of the grid → the triode has extremely high Z_{IN}. The graph of anode current i_A against grid voltage v_G (Fig. 5.29) has a similar shape to the graph of drain current i_D against gate voltage v_{GS} for a JFET (Fig. C.17). From this we can find the transconductance of the triode:

$$g_m = \Delta v_G / \Delta i_A$$

For the triode tested in Fig. 5.29 (using a generic model), $g_m \approx 20\,\text{mA}/4\,\text{V} = 5\,\text{mS}$ when $i_A = 4\,\text{mA}$ and $V_{AA} = 200\,\text{V}$.

The response of a triode to high frequency signals is attenuated by the capacitance between the grid and anode, which corresponds to the base-collector capacitance of a BJT. The effect of capacitance is reduced by another grid, placed between the control grid, and known as the screen grid. This is held at a potential approximately equal to that of the anode but is coupled to the 0 V line through a capacitor. Only a small proportion of electrons strike the screen grid. The majority, being strongly attracted by the high voltage on the screen grid, are accelerated through the meshes of the grid and pass to the anode. Thus the anode is screened from the control grid by the steady electrical field around the screen grid, reducing the capacitance between them. A valve which has these two grids is called a *tetrode*.

The triode and tetrode are the simplest amplifying valves. There are valves with three grids (pentodes) and even more.

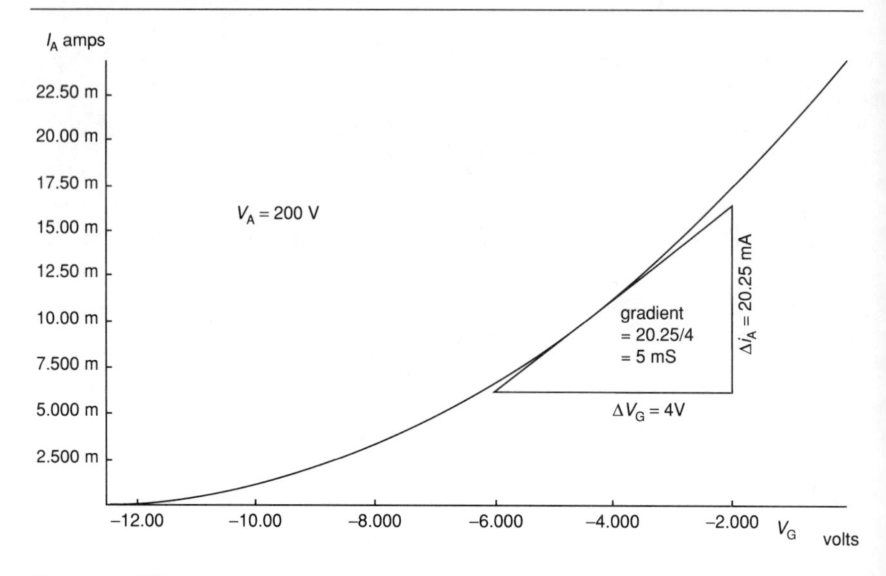

Figure 5.29

5.3.3 Valve amplifier

The amplifier circuit of Fig. 5.30 has almost the same configuration as the common-source amplifier (Fig. 5.1). Capacitors couple it to preceding and subsequent stages. Current i_A (a few milliamps, as in a transistor amplifier) flows through the triode from the supply (V_{AA}, usually 200 V–400 V). Passing through R_A, it generates a pd and R_A is usually selected to make $v_{OUT} = V_{AA}/2$. The same current passes through R_K to hold the cathode a few volts above zero. R_K is bypassed by a capacitor to give stability (compare 4.1.8). The grid resistor R_G holds the grid close to 0 V, so giving it negative bias. The amplifying action is:

$$v_{in} = v_g$$

$$i_a = g_m \times v_g$$

$$v_{out} = -R_A \times i_a$$

$$\Rightarrow \quad v_{out} = -v_{in} R_A g_m$$

This is an inverting amplifier. Because g_m varies with i_A (Fig. 5.29), there is distortion unless signals have small amplitude. The input impedance of this amplifier is R_G (usually several megohms) since the Z_{IN} of the triode itself is extremely high.

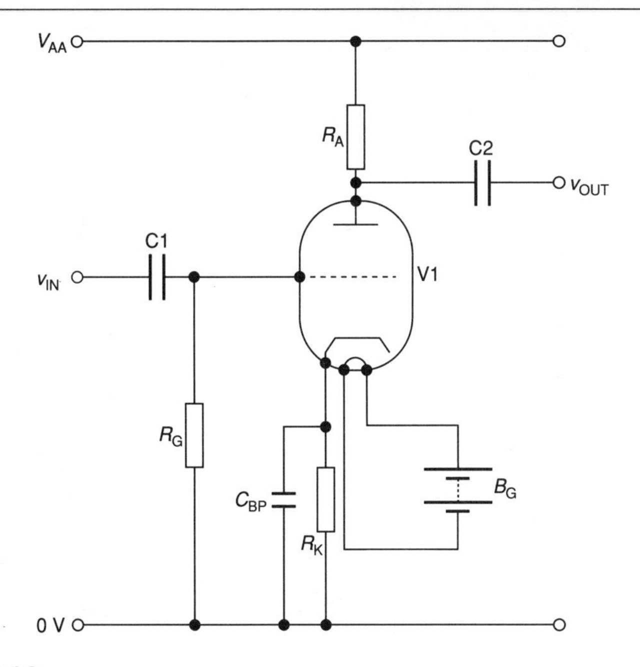

Figure 5.30

5.4 Parametric amplifiers

The pd V across a capacitor is related to the charge q and the capacitance C:

$$V = q/C$$

The capacitance depends on three parameters: the area of overlap of the plates, the distance between the plates, and the dielectric constant of the material between the plates. In particular, capacitance varies inversely with the distance between the plates. In Fig. 5.31, the capacitance between the plates decreases as the plates are moved further apart. But the charged plate is not able to lose or gain charge; q remains constant. Therefore v must increase. The additional energy to create the increased pd comes from the energy required to separate the plates against the attractive force between the positive and negative charges on the two plates. We have altered the pd across the capacitor by varying one of the parameters. We could also do this by moving one plate sideways, keeping its distance constant; this varies the area of overlap of the plates, and is the method used in the variable capacitors used in tuning circuits. Or we could insert sheets of different dielectrics between the plates. In all cases we are changing the pd across the capacitor by altering one of the parameters that

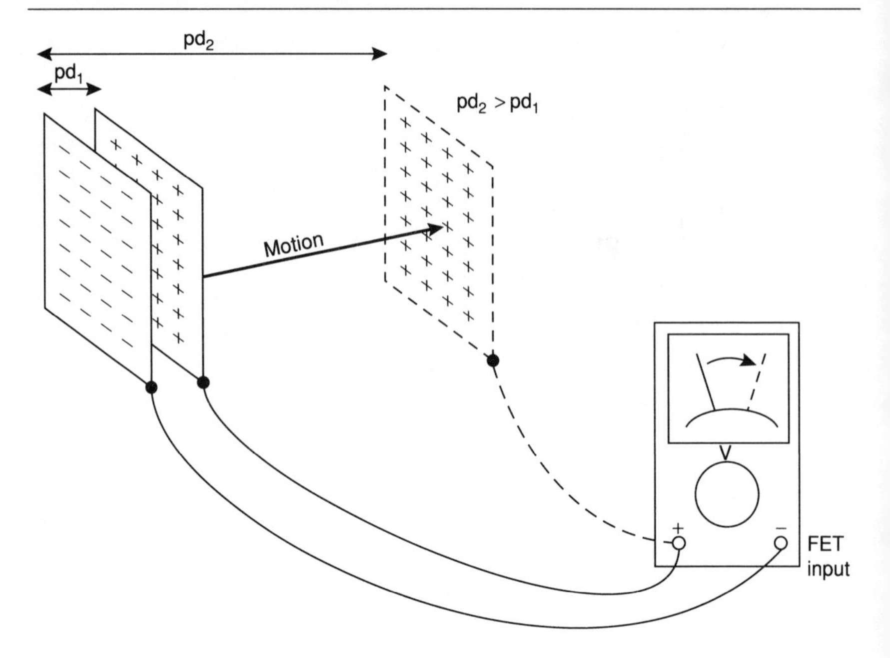

Figure 5.31

determine capacitance. We do not need to add or remove charge from the capacitor.

In a parametric amplifier, we vary the pd across a capacitance, not by using the gain of a transistor to amplify current or voltage, but by altering one of the parameters that determine capacitance. The capacitance is that of a reverse-biased varactor diode and we alter it by varying the pd across it. In the basic circuit (Fig. 5.32) the tuned *LC* network includes a varactor diode. The pd across it, and consequently its capacitance, is determined by a pulsed *pump signal*, which (in this mode of operation, there are other modes) is double that of the signal to be amplified. The pumping signal is timed so that it switches rapidly from positive to negative just as the sinusoidal input signal reaches its maximum, in either direction. The change of the pump signal from high to low instantly decreases the capacitance of the varactor. It is the equivalent of separating the plates in Fig. 5.31. Decreasing the capacitance raises the pd across the capacitor by an amount proportionate to its present value (Fig. 5.33), the energy to do this coming from the pumping circuit. At the peak of each half-cycle the amplitude of the input signal increases.

When the pumping waveform changes from negative to positive, the capacitance is restored to its former value. But this does not reduce the pd again because this action is timed to occur when the signal voltage is zero. There is

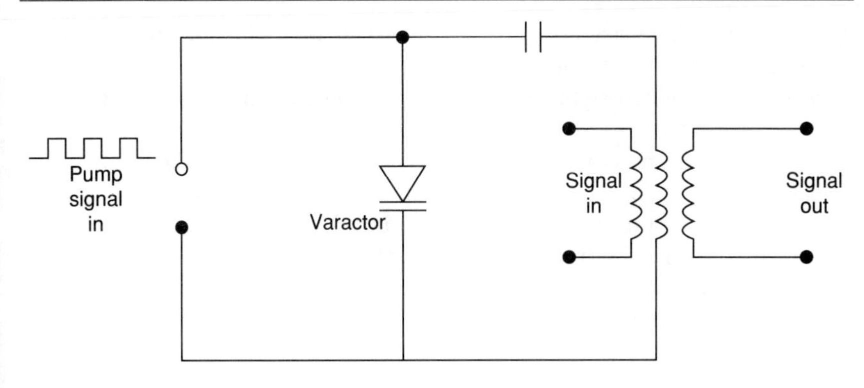

Figure 5.32

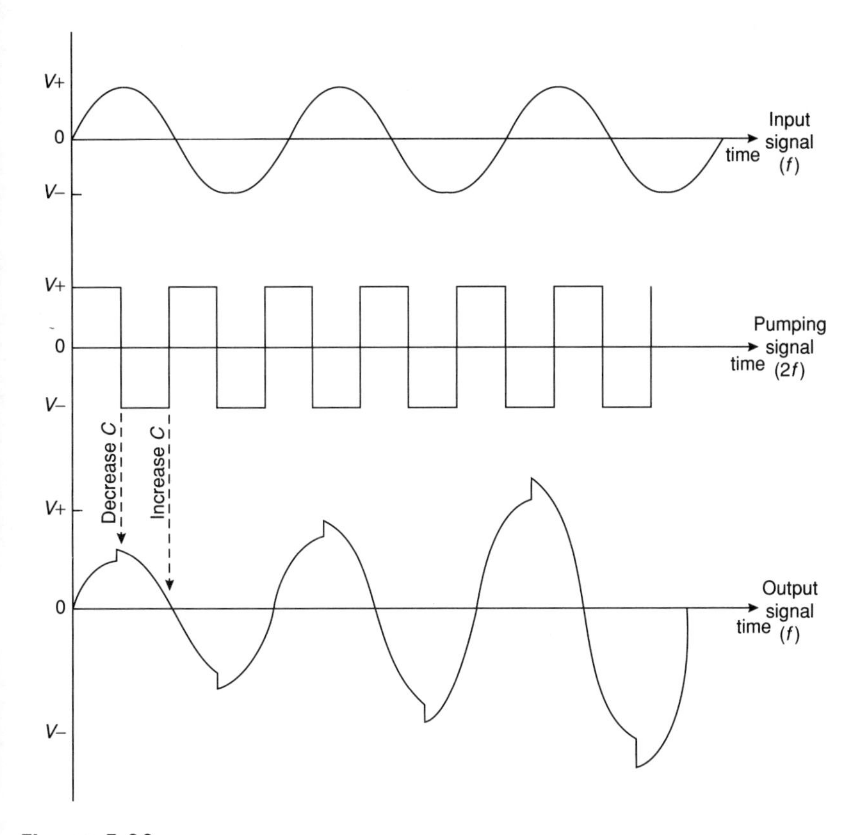

Figure 5.33

zero charge on the capacitor so there is zero change. The amount of amplification is limited by the ability of the pumping circuit to supply the required energy. The amplified signal is coupled to the output terminals by a third coil in the transformer.

One of the prime advantages of parametric amplifiers is their low noise. This is because there is no resistance in the amplifier and therefore no Johnson noise (7.2.1). Cooling the amplifier may, but does not necessarily, decrease noise still further. The amplifiers have applications in radio-astronomy and in satellite and space communications, where signals are weak and low noise levels in the amplifier are imperative.

6 Filtering analogs

An electrical signal, except for a pure sine wave, can be considered to be the sum of two or more sinusoidal signals of different frequencies and usually of different amplitudes. The function of a filter is to change the relative amplitudes of these component signals, and possibly change their phase relationships as well. Analog filters all rely on having at least one reactive component (a capacitor or inductor) in them, for these are the devices with frequency dependent behaviour. There are very many ways of putting together reactances to build a filter but in this chapter we concentrate on the essentials of filtering, using just a few of the many types of filter as illustrations. Refer to E.7–E.9 for background information.

6.1 Passive filters

The simplest types of filter are built from passive components, such as resistors, capacitors and inductors. All the energy in the output signal is derived from the energy present in the input signal. Some energy is lost during filtering because of the resistance of the filter components, particularly the resistors.

6.1.1 Resistor-capacitor low-pass filter

Figure 6.1 is the same network as in Fig. E.8, but redrawn to emphasize its function as a *low-pass filter*. Comparing the two figures:

$$v = \sin 2000\pi t \qquad \Leftrightarrow \qquad v_{\mathrm{IN}} = V_0 \sin \omega t$$
$$v_{\mathrm{c}} = 0.82 \sin(2000\pi t - 34.7^{\circ}) \qquad \Leftrightarrow \qquad v_{\mathrm{OUT}} = v_{\mathrm{IN}} H(s)$$

On the left are the values for Fig. E.8 (results plotted in Fig. E.9) and on the right are the corresponding symbolic equations relevant to Fig. 6.1 The output of this filter is the voltage across the capacitor, determined by the

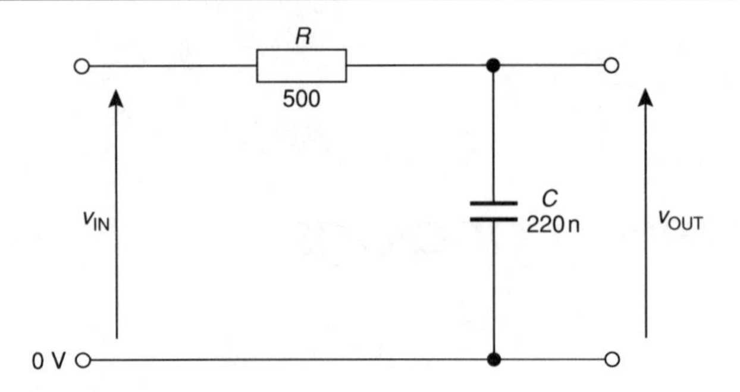

Figure 6.1

value of $H(s)$, the transfer function. In complex form, the impedance of the capacitor is:

$$z_C = 1/sC$$

and the total impedance is:

$$z_{IN} = 1/sC + R$$

The network of Fig. 6.1 may be thought of as a potential divider:

$$v_{OUT} = v_{IN} \times \frac{1/sC}{1/sC + R}$$

$$\Rightarrow \qquad H(s) = \frac{v_{OUT}}{v_{IN}} = \frac{1/sC}{1/sC + R} = \frac{1}{1 + sRC}$$

Because $v_{IN} = V_0 \sin \omega t$, it has constant amplitude V_0. Thus $\sigma = 0$, so we may replace s by $j\omega$. $H(s)$ is now symbolized by $H(j\omega)$:

$$H(j\omega) = \frac{1}{1 + j\omega RC}$$

Using the values of Fig. E.8, in which $\omega = 2000\pi = 6283.2$:

$$\omega RC = 0.69115$$

and

$$H(j\omega) = \frac{1}{1 + j0.69115}$$

Multiplying top and bottom by the conjugate, $1 - j0.69115$, then simplifying:

$$H(j\omega) = 0.6767 - j0.4677$$

In polar form:

$$H(j\omega) = 0.82\angle -34.7^\circ$$

The output amplitude is 0.82 of the input amplitude and v_{OUT} lags 34.7° behind v_{IN} (Fig. E.8).

This calculation can be repeated over a given range of frequencies and a graph plotted to show the relationship between frequency and $H(j\omega)$ and between frequency and phase, Fig. 6.2. When frequency is low, the amplitude of v_{OUT} is very close to 1 V, but falls steeply, passing through 820 mV at 1 kHz. The network is a low-pass filter. Phase lag is close to zero for very low frequencies, becomes −34.7° at 1 kHz, and continues falling at higher frequencies. Figure 6.2 is a linear plot, but the frequency range may be extended by plotting frequency on a logarithmic scale, Fig. 6.3.

The amplitude scale also covers a correspondingly greater range by being plotted on a decibel scale, which is also logarithmic. Figure 6.3 is known as a *Bode plot* and, in this example, shows the frequency response up to 1 MHz.

The response is divided into three regions:

- Pass band: v_{OUT} is within 3 dB of 1 V. The stop band ranges from 0 Hz to the corner frequency f_c (about 1.5 kHz).
- Transition region: v_{OUT} begins to fall more steeply as the curve turns gradually down.
- Stop band: above about 5 kHz, v_{OUT} falls at a regular rate, known as *roll-off*.

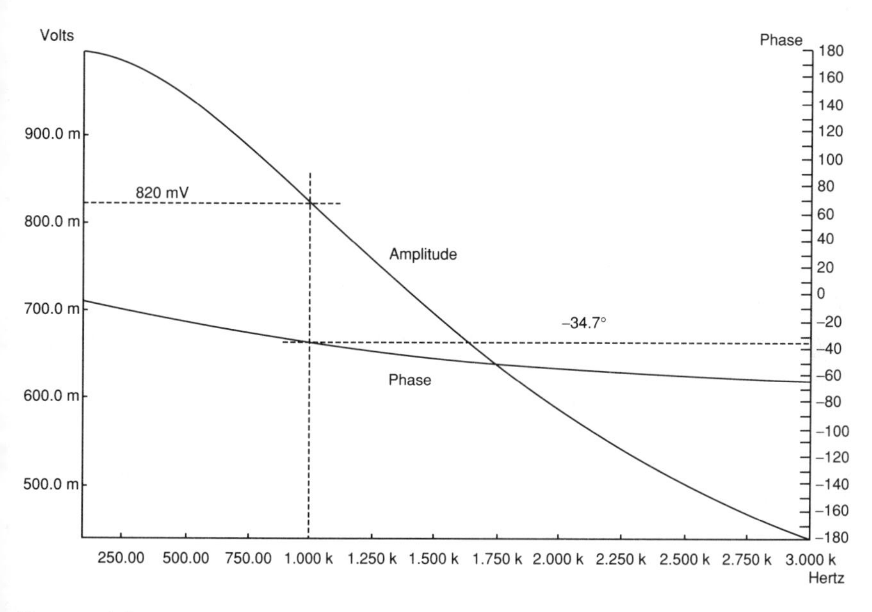

Figure 6.2

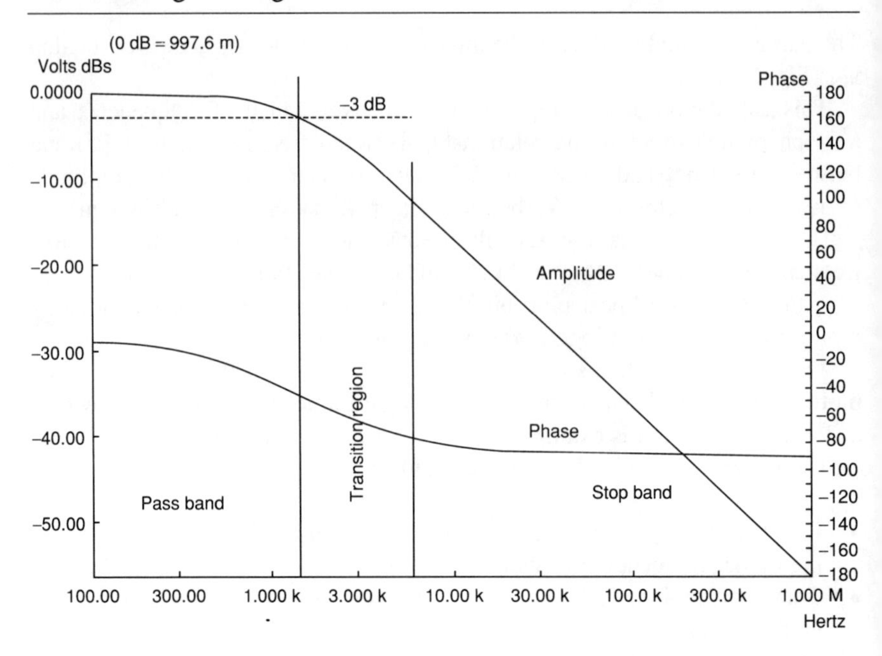

Figure 6.3

The most useful way of characterizing a frequency plot is to determine f_c and the roll-off in the stop band. The -3 dB level is chosen to delimit the pass band because this is the *half-power level*, when:

$$\frac{\text{Power out}}{\text{Power in}} = 0.5$$

Expressed in decibels:

$$10 \log(0.5) = 10 \times -0.3010 = -3.010\,\text{dB}$$

At the -3 dB point, the power developed across the capacitor is half that developed in the whole network. The other half of the power is developed across the resistor. For this to occur, the resistor and capacitor must have equal impedances. The impedance of the resistor is independent of frequency but f_c is the frequency at which the impedance of the capacitor is exactly equal to that of the resistor. Expressing f_c in terms of angular frequency:

$$\omega_c = 2\pi f_c$$

Ignoring phase, the equality of impedances gives:

$$R = 1/\omega_c C$$

$$\Rightarrow \qquad \omega_c = 1/RC$$

or

$$f_c = 1/2\pi RC$$

Example: In Fig. 6.1: $f_c = 1/(2\pi \times 500 \times 220 \times 10^{-9}) = 1447\,\text{Hz}$.

To find the phase relationship, recall that, at the $-3\,\text{dB}$ point:

$$\frac{\text{Power out}}{\text{Power in}} = 0.5$$

$$\Rightarrow \qquad v_{\text{OUT}}^2/v_{\text{IN}}^2 = 0.5$$

$$\Rightarrow \qquad v_{\text{OUT}}/v_{\text{IN}} = \sqrt{0.5} = 1/\sqrt{2}$$

Figure 6.4 demonstrates that this ratio between V_{IN} and V_{OUT} occurs when the phase diagram is a 45° triangle. Consequently, the phase lag at the $-3\,\text{dB}$ point is 45°.
In the pass band, where $\omega RC \approx 1$, and being concerned only with amplitude, not phase, the equation for $H(\text{j}\omega)$ may be rewritten:

$$H(\omega) = \frac{1}{\omega RC}$$

This indicates that the amplitude of V_{OUT} is inversely proportional to the frequency, the resistance and the capacitance. In particular, if frequency is doubled, amplitude is halved. A doubling of frequency is referred to as an *octave*. A halving of V_{OUT} from V to $V/2$ as a power ratio is equal to $V^2/(V/2)^2 = 0.5^2$. Therefore a halving of v_{OUT} in terms of the resulting power ratio is $10\log(0.5^2) = -6\,\text{dB}$.

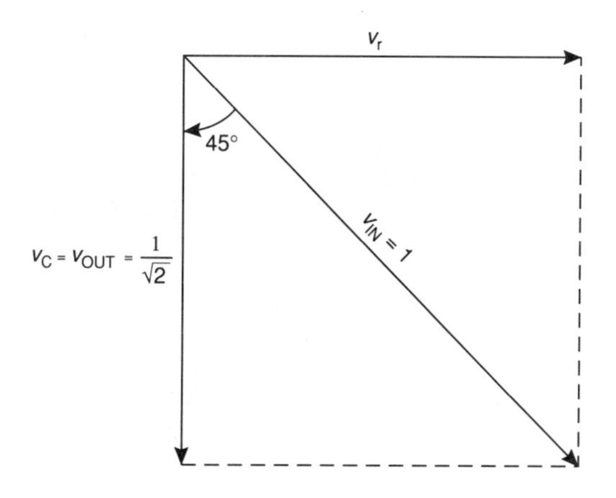

Figure 6.4

Example: In Fig. 6.3, amplitude falls from −16.8 dB at 10 kHz to −22.8 dB at 20 kHz. See 6.1.4 for a method of calculating roll-off.

The equation for $H(s)$ includes R and C as separate variables, but:

$$RC = 1/\omega_c$$

$$\Rightarrow \qquad H(s) = \frac{1}{s/\omega_c + 1}$$

This form of the transfer function applies to any combination of R and C that produces a given value of ω_c. Figure 6.5 is a three-dimensional plot of $H(s)$ against the components of s. This is for a low-pass filter for which $\omega_c = 10$. The horizontal plane of the plot is the *s-plane*. The highest plotted value for the transfer function is 2, when $\sigma = -5$ and $\omega = 0$. This is for a non-periodic signal decreasing exponentially at the rate e^{-5t}. The shape of the curve on the rear left surface (for $\sigma = -5$, $\omega = 0$ to 20) shows the response of the filter to increasing frequency, with an exponentially decreasing amplitude. The shape is very similar to that of Fig. 6.2, and is indicative of a low-pass filter. In Fig. 6.6 we have sliced vertically through the plot of Fig. 6.5 along the vertical plane for which $\sigma = 0$. We have also rotated the plot for a better view of the $\sigma = 0$ plane. The intersection between the plane for $\sigma = 0$ and the plotted surface is the graph for the transfer function for sinusoidal signals of fixed amplitude. It has the same form and the same interpretation as the response curve plotted in Fig. 6.2. At the left-hand end

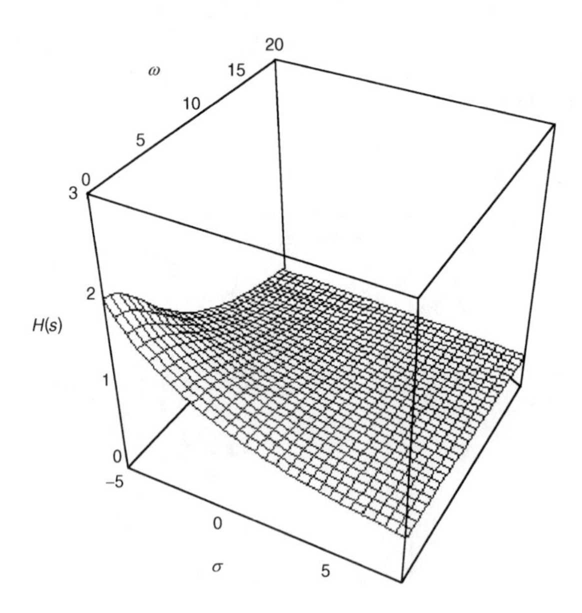

Figure 6.5

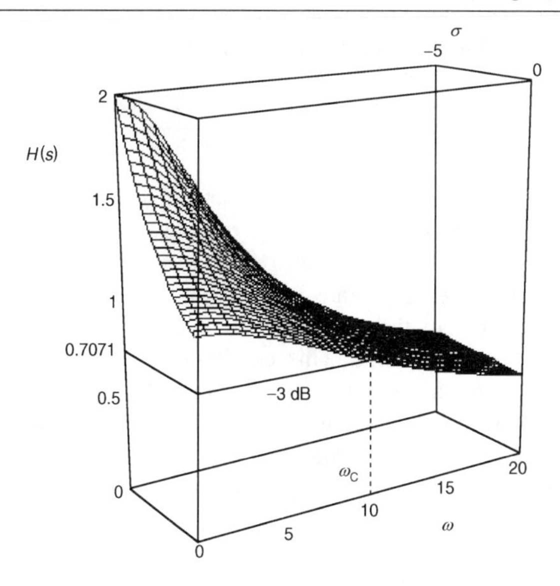

Figure 6.6

of this curve, $\sigma = 0$ and $\omega = 0$, which makes $s = 0$ and we have a constant DC level. At this point the term s/ω_c disappears from $H(s)$, the value of which becomes 1. As ω increases $\rightarrow$ s increases $\rightarrow$ s/ω_c increases $\rightarrow$ $H(s)$ decreases. When $\omega = 10 = \omega_c$, $H(s) = 0.7071$ (or $1/\sqrt{2}$), the -3 dB value. Returning to Fig. 6.5, we see that a similar low-pass action occurs when σ is negative, that is, for signals which are decreasing in amplitude. On the other hand, if σ is positive, meaning that amplitude is increasing exponentially, the filter response becomes much flatter and has much reduced overall gain.

6.1.2 RC high-pass filter

If the resistor and capacitor of Fig. 6.1 are transposed, lines of argument similar to those above show that the network is a high-pass filter. Its characteristics are:

$$H(s) = \frac{R}{R + 1/sC} = \frac{sRC}{sRC + 1}$$

$$H(j\omega) = \frac{R}{R + 1/j\omega C} = \frac{j\omega RC}{j\omega RC + 1}$$

$$f_c = 1/2\pi RC \text{ (same equation)}$$

a phase lead of 45° at f_c
roll-off of 6 dB per octave (positive) in the stop band
(from 0 Hz to about 500 Hz).

Figure 6.7 is the s-plane plot of $H(s)$ for a high-pass RC filter. Given $RC = 1/\omega_c$:

$$H(s) = \frac{sRC}{sRC + 1} = \frac{s}{s + \omega_c}$$

The intersection of the surface with the near right plane (for $\sigma = 8$, $\omega = 0$ to 30) shows a high-pass response. This response is stronger when $\sigma = 0$, for then $H(s)$ drops all the way to zero when $\omega = 0$, that is for a DC signal with constant amplitude. The transfer function predicts this result because substituting zero for s in the numerator of a quotient such as the transfer function makes it equal to zero. This condition is said to be a *zero* of the transfer function. We can also infer the response to high frequencies. As ω increases, so does s, but ω_c remains constant. Consequently, large values of s mean that ω_c plays a relatively smaller part in determining the value of $H(s)$. If ω increases $\rightarrow$ s increases $\rightarrow$ $H(s)$ approaches s/s $\rightarrow$ $H(s)$ approaches 1 $\rightarrow$ no attenuation at high frequencies.

If we expand the range of the plot of Fig. 6.7 to cover more negative values of σ, another feature of $H(s)$ appears (Fig. 6.8). There is a prominent upward sweep, known as a *pole*. Here we see only the base of the pole for it has been truncated by plotting only as far up as $H(s) = 2$. The axis of this pole is at the point where $\sigma = -10$ and $\omega = 0$. At these values, $s = -10$, and equals $-\omega_c$. Substituting $s = -10$ and $\omega_c = 10$ in the transfer function makes the expression $(s + \omega_c)$ equal to zero $\rightarrow$ the denominator $= 0 \rightarrow H(s)$ approaches infinity. A pole in this position means that, when the input is a non-periodic signal decreasing exponentially at the rate e^{-10t} the output at any instant is infinite. In practice there is always some damping due to the resistance of connections and terminal wires and also to power losses in dielectric of the capacitor, but at least the output of the filter is many times

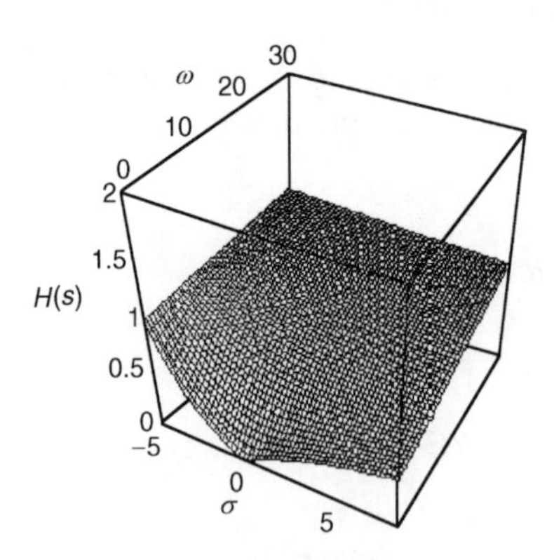

Figure 6.7

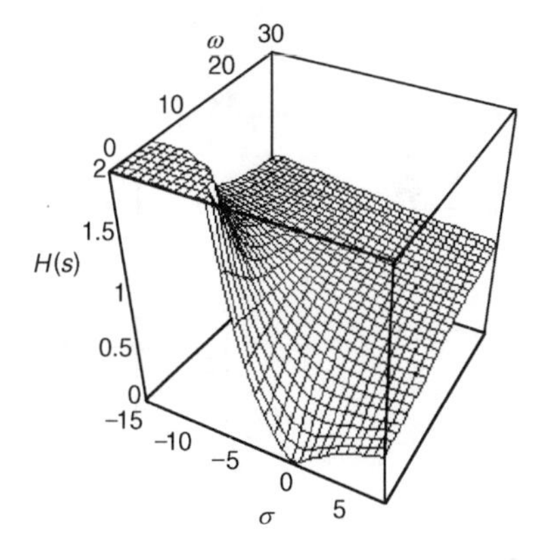

Figure 6.8

the input. This high gain is explainable. The input is falling so rapidly that the capacitor is unable to discharge through the resistor fast enough to keep up with it. Output (pd across the capacitor) exceeds input $\rightarrow H(s)$ is high. As the figure shows, the same applies to periodic signals decreasing at similar rates and simultaneously oscillating at low frequencies. The effect of rapid exponential decrease swamps signal changes caused by its periodicity.

Locating poles and zeros is important in filter design as we are able to predict regions of instability and regions of high attenuation. The pole in this filter is not on the $s = 0$ axis, so will not interfere with the action of the filter for constant-amplitude signals.

6.1.3 Other first-order passive filters

A first-order filter has just one reactive component, either a capacitor or an inductor. The reactance of a capacitor decreases with increasing frequency but the reactance of an inductance increases:

$$X_{\mathrm{L}} = sL$$

or, for constant amplitude signals:

$$X_{\mathrm{L}} = \mathrm{j}\omega L$$

We characterize resistor–inductor filters in the same way as *RC* filters. Figure 6.9 summarizes the properties of all four types of first-order passive filter.

The disadvantage of *RL* filters is that inductors tend to be heavy, large and expensive compared with capacitors. They are subject to electromagnetic interference and they also generate it. This in practice restricts inductors to

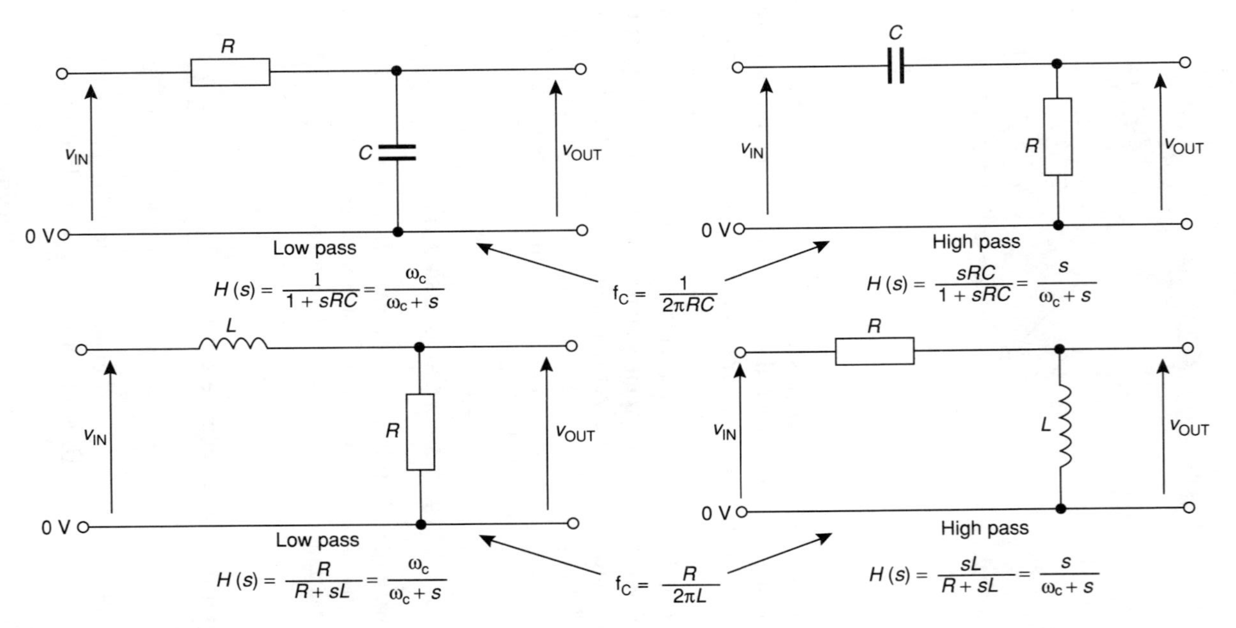
R
C
v_{IN}
v_{OUT}
0 V
Low pass
$H(s) = \frac{1}{1 + sRC} = \frac{\omega_c}{\omega_c + s}$
C
R
v_{IN}
v_{OUT}
0 V
High pass
$H(s) = \frac{sRC}{1 + sRC} = \frac{s}{\omega_c + s}$
$f_C = \frac{1}{2\pi RC}$
L
R
v_{IN}
v_{OUT}
0 V
Low pass
$H(s) = \frac{R}{R + sL} = \frac{\omega_c}{\omega_c + s}$
R
L
v_{IN}
v_{OUT}
0 V
High pass
$H(s) = \frac{sL}{R + sL} = \frac{s}{\omega_c + s}$
$f_C = \frac{R}{2\pi L}$

Figure 6.9

high-frequency (e.g. radio frequency) applications, which require only small inductors, with often only 1 or 2 turns of wire.

6.1.4 Second-order passive filters

Second-order filters contain two reactive components:

- Two cascaded low-pass *RC* or *RL* networks produce a low-pass filter with steeper roll-off.
- Two cascaded high-pass *RC* or *RL* networks produce a high-pass filter with steeper roll-off.
- An *RC* or *RL* low-pass filter cascaded with a high-pass filter of the same type produces a *band-pass filter* or a *band-stop (notch) filter*, depending on the corner frequencies of the two sections.

The response of such filters is complicated by the fact that the second filter draws current from the first filter. In the calculations above, for a single *RC* or *RL* network, it has been assumed that the output impedance of the source (v_{IN}) is zero and that no current is being drawn from the network. Neither assumption applies to cascaded filters. Design of cascaded filters usually relies on ready-compiled tables or on filter-design software.

When a second-order filter contains one capacitor and one inductor, their contrary responses to frequency confer advantageous properties. For example, the frequency response of the *LC* network in Fig. 6.10 has a notable peak at f_c (Fig. 6.11). This in itself is not advantageous but is put to service as explained later.

The explanation of the high peak in Fig. 6.11 is that f_c (about 1 kHz) is the *resonant frequency* of the *LC* network. The network resonates when the impedances of the capacitor and inductor are equal:

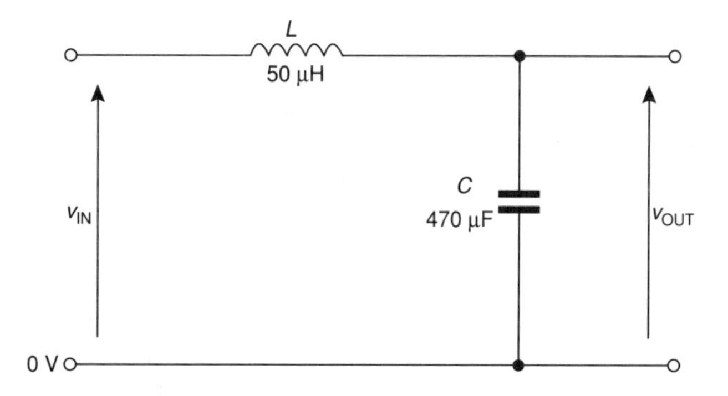

Figure 6.10

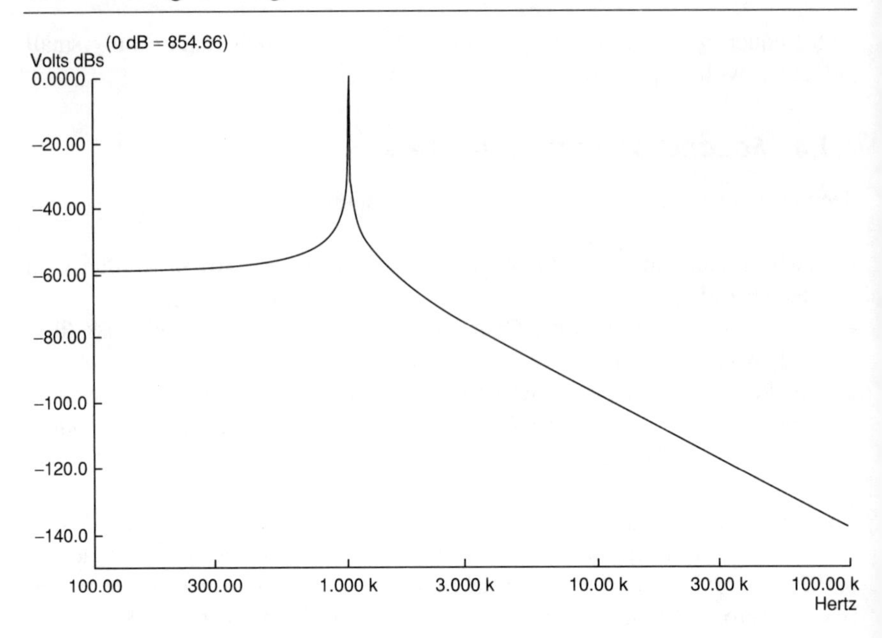

Figure 6.11

$$1/\omega_c C = \omega_c L$$

$$\Rightarrow \qquad \omega_c^2 = 1/LC$$

$$\Rightarrow \qquad \omega_c = \frac{1}{\sqrt{LC}}$$

During each cycle of v_{IN}, energy stored in the form of an electrical field in the capacitor is transferred to the inductor and reappears as a magnetic field. Then the magnetic field decays, transferring the energy back to the capacitor. Because v_{IN} is oscillating at the resonant frequency, it adds just a little more energy to the system at each cycle of energy transfer. Cycle by cycle, the amount of energy increases and, with it, the amplitude of v_{OUT}. Because the signal across the capacitor is 180° out of phase with that across the inductor, the total voltage across the *LC* pair alternates by only as much as v_{IN}, even though the signals across the capacitor or inductor may individually be of high (and equal) amplitudes. Figure 6.11 reveals that the simulator has calculated these amplitudes to be as much as 850 V, which is unrealistic. Energy builds up in the electrical and magnetic fields but, as time goes on, more and more energy is being dissipated by heating in the coil of the inductor and in the dielectric of the capacitor. These energy dissipating effects are not modelled in the simulation.

The high peak in Fig. 6.11 is predictable from the equation for $H(s)$. Using the same potential-divider approach as for the low-pass filter (6.1.1), we find at the resonant frequency ω_c:

$$H(s) = \frac{1/sC}{sL + 1/sC} = \frac{1}{s^2LC + 1} = \frac{1}{j^2\omega_c^2LC + 1} = \frac{1}{-\omega_c^2LC + 1}$$

assuming that $\sigma = 0$ and substituting $\omega_c^2 = 1/LC$:

$$\Rightarrow \qquad H(s) = \frac{1}{-1 + 1} = \frac{1}{0}$$

v_{OUT} becomes infinitely large at ω_c. There is a pole at $\sigma = 0$, $\omega = \omega_c$. As with the simulator result, this does not allow for energy dissipation in the components.

The resonance is damped by including a resistor in the network (Fig. 6.12). The transfer function becomes:

$$H(s) = \frac{1/sC}{sL + R + 1/sC} = \frac{1}{s^2LC + sRC + 1}$$

As oscillations build up, there comes a point at which the amount of energy being supplied to the system by v_{IN} is equal to the energy lost by dissipation in the resistor. Figure 6.13 shows the frequency response as the resistance is swept from 0.1 Ω to 1.1 Ω in steps of 0.2 Ω. When $R = 0.1\,\Omega$, there is still a pronounced peak at the corner frequency. This peak diminishes as R is increased, the response gradually becoming more rounded. At $R = 0.1\,\Omega$ damping is insufficient to prevent a certain amount of resonance; the filter is *under-damped.* At $R = 1.1\,\Omega$, there is no resonance; the response is more like that of the *RC* filter of Fig. 6.3, with a wide transition region between the pass and stop bands. The filter is *over-damped.* At an intermediate value of R, fairly close to $R = 0.3\,\Omega$, there is a fairly sharp knee to the curve. There is a relatively sharp distinction between the pass band and the stop band. This is ideal behaviour for a filter; it is *critically damped.*

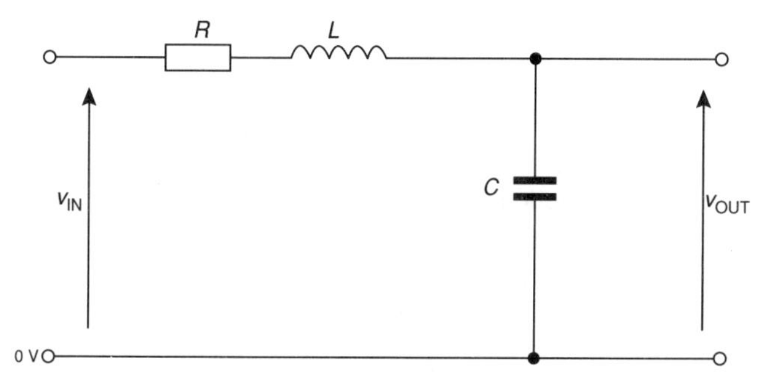

Figure 6.12

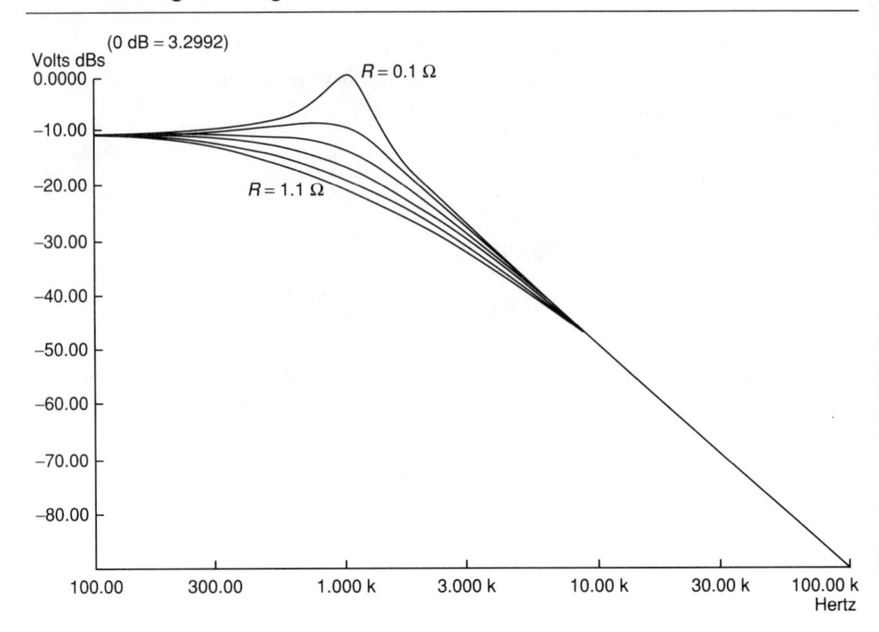

Figure 6.13

Roll-off in the stop band may be calculated from the equation above:

$$H(s) = \frac{1/sC}{sL + R + 1/sC} = \frac{1}{s^2LC + sRC + 1} = \frac{1}{(\mathrm{j}\omega)^2LC + \mathrm{j}\omega RC + 1}$$

Ignoring phase:

$$H(\omega) = \frac{1}{-\omega^2LC + \omega RC + 1}$$

At frequencies in the stop band:

$$H(\omega) \approx \frac{1}{-\omega^2LC + \omega RC}$$

If ω_1 and ω_2 are two frequencies an octave apart so that $\omega_2 = 2\omega_1$

$$\text{Roll-off} = \frac{H(\omega_2)}{H(\omega_1)} = \frac{-\omega_1^2LC + \omega_1RC}{-\omega_2^2LC + \omega_2RC}$$

$$= \frac{-\omega_1^2L + \omega_1R}{-\omega_2^2 + \omega_2R}$$

Substituting for ω_2, and ignoring the term in R, since R is small:

$$\text{Roll-off} = \frac{\omega_1^2 L}{(2\omega_1)^2 L} = 0.25$$

$$\text{Roll-off (dB)} = 10 \times \log(0.25^2) = -12\,\text{dB}$$

This is double the roll-off in the first-order filters. In practice, damping requires only a small resistor, for only a small amount of energy is injected into the system at each cycle. R could be present as the resistance of the wire of the inductor coil, no separate resistor being required. The inductance of an inductor depends on several factors: the number of turns, the diameter of the turns, the length/diameter ratio, the magnetic permeability of the core. Consequently it is possible to produce a range of inductors of different sizes and shapes, all having the same inductance. The resistance of the wire of their coils will vary and, within limits, it is possible to wind an inductor so that it has a required resistance. For instance, an inductor could be produced which has the resistance needed for critical damping. The property of the inductor which relates to its damping ability is known as the *quality factor*, or Q, where:

$$Q = \omega_c L/R$$

Example: If $\omega_c = 1\,\text{kHz}$, an inductor with $L = 50\,\mu\text{H}$ and $R = 0.2\,\Omega$, has $Q = 0.25$.

Looking again at the equation for $H(s)$, although the pole at $\sigma = 0$, $\omega = \omega_c$ has been damped by including resistance, there may be other poles to be discovered. To locate these, put the denominator equal to zero:

$$s^2LC + sRC + 1 = 0$$

This is a quadratic equation and therefore has two roots. Substituting the known values of R, L and C and using the quadratic formula we find that the roots of the equation are both imaginary:

$$s = -2000 \pm \text{j}6209$$

That is to say, poles exist where $\sigma = -2000$ and $\omega = \pm 6209$. Figure 6.14 shows the pole at $\sigma = -2000$ and $\omega = +6209$. The right-hand surface of the plot displays the frequency response for constant-amplitude signals. The response is unity for DC and low-frequency signals. There is a moderate upward swing when $\omega = \omega_c$ when ($f = 1038\,\text{Hz}$), then a downward slope. This is equivalent to the curve for $R = 0.3\,\Omega$ in Fig. 6.13, except that it is plotted on linear scales. The slight upward knee of the curve at $\sigma = 0$ expands into a pole at $\sigma = -2000$. Although the filter is damped for signals of constant amplitude, it is far from damped for signals of exponentially decreasing amplitude. Figure 6.15 analyses the behaviour of the filter when the amplitude of the input signal is decreasing rapidly:

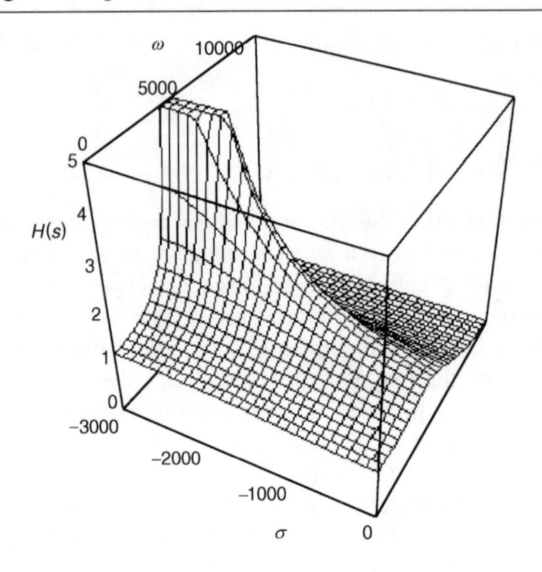

Figure 6.14

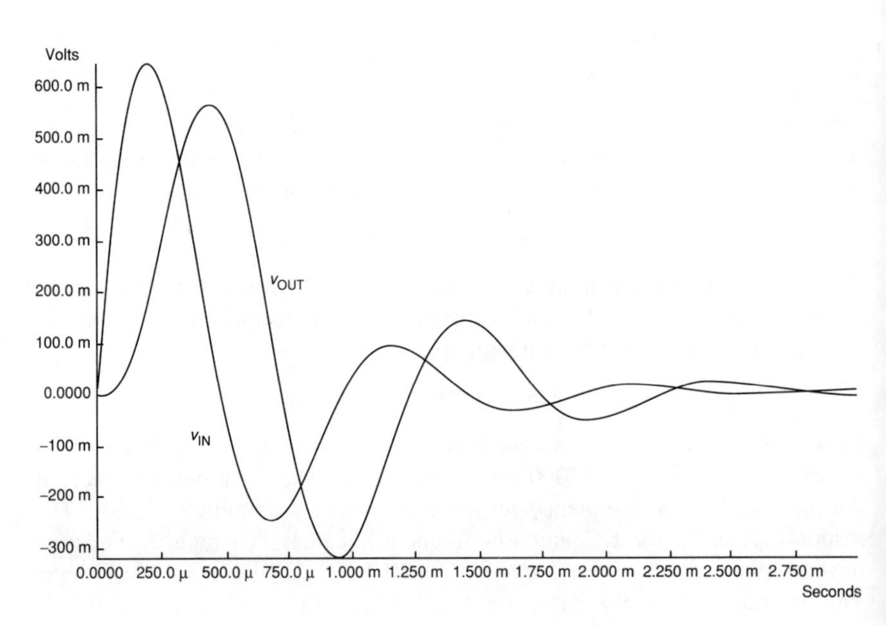

Figure 6.15

$$v_{IN} = e^{-2000t} \sin 6522t$$

The amplitude of v_{OUT} is greater than that of v_{IN}. But σ is negative, and the signal is decreasing, so the situation is short-lived.

6.2 Active filters

The active stage may be a transistor but most often it is an operational amplifier. The amplifier draws power from an external source and uses this to increase the power of the output signal. Thus the attenuation associated with passive filters is avoided. An equally important outcome of introducing operational amplifiers is that they can be used to shape the response of the filter, particularly to sharpen the knee of the response curve, so making it possible to dispense with weighty and bulky inductors.

6.2.1 Two-pole filters

It is possible simply to follow any passive filter with an op amp to restore the signal to its original power or even to exceed it. But this is not making full use of the op amp, particularly its feedback facility. The low-pass filter in Fig. 6.16 is referred to as a *two-pole filter* because it comprises two *RC* stages. It is also known as a *VCVS filter* because the op amp is used as a voltage-controlled voltage source. Figure 6.17 is a 3-pole filter constructed on the same principles. Filters of higher order may be built by cascading several 2-pole or 3-pole filters.

There are two types of feedback in this circuit. There is negative feedback from the output, making the op amp a voltage follower (D.2.2); some filters have potential divider R_A and R_F as in Fig. D.2a, to set the gain of the filter to other values. In addition, there is positive feedback through C1. The amount of feedback is frequency dependent, being greatest at f_c. The effect of this is

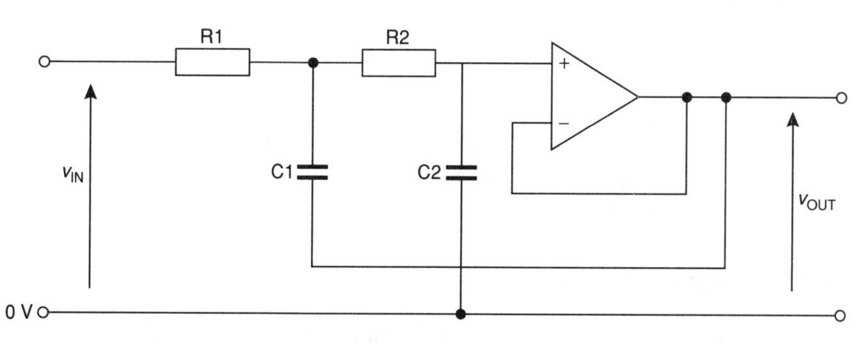

Figure 6.16

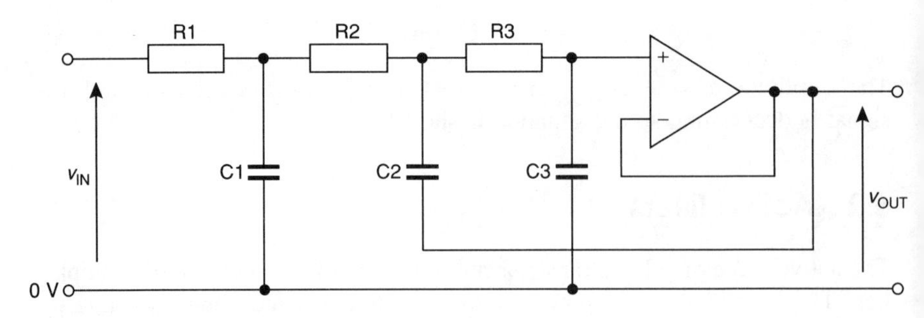

Figure 6.17

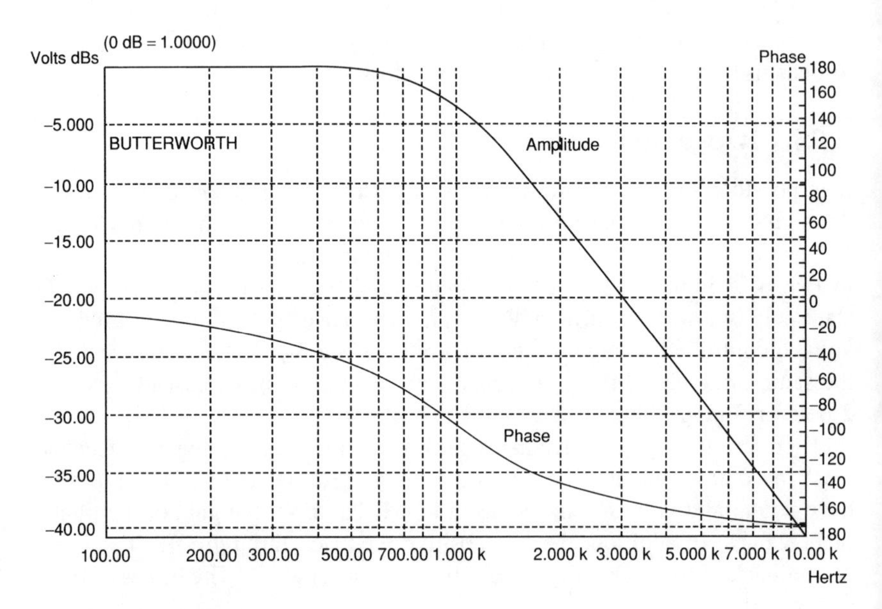

Figure 6.18

to sharpen the knee of the response curve, reducing the width of the transition region. The improved response (Fig. 6.18) is very similar to that obtained by using an inductor. There being two reactive devices in the network, roll-off is −12 dB per octave.

6.2.2 Filter responses

By using op amps we can do more than mimic the behaviour of an inductor. We can exercise much more precise control over the shape of the response curve. Given the networks of Figs. 6.16 and 6.17, we can produce an infinite number of filters with a low-pass response, simply by altering the values of the capacitors and resistors. This is also possible for high-pass and other categories of filter. Responses are classified under a few main headings, based on certain characteristics:

- Pass band flatness: Ideally, the pass band should be flat, but in certain types of filter the pass band has ripples, that is, has regions of slightly increased or decreased response.
- Phase response: Filters differ in the degree to which signals of different frequencies are changed in phase as they pass through the filter. A change in phase is equivalent to a delay and, if the different components of a signal are delayed by different amounts, the result is distortion of the signal. Ideally, though seldom in practice, all frequencies should have the same phase change.
- Roll-off: The steeper the curve in this region, the more effectively the filter separates out the frequencies that are to be passed from those that are to be stopped.
- Overshoot: When the input level changes abruptly (especially with square-wave or pulsed signals) the output may swing too far, oscillating for a while before settling to the correct new level.

All filter responses are a matter of compromise. Figure 6.18 to 6.23 show the responses of three types of 2-pole filter, all based on the network of Fig. 6.16. The component values were obtained by using sets of ready calculated tables, but could also have been obtained by using filter-design software. In all networks $R1 = R2 = 1\,k\Omega$. Only capacitor values are altered.

The Butterworth response (Fig. 6.18) with $C1 = 225\,nF$ and $C2 = 112\,nF$, shows a flat pass band and moderate roll-off (12 dB/octave). The phase response is poor ranging from $-22°$ at 100 Hz to $-172°$ at 10 kHz. The Butterworth filter has the flattest pass band of all types, making this a popular filter response.

The transfer function for the Butterworth response is:

$$H(s) = \frac{1}{1 - s^{2n}}$$

where n is the order of the filter. Thus the functions for the first three orders are:

$$H(s) = \frac{1}{s+1} \qquad H(s) = \frac{1}{s^2 + \sqrt{2}s + 1} \qquad H(s) = \frac{1}{s^3 + 2s^2 + 2s + 1}$$

Figure 6.19 demonstrates the way in which the response varies with filter order. This is for a normalized filter, a filter with amplitude 1 in the pass band

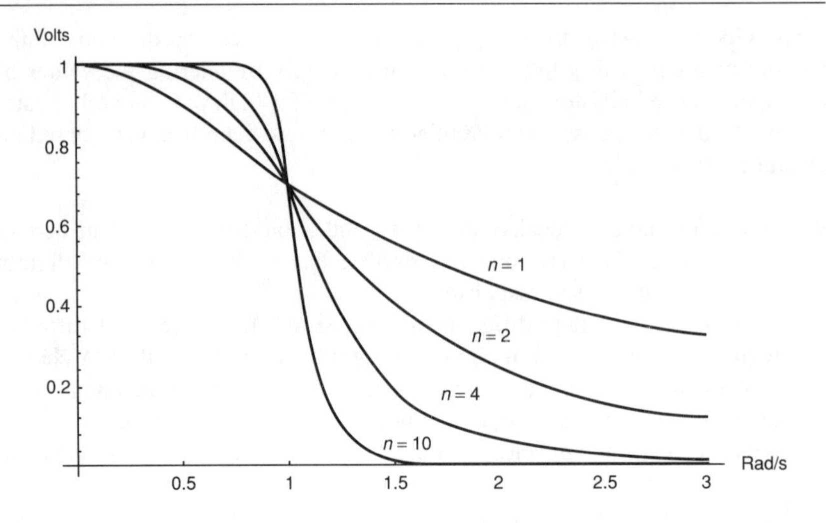

Figure 6.19

and $\omega_c = 1$ rad/sec ($f_c = 0.159$ Hz), and is representative of the response of all Butterworth filters. The curves plot the response for orders 1, 2, 4 and 10. The response sharpens with increasing order but all have a flat pass band except for order 1. All have the -3 dB point (magnitude $= 0.7071$) at $\omega = 1$. The roll-off is $12n$ dB per octave ($20n$ dB per decade), where n is the order.

The Chebyshev response (Fig. 6.20) with C1 $= 353n$, C2 $= 96n$ (all other components unchanged) shows a pass band with a ripple. There is only one ripple here, in that the curve swings up slightly in the pass band before it curves down to the stop band. The maximum amplitude of the ripple is 1 dB, which is a relatively small drawback to set against the sharp 'knee' which causes roll-off to begin at a lower frequency than in the Butterworth filter. Although the roll-off is the same (-12 dB per octave), amplitude is down to -15 dB at $2f_c$ (compared with -13 dB for Butterworth). The phase response in the 2nd order filter is similar to that of the Butterworth response, but slightly steeper around f_c, which increases distortion in frequencies in that region.

The shape of the Chebyshev response is similar to that of the capacitor–inductor filter (Fig. 6.13), illustrating the point that, by using op amps, we are able to dispense with inductors, with all their disadvantages.

The transfer function for the Chebyshev filter is based on the Chebyshev polynomials, which are evaluated as follows:

$$C_0 = 1$$

$$C_1 = \omega$$

$$C_2 = 2\omega^2 - 1$$

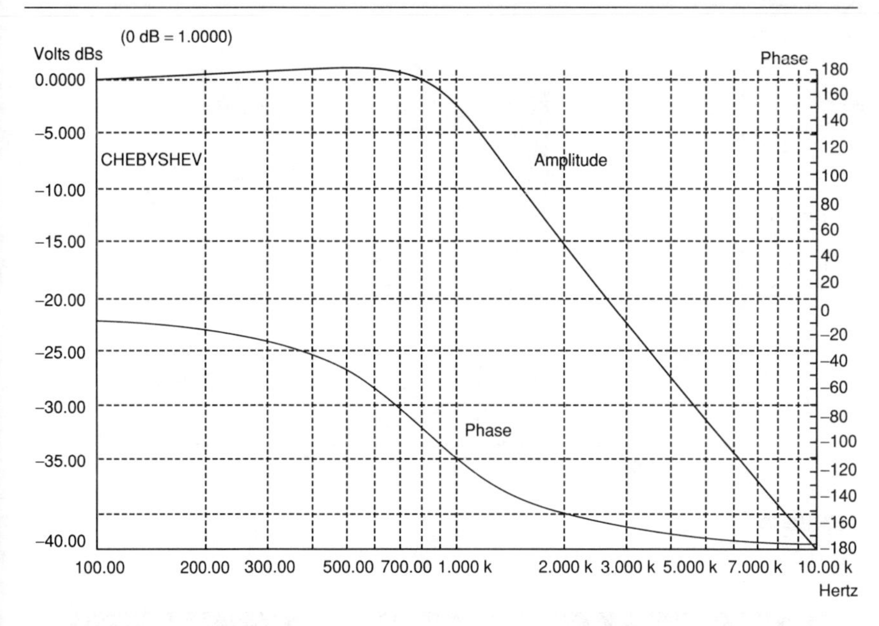

Figure 6.20

$$C_3 = 4\omega^3 - 3\omega$$

$$\Rightarrow \qquad C_n = 2\omega C_{n-1} - C_{n-2}$$

The magnitude of transfer function is given by:

$$\left|H^2(\omega)\right| = \frac{1}{1 + \varepsilon^2 C_n^2}$$

where ε determines the depth of the ripples. In the normalized filter, the magnitude of the ripple maxima is 1 and that of the ripple minima is:

$$\frac{1}{\sqrt{(1 + \varepsilon^2)}}$$

Normalized Chebyshev response curves for orders 1, 2, 4 and 10 are plotted in Fig. 6.21. In these curves, $e = 0.4$ so the ripple minima are 0.928. This corresponds with a ripple of 0.65 dB. The roll-off in a Chebyshev filter increases markedly with increasing order. Comparison of Fig. 6.21 with Fig. 6.19 demonstrates that, for a filter of given order, the Chebyshev filter has appreciably greater attenuation than the Butterworth filter. If roll-off is the prime consideration, a Chebyshev filter requires fewer stages than the Butterworth filter.

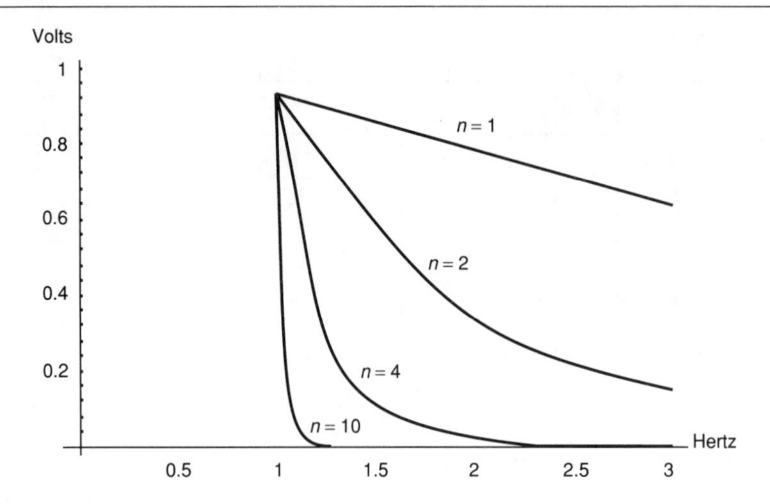

Figure 6.21

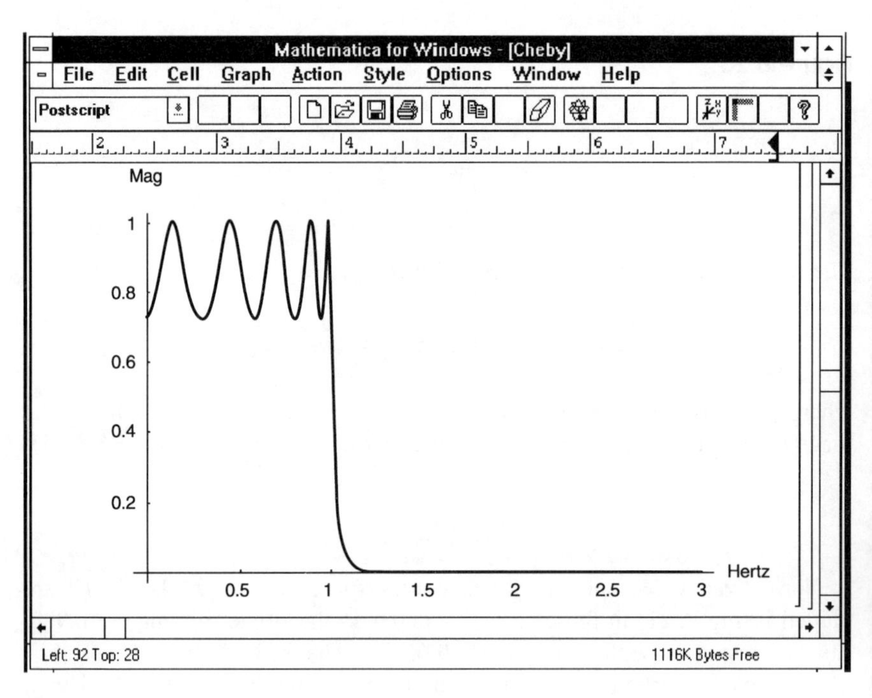

Figure 6.22

In the Chebyshev filter, the roll-off is related to ε. Figure 6.22 shows the response of a 10th order Chebyshev filter with $\varepsilon = 0.95$ (ε must be < 1). Now the ripple minima are at 0.725, corresponding to ripple depth of 2.8 dB. Compromise is required; the steeper the roll-off, the deeper the ripples. Note that the number of maxima and minima in the pass band (excluding the point at which the curve passes through ω_c) equals the order of the filter. The higher the order, the more ripples. In Fig. 6.20, a 2nd order filter, there is one minimum (at $\omega = 0$) and one maximum, total 2. In Fig. 6.22 there are 5 minima and 5 maxima, total 10. The ripples always end with a maximum before the response dips down to the stop band so, if the order is even, magnitude is 1 at $\omega = 0$; if the order is odd, there is a minimum when $\omega = 0$.

The Bessel response (Fig. 6.23), with $C1 = 215\,\text{nF}$ and $C2 = 107\,\text{nF}$, produces a more rounded knee and a less steep roll-off than either Butterworth or Chebyshev, but has the advantage that the phase response is flatter in the pass band. The effect is more marked in filters of higher order. Another advantage of the Bessel filter is that it shows less overshoot, another cause of signal distortion, than the other two types. Figure 6.24 illustrates what happens when the input to a 2nd order Bessel filter is suddenly increased from 0 V to 1 V. The output reaches 1 V in 0.5 ms, overshoots by 44 mV, finally settling to 1 V in a total time of 1.6 ms. By contrast, the Chebyshev filter (Fig. 6.25) reaches 1 V slightly earlier (0.46 ms), but overshoots much

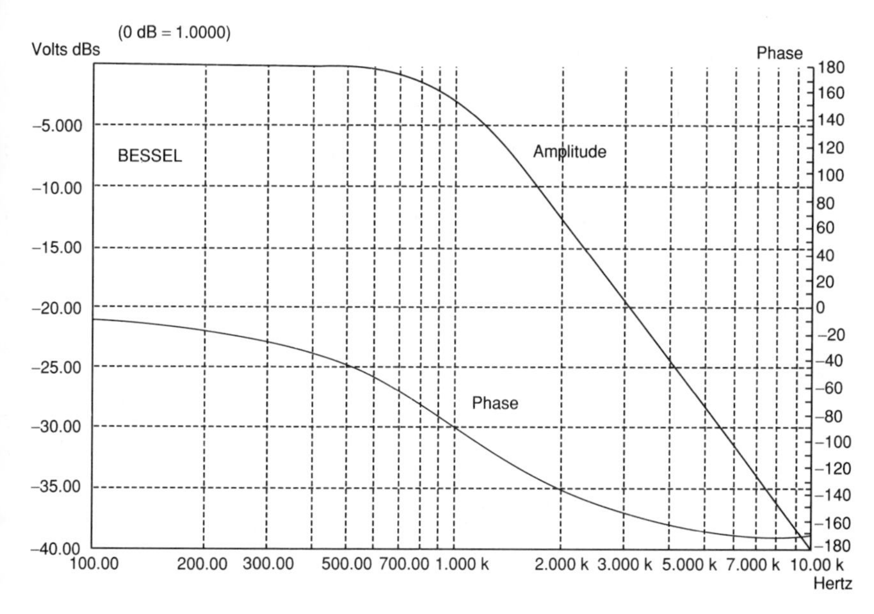

Figure 6.23

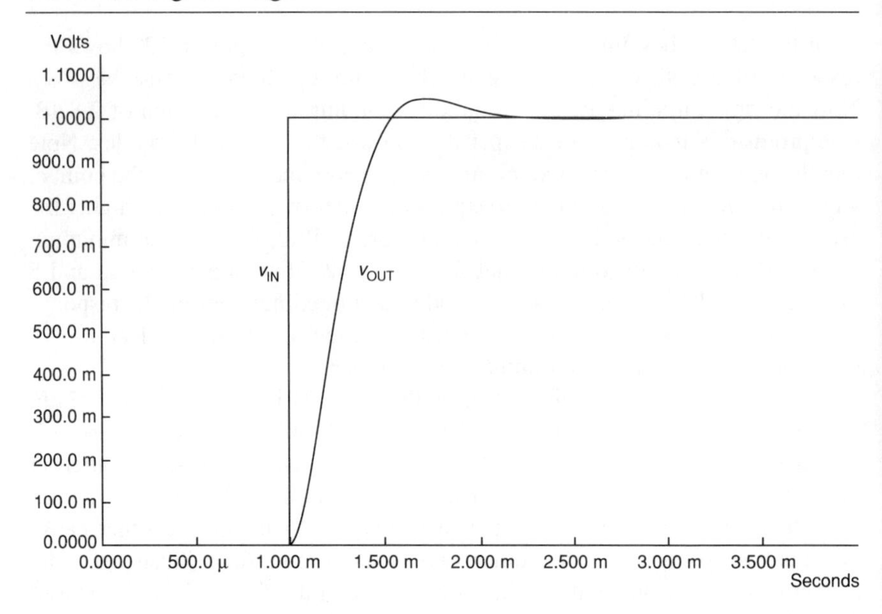

Figure 6.24

more (150 mV) and oscillates about the 1 V level for 2.6 ms before settling. We have demonstrated overshoots with a step wave but it occurs with all waveforms, including sinusoids, even if it is less obvious.

The Cauer or elliptic filter is a relative of the Chebyshev filter and has a very steep roll-off. Like the Chebyshev filter, it has ripples in the pass band. Also, because its transfer function contains zeros, which the other types do not, it has ripples in the stop band. This means that the low-pass Cauer filter may pass certain frequencies higher than f_c, but these frequencies are too high to be significant in the applications for which the elliptic filter is used. Although the Cauer filter has the benefit of steep roll-off, its phase response is very irregular and it cannot be used in situations in which phase is important.

6.2.3 Other active filters

The low-pass filters described above are readily convertible to high-pass filters by exchanging the resistors with the capacitors and by looking up their values in the tables. Their characteristics are the high-pass equivalents of the low-pass prototypes.

One way to produce an active band-pass filter is to cascade a low-pass active filter with a high-pass active filter (Fig. 6.26). The cut-off frequency of the low-pass filter f_H is set to be higher than that of the high-pass filter f_L

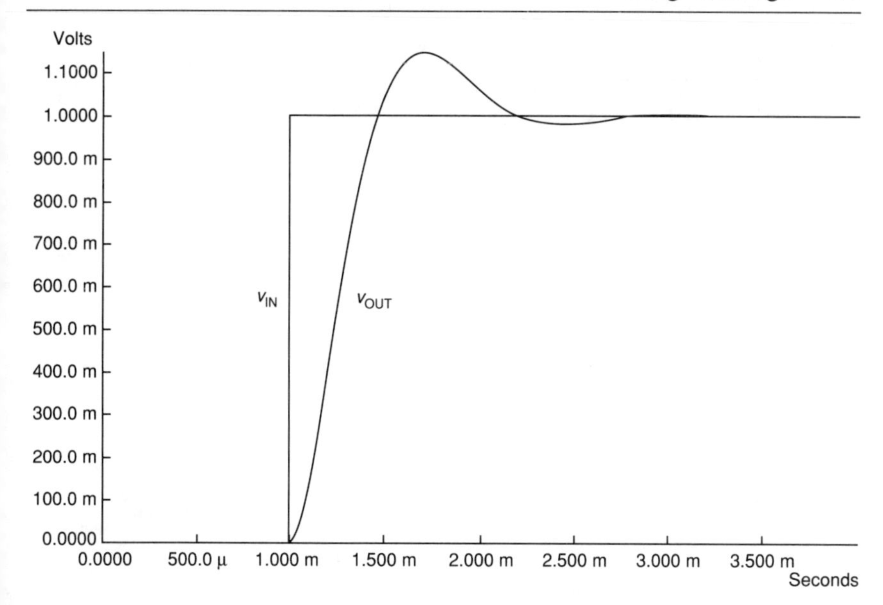

Figure 6.25

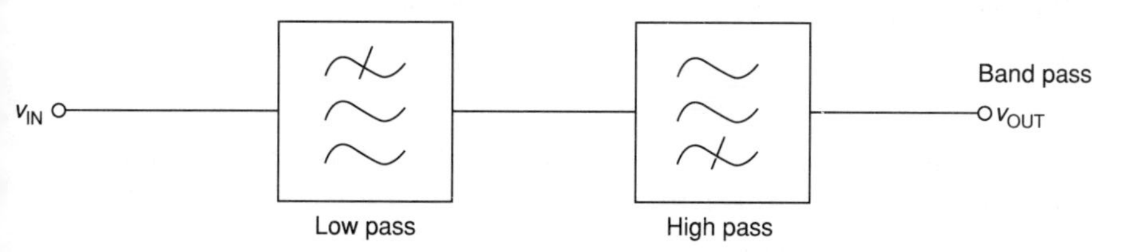

Figure 6.26

(Fig. 6.27). The *bandwidth* is the difference between the two cut-off frequencies, $f_H - f_L$. The central frequency of the pass band is the geometric mean of the two -3 dB frequencies:

$$f_c = \sqrt{(f_H \times f_L)}$$

Example: If $f_H = 1050$ Hz and $f_L = 950$ Hz, the bandwidth is 100 Hz and the central frequency is 998.7 Hz. An identical bandwidth when the central frequency is, say, 100 kHz represents a very precisely tuned filter with a narrow pass band. By contrast, a bandwidth of 100 Hz represents a very broad pass band if the central frequency is only 300 Hz.

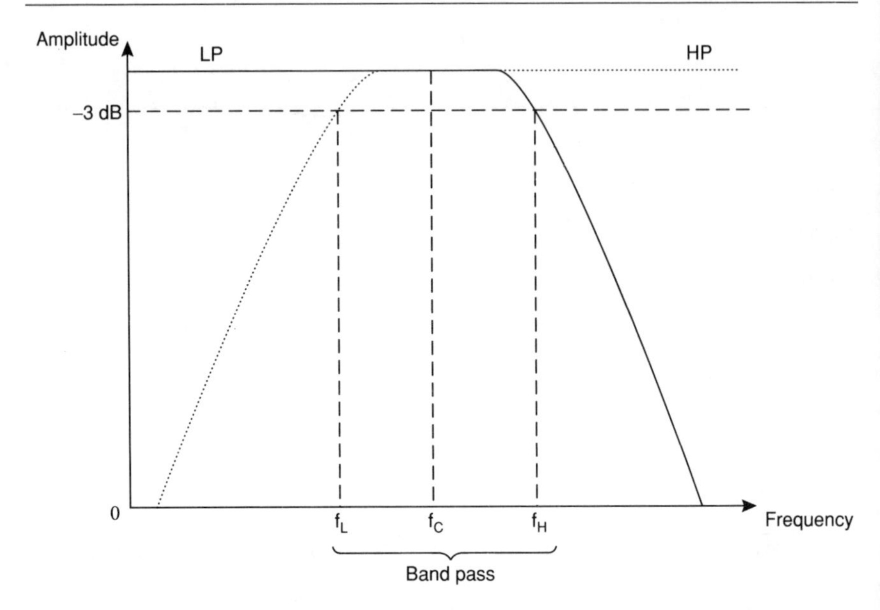

Figure 6.27

We need to quote the bandwidth relative to the central frequency:

$$\frac{\text{Bandwidth}}{f_C} = \text{fractional bandwidth}$$

For the example above, fractional bandwidth $= 100/998.7 = 0.100$. The quality factor Q (6.1.4) is a measure of the sharpness of response and, in the case of band-pass filters, of the width of the pass band (bandwidth) relative to the central frequency. The greater the bandwidth, the lower the quality factor. This gives another definition for Q:

$$Q = \frac{1}{\text{fractional bandwidth}}$$

For the example above, $Q = 1/0.100 = 10$.

Cascaded high-pass and low-pass filters are suitable if the pass band is reasonably wide. As we move f_H and f_L closer together to obtain a narrow pass band, the region of overlap includes more and more of the knees, so that the amplitude at the centre frequency is attenuated.

For a narrow pass band with no attenuation we make use of the feedback feature of op amps. An inverting op amp active filter (Fig. 6.28) has a *twin-T frequency rejection network* in the negative feedback loop. One of the Ts consists of two resistors, value R and a capacitor value $2C$. This acts as a low-pass filter (compare Fig. 6.1). The other T, consisting of two capacitors value C and a resistor value $R/2$ is a high-pass filter. The filters are connected in

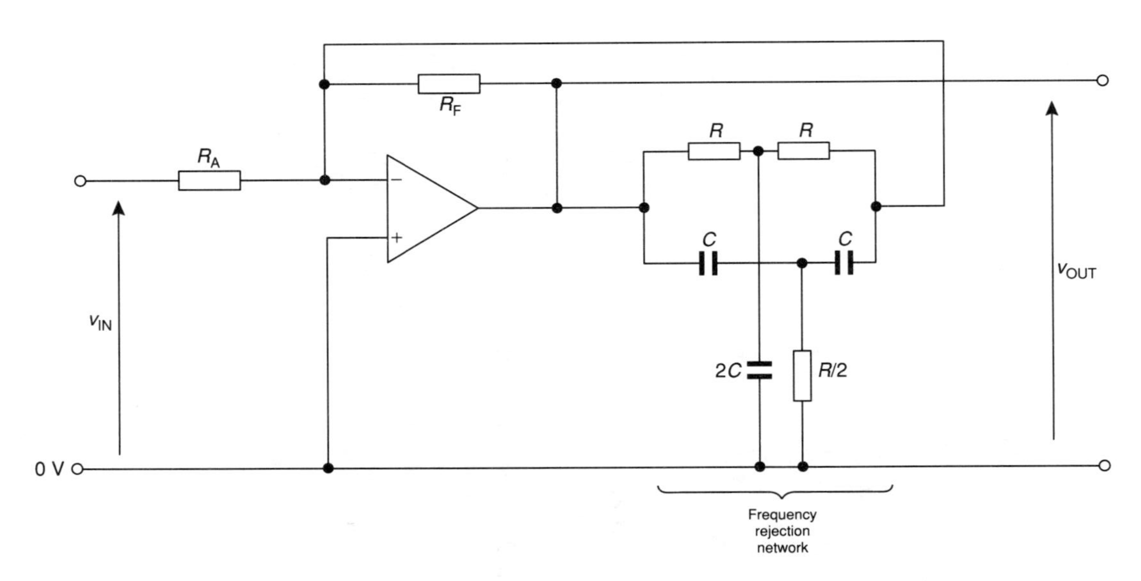

Figure 6.28

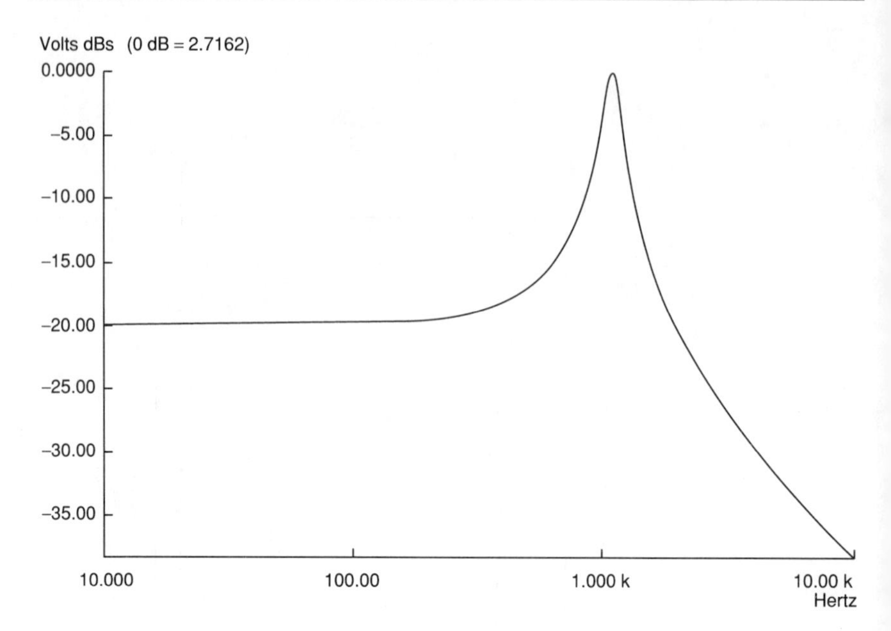

Figure 6.29

parallel so that one passes low frequencies, the other passes high frequencies and only intermediate frequencies are attenuated by the network. The signal from the network, consisting of low and high frequencies, but not intermediate frequencies, is fed back to the negative input of the op amp. This cancels out the low and high frequencies present in v_{IN}, allowing only intermediate frequencies to appear in v_{OUT}. Thus we have a band-pass filter. The frequency response (Fig. 6.29) has a sharp peak at 1 kHz ($f_c = 1/(2\pi RC)$, where $R = 1.592\,k\Omega$ and $C = 100\,nF$. The stop band below the peak is at $-20\,dB$. Above the peak the response drops at $-6\,dB$ per octave. Note that v_{IN} goes to the inverting input, so the filtered signal is the inverse of the input signal. In other words, there is a phase difference of 180° in addition to the phase changes associated with the capacitors.

The response shown in Fig. 6.23 is that obtained when $R_F = 20\,k\Omega$. This feedback resistor is in parallel with the rejection network and acts to dampen the response. If R_F is omitted, the output peaks very sharply to a high level, as there is a pole when $f = 1\,kHz$. The smaller the value of R_F, the greater the damping and the broader the pass band.

A *multiple feedback* band-pass filter (Fig. 6.30) has two feedback loops. The loop through C1 has a high-pass filter consisting of C1 and R1. High-frequency signals enter the circuit through R1, then pass easily through C2 to

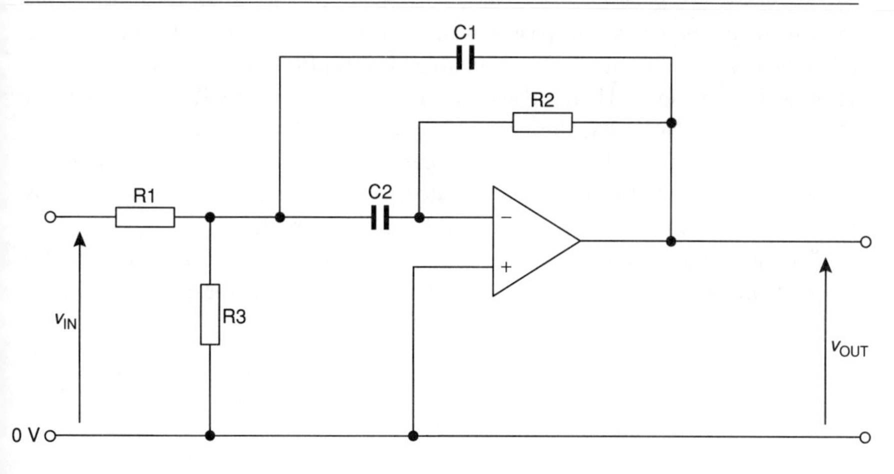

Figure 6.30

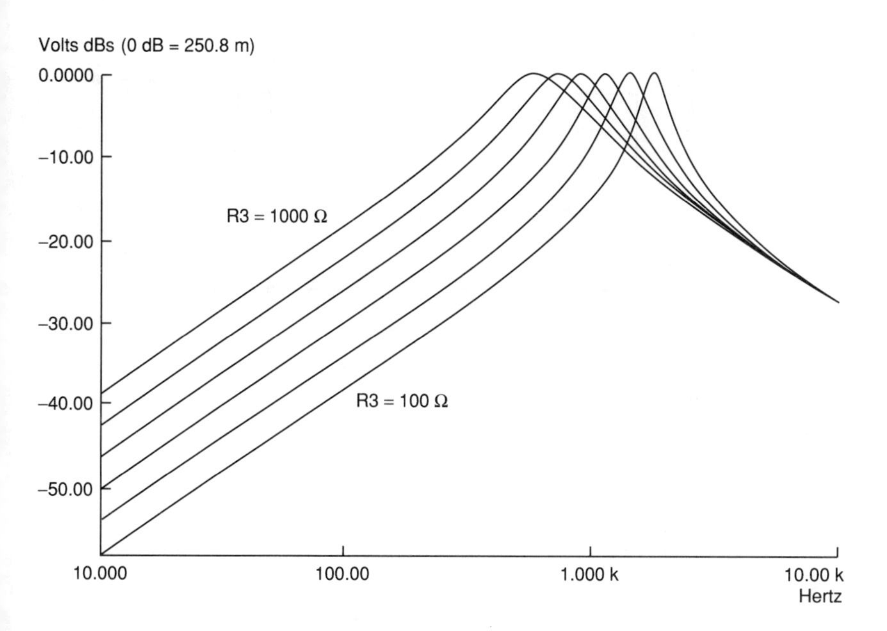

Figure 6.31

the op amp; they are inverted and then fed back through the C1/R1 high-pass filter, so that they cancel out. The loop through R2 is a low-pass filter consisting of R2 and C2. Any low frequency signals that manage to pass through C2 on their way to the op amp are inverted and fed back through the R2/C2 low-pass filter, to cancel out low-frequency signals. Only signals of intermediate frequency can pass through the circuit without attenuation. The advantage of this filter is that it is tunable. The centre frequency is adjusted by varying R3. In Fig. 6.31, the response curves are shown as R3 is swept from 100 Ω to 1 kΩ. Note that the centre frequency is changed but the amplitude of the output and the sharpness of the pass band (Q) are unaltered. Although in Fig. 6.31 the bandwidth appears to decrease as frequency increases, this is because the responses are plotted on a logarithmic scale.

The fact that op amps have a built-in roll-off (see D.1, gain bandwidth product) must be taken into account when designing high-pass and band-pass

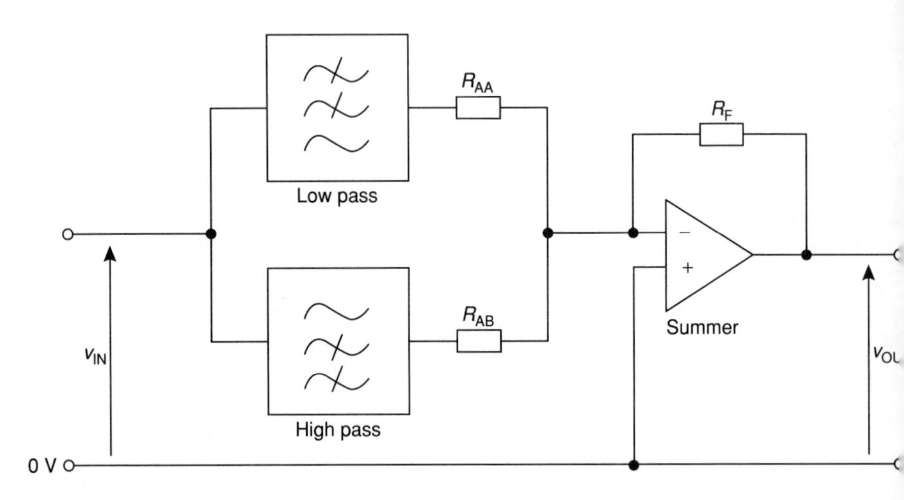

Figure 6.32

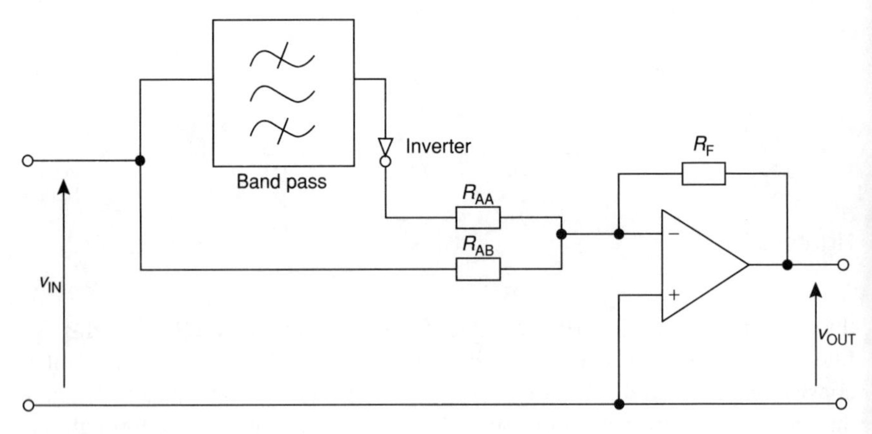

Figure 6.33

filters for frequencies over about 10 kHz. This feature can simplify filter design as a low-pass filter takes on the properties of a band-pass filter. It can cause problems when designing high-pass filters.

Band-reject or notch filters are useful for removing a particular frequency (e.g. 50 Hz mains interference) from a signal. Figure 6.32 shows the outputs of paralleled low-pass and high-pass filters summed by an op amp. The filters may be active filters of two or more poles (compare Fig. 6.26). A similar principle is used in Fig. 6.33, where the inverted output of a band-pass filter is mixed with the original signal. The inverted signal cancels out the intermediate frequencies, producing a notch. The depth of the notch is determined by the relative values of R_{AA} and R_{AB}.

6.3 Switched-capacitor filter

This uses an entirely different technique for filtering. The filter consists of two capacitors, two analog switches and an operational amplifier connected as a unity gain buffer (Fig. 6.34). The complete filter is usually built as an integrated circuit, including the capacitors (about 1 pF) and the clock, which requires an external timing capacitor and resistor. The clock produces a square-wave output at high frequency, usually several hundred kilohertz or even a few megahertz. Because of the inverter, the switches are opened and closed alternately at high frequency. Typically the clock rate is 100 times the cut-off frequency so that, although the switching action introduces noise (7.2.4) at clock rate, this has too high a frequency to corrupt the output.

When S1 turns on and S2 turns off, current flows from the input and charges C1. The action is similar to that of the sample-and-hold network (9.1). The clock frequency is about 100 times greater than the maximum frequency that the filter is required to work with, so there is insufficient time for v_{IN} to alter significantly while charging takes place. When S1 is turned off and S2 is turned on, the charge on C1 is shared with that already on C2 from the previous cycle. Since the capacitors are fabricated with high precision and are close together on the same chip, they are practically equal in capacitance and share the charge equally. A current flows either way through S2 until the potential across them is the mean of the charges prior to S2 being turned on. Then S2 is turned off and C2 is isolated while C1 is being charged to the new value of v_{IN}. The pd across C2 is detected by the unity gain voltage follower, the high input impedance of the op amp preventing significant leakage of charge from C2.

The overall effect of one switching cycle is to average the present value of v_{IN} with its previous value. Short-term variations in v_{IN} are evened out. Long-term trends are passed through to the output virtually unaffected. The circuit has the action of a low-pass filter.

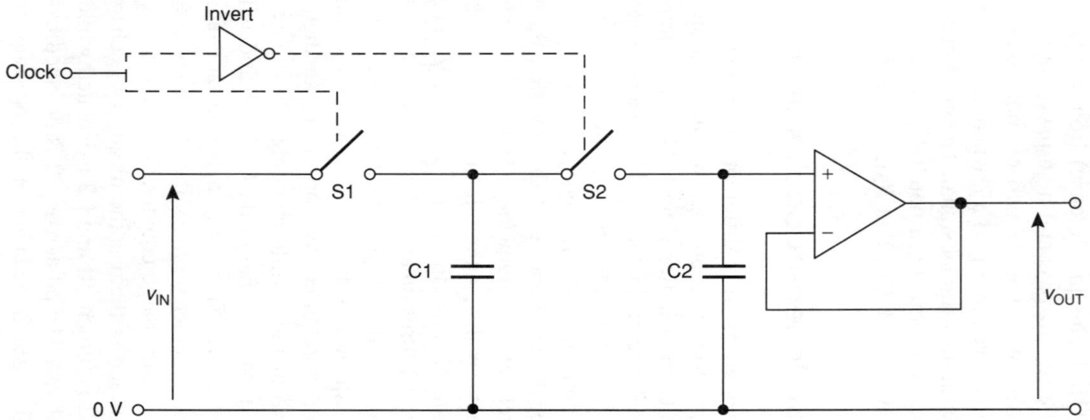

Figure 6.34

The action of this filter is affected by the clock rate. If the clock rate is made faster, there is insufficient time for C1 to charge fully, or for the charge to equilibrate between C1 and C2. This has the effect of increasing the apparent resistance of the filter and affects its response. If the clock rate is voltage controlled, the circuit becomes a voltage-controlled filter.

This principle is extendible to other types of filter, including high-pass, band-pass and band-reject filters. The necessary building blocks are present on a single chip and can be connected externally to produce the required filter type, which can be configured to produce Butterworth, Chebyshev, Bessel and other responses. There are also integrated circuits for building up to 6th order filters using the switched capacitor technique.

7 Noisy analogs

Often an analog signal is corrupted by superimposed signals, which we call *noise*. The term noise covers two distinctive kinds of signal. One type is another analog signal which has somehow found its way into the system, or from one part of the system to another part.

Examples: The signal from a powerful local radio transmitter is heard in the background on a tape player; a hum at mains frequency is heard from the loudspeaker of a stereo system; the output signal feeds back to the input stage of an amplifier, causing distortion and possibly oscillation. This type of noise is better known as *interference*.

In the other sense of the term, noise consists of voltage or current fluctuations and spikes occurring randomly, and originating within the circuit, or entering the circuit from a signal source, superimposed on the input signal. Noise may manifest itself as audible noise, as when we hear a hissing or rasping sound coming from a loudspeaker, but it may also appear as unpredictable peaks on the trace of an electronic chart recorder or oscilloscope. It may also take the form of erratic motion of electronically controlled machinery, such as a robot. This chapter describes different types of interference and noise, and how they may be reduced or eliminated.

7.1 Interference

Interference has a minimal effect if the signal itself is large by comparison. The effects of interference can be reduced by operating with a large signal amplitude. But there are many situations in which this is not practicable. For example, if the interference enters the system at one of the early stages when signal amplitude is low, amplification will merely amplify the interference along with the signal. Another precaution is to avoid sources of interference by not operating in noisy environments, such as close to radio or TV transmitters, or near mains power lines.

Mains-powered equipment may suffer interference caused by spikes on the supply lines, produced by heavy switching on nearby circuits. Supply line filters and transient suppressors reduce these.

One of the most frequent types of interference is *electro-magnetic interference*. This includes radio and TV transmissions as well as electromagnetic waves generated by sparking, such as from vehicle ignition systems, welding equipment, or electric locomotives. The principal precaution is to shield equipment in a metal enclosure which is connected to the 0 V or earthed line of the circuit. This must be done at a single point, otherwise a loop of conductors is formed. Magnetic fields threading through this loop generate interfering currents in the conductors and some of these currents may find their way into critical parts of the circuit. This applies also to connections to the mains power line. If several pieces of equipment, each separately mains powered, are plugged into different mains sockets, the loops so formed can have mains hum currents generated in them. The shielding must be extended to include input and (sometimes) output cables. It is particularly necessary to shield input cables as the signals in these are usually small. Shielding consists of a braided wire layer, electrically connected at one end to the metal circuit enclosure.

Interference may occur between one part of the circuit and another. This is sometimes difficult to anticipate when designing a circuit but general principles for reducing it include:

- Make wires or tracks as short as possible (avoids inductive and capacitative pickup).
- Lay wires or tracks as far apart as possible (avoids inductive and capacitative pickup).
- Use twisted pairs of wires for send and return leads (avoids inductive pickup), or use balanced cables for long-distance transmission of signals (Fig. 8.14).
- Make no unnecessary crossings of wires or tracks.
- Screen transformers, inductors and tuning coils to minimize stray magnetic fields; toroidal power transformers generate the smallest fields externally.
- Mount power supply and output transformers as far as possible from the input side of a circuit.
- The shielding around transformers or between parts of a circuit should be of high magnetic permeability (high mu), otherwise currents induced in the shielding itself by magnetic fields in one part of the circuit may generate magnetic fields on the opposite side of the shielding.

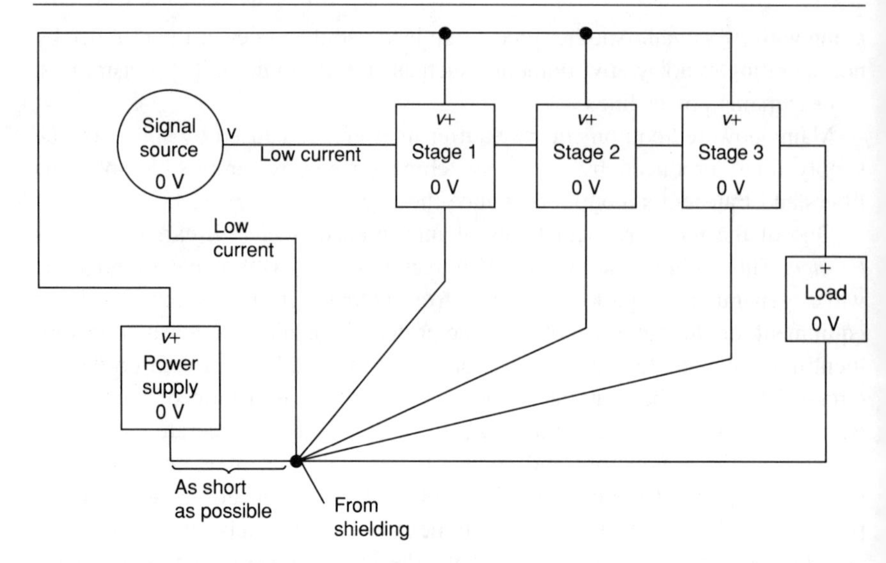

Figure 7.1

- Keep wires close to a ground plane to avoid capacitance effects.
- Thread signal wires though ferrite beads to reduce radio-frequency coupling.
- Return the load current directly to the 0 V terminal of the power supply along a separate wire (Fig. 7.1). If heavy, the load current generates a small pd across the return line; if the same line is used for the 0 V input or the 0 V line of amplifiers, this pd could be reintroduced into earlier stages.
- In general, return all 0 V lines and screening connections directly to the same point in the system.

7.2 Electrical noise

7.2.1 Johnson noise

Alternatively known as *thermal noise*, this is due to the random motion of charge carriers when they are excited by thermal energy. It is generated in resistive materials, such as semiconductors and conductors, which have free charge carriers. It appears as a fluctuating voltage:

$$e_{\mathrm{n}} = (4kTRB)^{1/2}$$

where k is Boltzmann's constant ($1.38 \times 10^{-23}\,\mathrm{JK}^{-1}$)
T is absolute temperature

B is bandwidth (Hz)
and R is the resistance (Ω)

Since all circuits have resistance in them, all circuits are liable to Johnson noise. The voltage at any instant is unpredictable, but e_n is an averaged value, the root mean square of all possible instantaneous values.

Noise is an aperiodic signal, which we can consider to be made up from signals of all frequencies in a given range (B). Frequency does not appear in the equation above $\rightarrow$ Johnson noise occurs with equal magnitude at all frequencies. If we consider only a narrow range, there is a relatively limited number of frequencies in that range so the total noise is likewise limited. If we extend the range to include lower and higher frequencies, we include more signals and the total noise level rises.

Example: Given a $10\,\mathrm{k}\Omega$ resistor, and a bandwidth of 100 kHz. At room temperature (20°C), $4kT = 1.62 \times 10^{-20}$

$$\begin{aligned} e_n &= (4kTRB)^{1/2} = 1.27 \times 10^{-10} R^{1/2} B^{1/2} \mathrm{V/Hz}^{1/2} \\ &= 4.02\,\mu\mathrm{V/Hz}^{1/2} \end{aligned}$$

The bandwidth may extend from 10 Hz to 100.001 kHz or from 1 MHz to 1.1 MHz, or any other width of 100 kHz, and all contain the same noise power.

As an example of noise in a circuit, Fig. 7.2 is a purely resistive attenuator network (3.1.1, Fig. 3.3a) with resistor values calculated to produce attenuation of $a = 1.2$. The *noise density*, the noise voltage per root Hertz, is plotted in Fig. 7.3, the results being taken from a simulation in which the Johnson noise of resistors is modelled. In addition to plotting noise against frequency, noise has been plotted for a number of temperatures in the range 0°C to 100°. It is shown by the horizontal curves that the noise is distributed equally across the frequency range; in other words, there is the same noise power in each hertz of the frequency range. Noise of this type is known as *white noise*.

As well as illustrating that Johnson noise is independent of frequency, Fig. 7.3 also demonstrates the increase of noise with increasing temperature. At 100°C the noise is 28.79 nV/Hz and it is 1.357 dB below this at 0°C.

The effect of resistance on noise is illustrated by reducing all resistances of the network to 1/1000 of the values shown in Fig. 7.3 and rerunning the simulation. Attenuation is as before since all resistors have changed proportionately. The plot obtained is a set of curves identical to Fig. 7.3, except that the 0 dB level (100°C) is now 910.9 pV. Comparing results:

$$\frac{910.9 \times 10^{-12}}{28.79 \times 10^{-9}} = 0.03164 = \frac{1}{\sqrt{1000}}$$

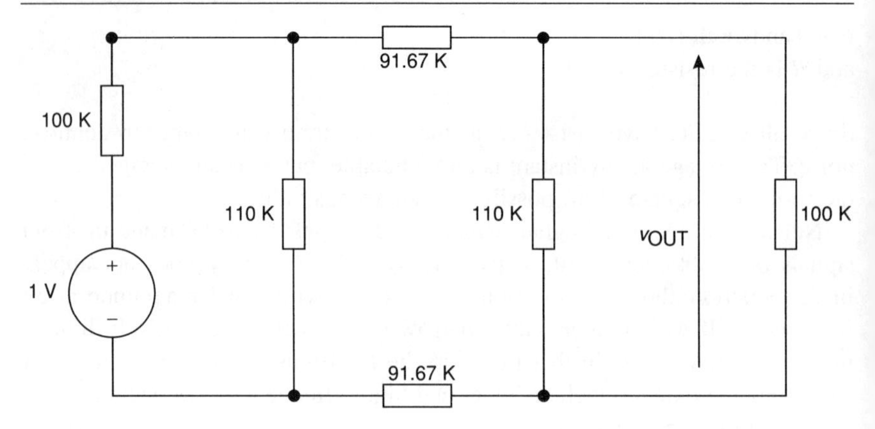

Figure 7.2

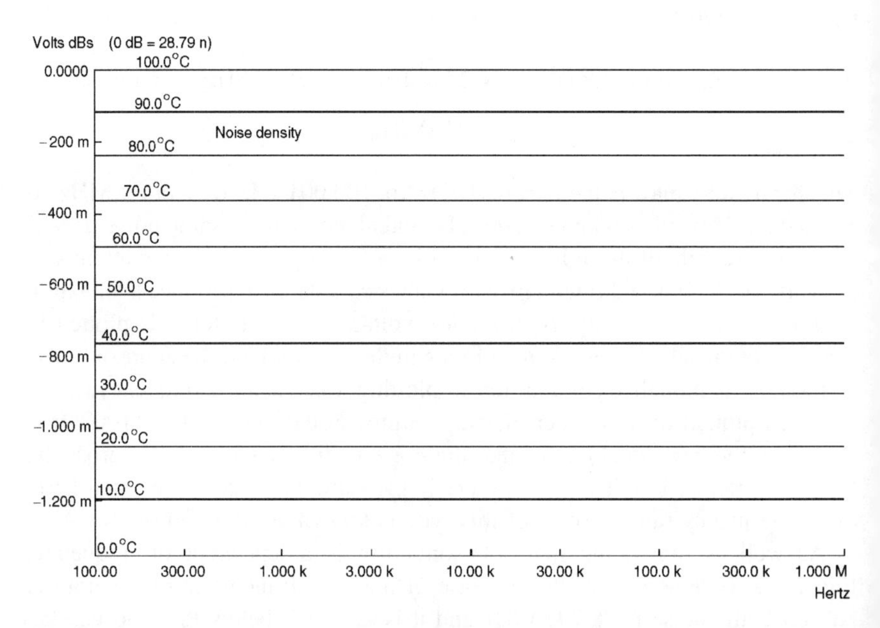

Figure 7.3

Dividing resistor values by 1000 has divided noise by $\sqrt{1000}$, as might be expected from the equation. This is why limiting resistor values as much as is feasible is an important technique for reducing Johnson noise.

It is in the nature of white noise that its amplitude at any instant is unpredictable. But there are further things that can be said about it. It is most often to be found at a mean value, often to be found at a value slightly positive or negative of this, and less often to be found at markedly positive or negative values. These statements can be summed up and made more precise in a diagram (Fig. 7.4). This shows the range of instantaneous values of the noise voltage against which are plotted the relative frequency with which each value occurs. Here the word frequency does not mean the frequency of a periodic signal in hertz, but how frequently (how often) we are likely to find that the signal has a particular value. The most frequent occurrence is for the voltage to be 0 V, for this is the value associated with the peak of the curve. The curve slopes symmetrically down on both sides, showing that values positive or negative of zero are equally likely to occur. The curve is bell shaped, showing that the voltage is more likely to be close to zero and, as we look for bigger and bigger differences from zero, they are less and less likely to occur.

This curve has a mathematically defined shape, known to statisticians as a *normal distribution*. It is also referred to as a *Gaussian distribution*. If the area under the curve (stretching to infinity in both directions) is taken to be 1, the probability of the noise voltage lying in any particular range v_a to v_b is equal to the area included between the two vertical lines drawn at v_a and v_b. In the figure, the shaded area represents the probability of the noise voltage being between v_a and v_b at any given instant, which a rough visual estimate shows to be about 0.1, or 10% probability. The spread of the curve is described by a parameter, the standard deviation σ. It is a property of the curve that 68% of the area lies between $-\sigma$ and $+\sigma$ and that 95% of the area lies between -2σ and $+2\sigma$.

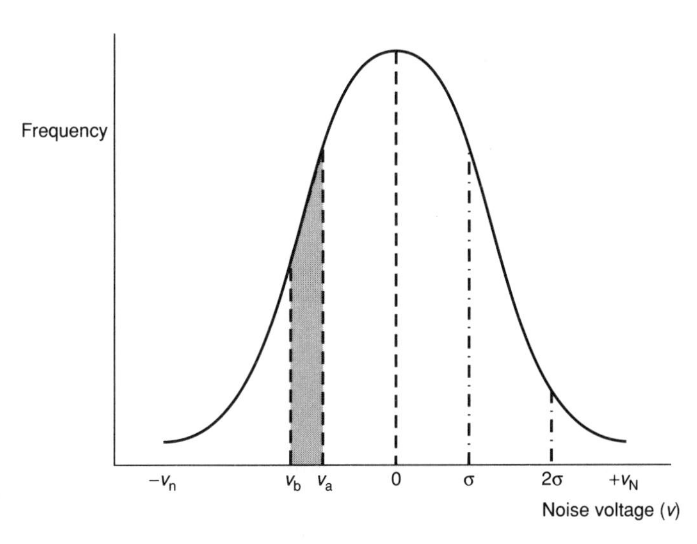

Figure 7.4

7.2.2 Shot noise

Although we think of an electric current as a smooth flowing of charge, current is actually carried by discrete charge carriers. With steady DC, the average number of carriers passing a given point in the circuit is constant, but the number passing at any instant varies randomly. This gives rise to the random variations in current known as shot noise. The effect is enhanced by the random combination of electrons and holes. Shot noise is characteristic of semiconductors, which includes the materials from which certain types of resistor are made.

$$i_n = (2qI_{dc}B)^{1/2}$$

q = electron charge (1.60×10^{-19} C)
B = bandwidth

Johnson noise is a fluctuation in voltage but shot noise is a fluctuation in *current*. However, the current, referred back to the input, generates a voltage $i_n R_s$ across the source resistance, so eventually appears as a voltage at the output of the circuit.

Shot noise is dependent on current magnitude, but not linearly. From the equation above:

Current	i_n in 100 kHz bandwidth	i_n as % of current
1 A	179 nA	0.0000179%
1 μA	179 pA	0.0179%
1 pA	179 fA	17.9%

The smaller the current, the relatively larger the variations due to individual charge carriers. Signals in the picoamp range are seriously degraded by shot noise.

Shot noise is independent of frequency, so it is white noise. Its amplitude has a normal (Gaussian) distribution. The equation above assumes that the carriers move independently of each other, as when they diffuse across a pn junction. In metallic conductors, mass flow predominates → carriers are less independent → much less shot noise.

7.2.3 Flicker noise (1/f noise)

These fluctuating currents are characteristic of semiconductors, and also occur in certain types of resistor made from semiconductor material. The cause is not known exactly, but it is thought to be due to fluctuations in electron and hole velocity due to imperfections in the semiconductor material. Flicker noise also occurs in thermionic valves and is probably due to irregularities in the surface of the cathode.

Flicker noise differs from Johnson noise and shot noise because it decreases with increasing frequency, which is why it is also known as $1/f$ *noise*. Its level falls 3 dB (half power) for every doubling in frequency. As a result, flicker noise is not significant above 1 kHz, and is important only below about 100 Hz. With this rate of fall against frequency, the $1/f$ spectrum has equal noise power per decade. That is to say, the power from 1 kHz to 10 kHz equals that from 100 Hz to 1 kHz, or from 1 MHz to 10 MHz.

Figure 7.5 is the analysis of noise density of a common-emitter amplifier similar to Fig. 4.1. Noise density is constant between 1 kHz and 300 kHz. This is a combination of Johnson noise, mainly from the resistors (and also from the source, since any source has a finite output resistance) and shot noise. There are two important features in this graph. One feature is that noise levels fall off by about 40 mdB between 300 kHz and 1 MHz. This is explained by the Bode plot which shows that signal gain falls off by about 40 mdB between 300 kHz and 1 MHz in this amplifier and, with it, the noise levels in that bandwidth.

In spite of the fact that the Bode plot (Fig. 7.6) also shows that low-frequency signals are reduced by about 35 mdB from 600 Hz down to 100 Hz, Fig. 7.5 shows noise levels increasing at low frequencies. This is flicker noise. Flicker noise is also evident at low frequencies in the Fourier analyses of Figs. 4.7 and 4.9.

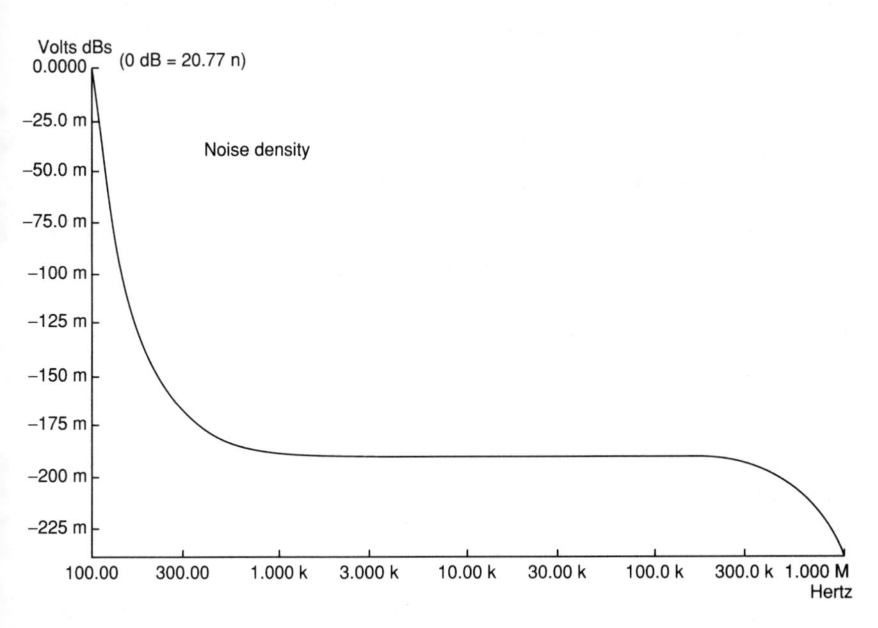

Figure 7.5

7.2.4 Quantization noise

When an analog signal is converted to digital form (9.2) there is inevitably a degree of approximation. For example, if an analog signal of 1 V amplitude (2 V peak-to-peak) is converted into a 12-digit binary number, the converted number can take any one of 2^{12} values. It increases from 0 to 2 V in 4096 steps, each of $2/4096 = 488\,\mu V$. If the analog value lies between one step and the next, it is rounded up or down to the nearest step. There is an error of up to 244 μV in evaluating the signal, which becomes apparent if the signal is converted back to analog form. This is equivalent to adding noise to the signal, and is known as quantization noise. This noise can be ignored in most applications with 12-bit resolution, but is significant when the resolution is only 8 or fewer bits.

The magnitude of quantization noise is reduced by increasing the number of bits in the conversion. For example, a 16-bit digital number has 2^{16} possible values, each 30.5 μV apart, giving a maximum noise error of 15 μV in a 2 V peak-to-peak signal. Unfortunately, in audio signals of very low amplitude, the effects of quantization noise sound like distortion. The greater the number of bits, the higher the resolution, but the greater the apparent distortion. The cure for this is adding white noise to the signal, a process known as *dithering*.

Another way in which signal processing may add noise occurs in switched-capacitor filters (6.3). It is inevitable that turning the analog switches on and off generates voltage peaks and current surges which appear in the signal as noise. This usually has a frequency that is appreciably higher than the frequencies present in the signal. In audio circuits it is too high to be heard, and in other applications it can be removed by low-pass filtering.

7.3 Noise levels

In noise analyses, the noise level at any part in a circuit is usually referred back to the input, that is to say, we find what level of noise at the input would produce that amount of noise at that part of the circuit. This allows us to make the essential comparisons between signal and noise at the same critical place, the input. Noise is generally measured in microvolts. The total noise voltage v_n, referred to the input, is obtained by adding together voltage noise (e_n) and the voltage produced by current noise ($i_n R_s$) from different parts of the circuit. When adding noises together or adding noise to a signal, add the squared amplitudes, then take the square root:

$$v = (v_s^2 + v_n^2)^{1/2}$$

This is because noise voltages are root-mean-squared quantities and we must square them before we add them, then square root their sum.

7.4 Signal to noise ratio

This is a way of expressing the amount of noise present in a system. It relates the power of the signal to that of the noise. Given a voltage V across a resistor R, producing a current I, we know that $V = IR$, and so $I = V/R$. Power $P = VI = V^2/R$. Power is proportional to voltage squared, and the ratio of the powers equals the ratio of the squares of the voltages. Expressing the ratio in decibels:

$$\mathrm{SNR} = 10\log_{10}(v_s^2/v_n^2)\,\mathrm{dB}$$

where v_s is the rms signal voltage and v_n is the rms noise voltage. Usually the aim is to design a system to have a high SNR.

7.5 Noise figure

Imagine two amplifiers each with unity gain, each with a resistor R_s connected across its input terminals. The resistor produces Johnson noise according to the equation in 7.2.1. If one amplifier is an ideal (noiseless) amplifier, the noise in its output signal is:

$$v_o^2 = 4kTR_s\ \mathrm{V/Hz}$$

If the amplifier is a real one, generating noise v_n^2 internally, the noise in its output signal is:

$$v_o^2 + v_n^2 = 4kTR_s + v_n^2\ \mathrm{V/Hz}$$

These are rms voltages squared, so they can be added. The noise figure is the ratio between the output voltages, in decibels:

$$\mathrm{NF} = 10\log_{10}\frac{(4kTR_s + v_n^2)}{4kTR_s} = 10\log_{10}\left(1 + \frac{v_n^2}{4kTR_s}\right)\ \mathrm{dB}$$

This is a measure of how much noise the amplifier adds to the signal. Note that v_n^2 comprises both e_n^2 and $i_n^2R_s$.

7.6 Low-noise amplifiers

In a BJT amplifier, thermal noise (voltage) is generated in the bias resistors, shot noise (current) from current flowing in the base-emitter region of the BJT, and flicker noise (current) in the region between the base connection and the base-emitter junction. All three contribute to the noise levels plotted in Fig. 7.5. Designing a low-noise amplifier is a matter of compromise. For example, if resistors are made small to minimize Johnson noise, currents are correspondingly larger, increasing shot noise. This is over-simplifying the issue but illustrates the complexity of the situation. One solution to low-noise design is to keep supply voltages low so that resistances can be low without passing excessively high currents. Another approach is to minimize base current (and hence the shot noise and flicker noise in the base-emitter junction) by using transistors with high h_{fe}. Johnson noise may be reduced by using low-noise BJTs and by operating with a large signal voltage to render Johnson noise low in comparison. JFETs have a low gate leakage current so are less subject to shot noise and flicker noise. MOSFETs have high noise, especially flicker noise at lower frequencies.

In general, the first stage of an amplifier is the one in which most effort should be made to reduce noise, because noise generated there is amplified by the gain of the whole amplifier. Noise occurring at later stages receives less amplification. Using a JFET, a low-noise BJT, or a BJT with low collector current, are ways of reducing first stage noise. Source resistance is important too. From the noise factor equation, it can be shown that the minimum value of NF is obtained when:

$$R_s = \frac{e_n}{i_n}$$

The input resistance should be matched to R_s. In addition, input noise can be reduced by keeping the leads from input devices as short as possible, to minimize the Johnson noise generated in them. In extreme cases a parametric amplifier (5.4) is used.

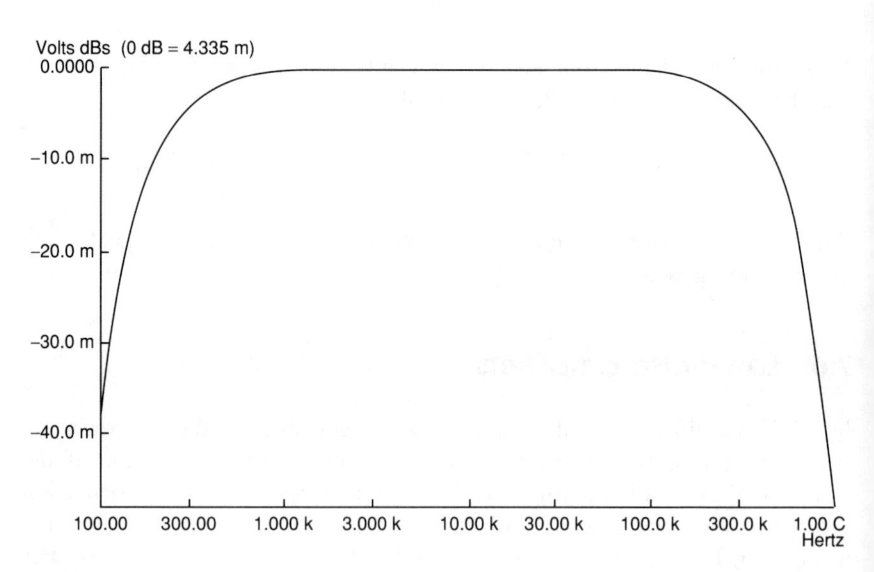

Figure 7.6

In most instances, noise may be further reduced by limiting the bandwidth of the circuit to include only those frequencies which are essential to its function. It may even be preferable to reduce bandwidth at the expense of distorting the output signal to a certain extent. Reducing bandwidth to its essential minimum reduces Johnson noise and shot noise proportionately. Filtering out the lowest frequencies, if they are not required for operation of the circuit, will eliminate almost all flicker noise.

8 Analog communications

Analogs generated in one location may be required somewhere else. DC and low-frequency signals may be carried for relatively short distances along ordinary wires or the tracks of a printed circuit board.

Example: Audio signals from amplifier to loudspeaker.

But when distances are greater or when frequencies are higher, simple wires become inefficient or impracticable, and other means are employed. These means include transmission lines, radio, and optical fibre dealt with in Sections 8.2 onward. Transmission of analogs by these involves modulation, which is the topic of the next section.

8.1 Modulation

Modulation is the technique of superimposing a low-frequency (e.g. AF) signal on a high-frequency (e.g. RF) *carrier* signal. Although it is possible to operate a radio transmitter at audio frequencies, it is impracticable. Even the highest audio frequency (20 kHz), transmitted as a radio wave, has a wavelength of 15 km, requiring transmitting and receiving antennas of at least a quarter this length for efficient communication.

Selectivity is another factor. Imagine a market-place full of vendors shouting their wares. No one is heard properly. The same applies to a number of radio transmitters in a restricted area. If they all transmit at audio frequencies, it is impossible to receive them individually. But, if they all transmit on different radio frequencies, we can tune our receiver to any one of them we choose. The fact that we are transmitting high (radio) frequencies means that problems with antenna size are eliminated. The matter of *bandwidth* is of even more consequence. It is shown in 8.1.1 that the bandwidth required for a modulated radio transmission of an analog signal up to, say, 20 kHz is 20 kHz on either side of the carrier frequency. Thus each transmission occupies only 40 kHz of bandwidth. The radio-frequency spectrum used for telecommunications extends roughly from 30 kHz to 30 GHz, a bandwidth of almost 30 GHz, so there is room for some 750 000 simultaneous radio transmissions with audio-frequency modulation.

A continuous transmission of a signal at a fixed frequency conveys no information, except that the transmitter is working. If we *modulate* the signal so that it carries, say, an audio signal, then information can be communicated. The oscillator of a radio transmitter produces a sine-wave carrier signal:

$$v_c = V_c \sin(\omega_c t + \phi)$$

where V_c is the amplitude, ω is the angular frequency ($= 2\pi f$, (see 2.2)), ϕ is the phase angle, and t is elapsed time. V_c, ω_c and ϕ are three (and the only three) parameters which together determine the value v_c at any given instant. This equation applies to the unmodulated carrier wave. Information may be added to the wave by altering the value (i.e. modulating) of any one of the three parameters, amplitude, frequency or phase, in conformity with variations in the magnitude of the modulating signal.

8.1.1 Amplitude modulation

In AM, the carrier signal is modulated by varying its amplitude. Given a carrier wave:

$$v_c = V_c \sin \omega_c t$$

We have omitted ϕ, since it is not affected in AM. Given also a modulating signal:

$$v_m = V_m \sin \omega_m t$$

where $\omega_m \ll \omega_c$. AM consists in making V_c proportional to the instantaneous value of v_m. When the modulating signal is superimposed on the carrier, the form of the resulting signal depends on the ratio:

$$\text{modulation factor, } m = V_m / V_c$$

m expresses the depth of modulation of the carrier. If, for example, $m = 0.4$, then the amplitude of the carrier ranges between $1.4V_m$ and $0.6V_m$. In Fig. 8.1 a 10 Hz carrier is modulated with a 1 Hz signal with $m = 0.5$. Modulation ranges from 0 (unmodulated carrier) to 1 (deepest modulation, V_c is periodically reduced to zero). In terms of the equations above, the amplitude of the carrier is the sum of V_c and the modulating wave:

$$\begin{aligned} v_c &= (V_c + V_m \sin \omega_m t) \sin \omega_c t \\ &= V_c \sin \omega_c t + mV_c \sin \omega_c t \times \sin \omega_m t \end{aligned}$$

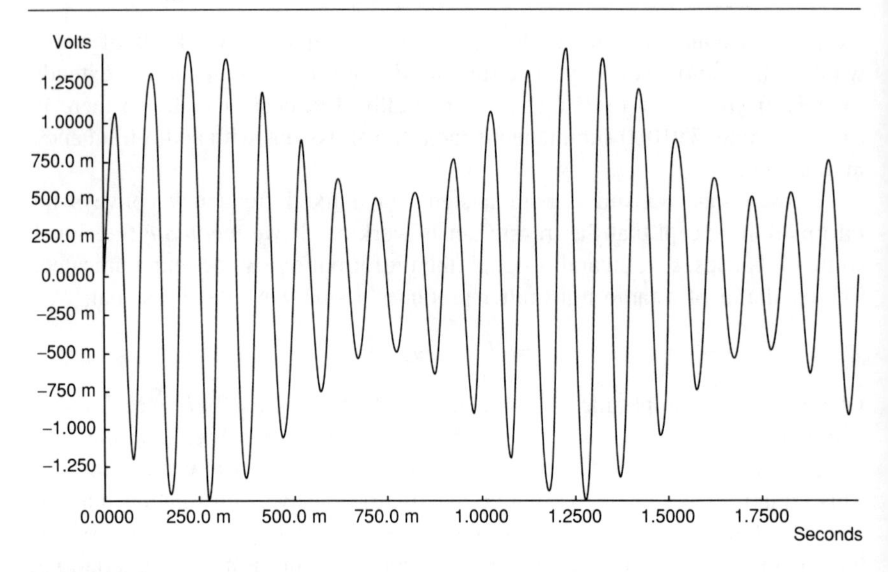

Figure 8.1

Using the trigonometrical identity for $\sin A \sin B$:

$$\sin \omega_c t \times \sin \omega_m t = \tfrac{1}{2}[\cos(\omega_c - \omega_m)t - \cos(\omega_c + \omega_m)t]$$

$$\Rightarrow \qquad v_c = V_c \sin \omega_c t + \tfrac{1}{2} m V_c \cos(\omega_c - \omega_m)t - \tfrac{1}{2} m V_c \cos(\omega_c + \omega_m)t$$

The modulated carrier consists of the sum of three sinusoids; Fig. 8.2 is its frequency spectrum. If the carrier is modulated by a saw tooth wave, for example, the fundamental and several harmonics are superimposed on it, giving a more complicated spectrum. For each harmonic there are two lines, representing the sum and difference frequencies. Taking this further, a complicated signal such as that representing speech or orchestral music is composed of signals of hundreds of frequencies and, for each of these, there are lines equally spaced above and below the carrier frequency. Figure 8.3 illustrates how these merge into two bands, known as *side bands*, on either side of the carrier frequency. The width of each side band is the difference between the carrier frequency and the highest frequency present in the modulating signal. In the case of an audio signal with frequencies ranging from 20 Hz to 20 kHz, each band extends to 20 kHz on either side of the carrier frequency → the total bandwidth required for transmitting this signal is 40 kHz.

Because of the symmetry of the side bands, all the information about the modulating signal is contained within one side band. There is no need to transmit two. This is the basis of *single side band* (SSB) techniques (8.3).

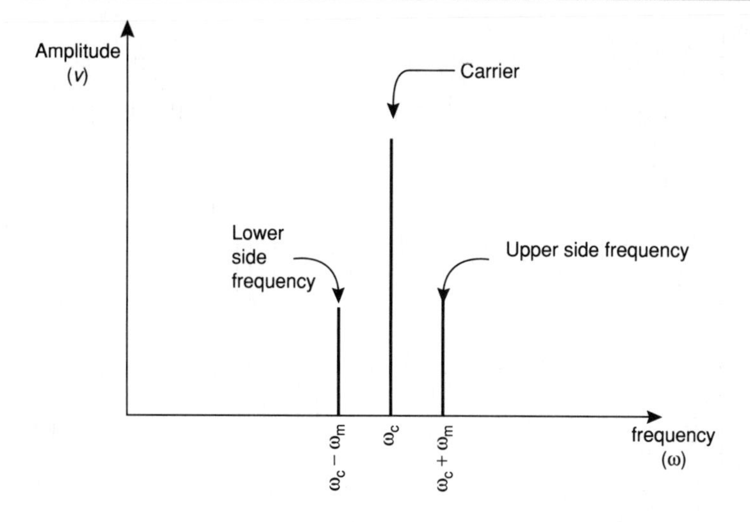

Figure 8.2

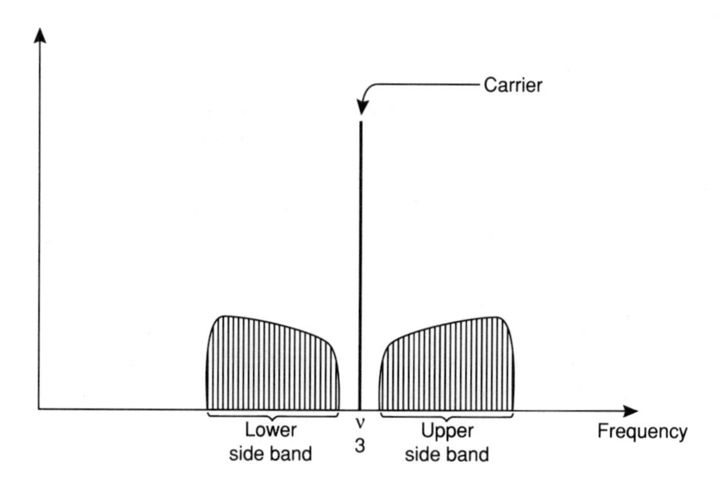

Figure 8.3

8.1.2 Frequency modulation

In FM the carrier amplitude is held constant but its frequency is modulated. In Fig. 8.4 a 10 kHz carrier is frequency modulated at 1 Hz. The instantaneous frequency f_i of the modulated carrier is given by:

$$f_i = f_c(1 + kV_m \cos \omega_m t)$$

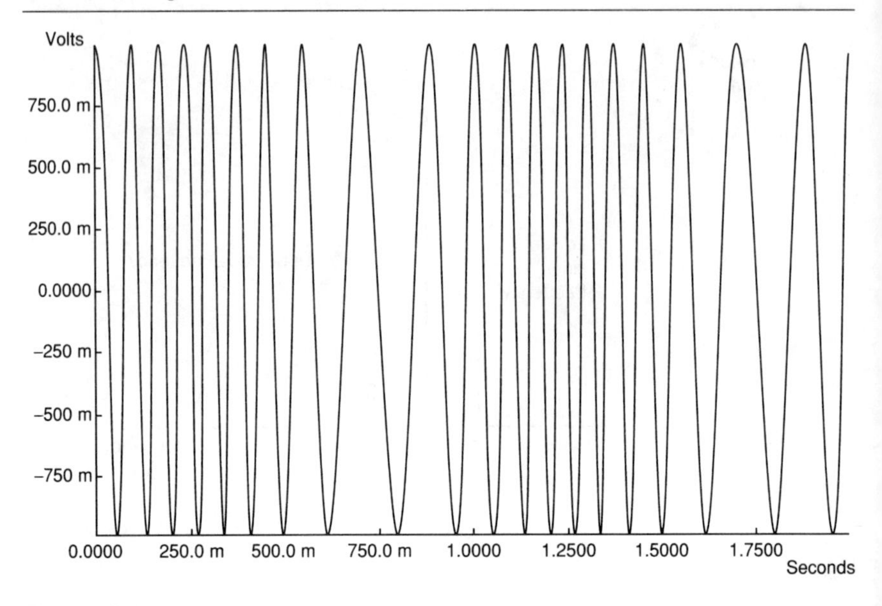

Figure 8.4

where f_c is the unmodulated carrier frequency, and k is a constant. $V_m \cos \omega_m t$ represents the modulating signal, except that we are relating it to cosines instead of to sines to make the calculation easier. Multiplied by k, it expresses the deviation of the modulated carrier from the unmodulated carrier at any instant t. The maximum deviation δ occurs when $\cos \omega_m t$ is either $+1$ or -1, when:

$$f_i = kV_m f_c$$

FM is not simply the summation of the carrier with the two sum and difference signals as in AM, but can be expressed as:

$$v_c = V_c \sin[F(\omega_c, \omega_m)]$$

where $F(\omega_c, \omega_m)$ is a function of the two frequencies, of a form described below. An unmodulated carrier wave can be represented by a rotating phasor, turning equal angular distances in equal intervals of time (Fig. 8.5), because its instantaneous angular frequency ω_i is constant. We calculate the phase angle θ at any given instant from the equation:

$$\theta = \omega_i t$$

With a frequency modulated carrier, ω_i is not constant so the phasor does not turn equal distances (Fig. 8.6). To find the phase angle at any instant we have

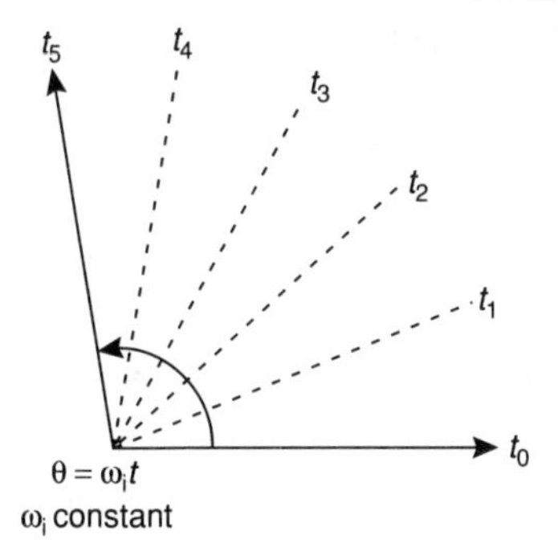

Figure 8.5

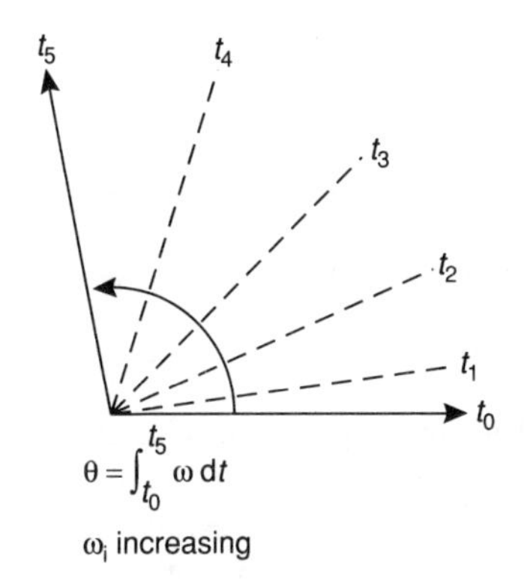

Figure 8.6

to integrate the varying ω_i with respect to time:

$$\theta = \int \omega_i \, dt$$

This is the angle represented by the function above. Note that if $f_i = f_c(1 + kV_m \cos \omega_m t)$, as previously defined, then multiplying both sides of the equation by 2π replaces frequencies with angular frequencies and $\omega_i = \omega_c(1 + kV_m \cos \omega_m t)$. We can now write:

$$\begin{aligned} F(\omega_c, \omega_m) = \int \omega_i \, dt &= \int \omega_c(1 + kV_m \cos \omega_m t) \, dt \\ &= \omega_c \int (1 + kV_m \cos \omega_m t) \, dt \end{aligned}$$

Integrating with respect to time:

$$F(\omega_c, \omega_m) = \omega_c \left(t + \frac{kV_m \sin \omega_m t}{\omega_m} \right)$$
$$= \omega_c t + \frac{kV_m \, \omega_c \sin \omega_m t}{\omega_m}$$

Dividing top and bottom of the fraction by 2π to return angular frequencies to frequencies:

$$F(\omega_c, \omega_m) = \omega_c t + \frac{kV_m f_c \sin \omega_m t}{f_m}$$
$$= \omega_c t + \frac{\delta}{f_m} \sin \omega_m t$$

Substituting this form of the function into the equation for the modulated signal:

$$v_c = V_c \sin \left(\omega_c t + \frac{\delta}{f_m} \sin \omega_m t \right)$$

In this expression, the quotient δ / f_m is referred to as the modulation index, m_f

$$\Rightarrow \qquad v_c = V_c \sin(\omega_c t + m_f \sin \omega_m t)$$

This is the expression for the frequency modulated carrier. It is more complicated than the AM expression since it contains the sine of a sine. To be able to plot the frequency spectrum we need to resolve this into the sum of a number of sines. This involves the use of Bessel functions, which are outside the scope of this book.

The results of such calculations show that when $m_f = 0.1$ the frequency spectrum of an FM signal (modulated by a single frequency) consists of the carrier frequency and a pair of sum and difference frequencies (Fig. 8.7). This appears to be the same as that for AM (Fig. 8.2) but phase relationships are different. When $m_f = 0.5$, there are two side bands. These are examples of *narrow-band* FM. Transmissions of this type can be received on an AM receiver. When $m_f > 1$ the number of side bands increases to three or more. Figure 8.8 illustrates the spectrum when $m_f = 2.5$. The exact relationships of the amplitudes varies according to m_f, depending on the coefficients of the Bessel function. In general, when $m_f > 1$ there are several frequencies spaced ω_m apart on either side of the carrier frequency. The amplitude of the carrier frequency is often much less than that of the side bands and may disappear altogether. This is an advantage because most of the energy of the signal is in the information-containing side bands and less in the carrier. In contrast to AM, in which most of the energy is transmitted as carrier, this is much more efficient.

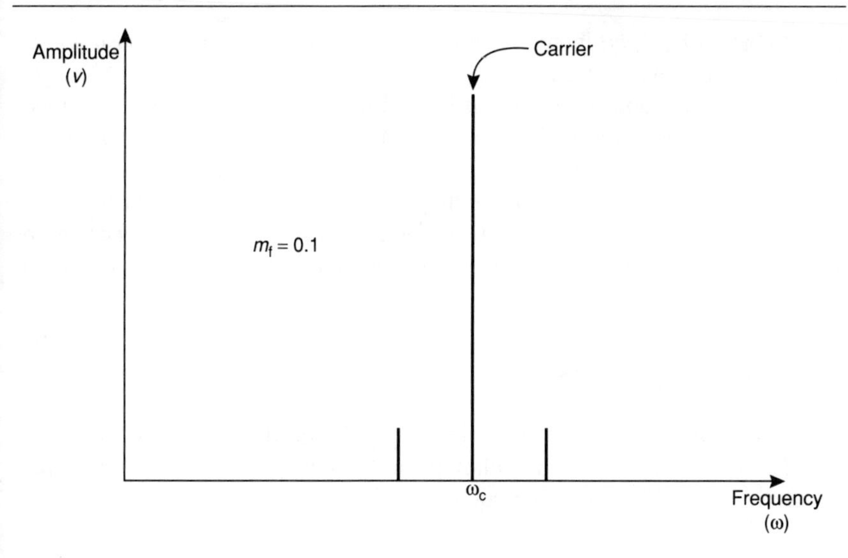

Figure 8.7

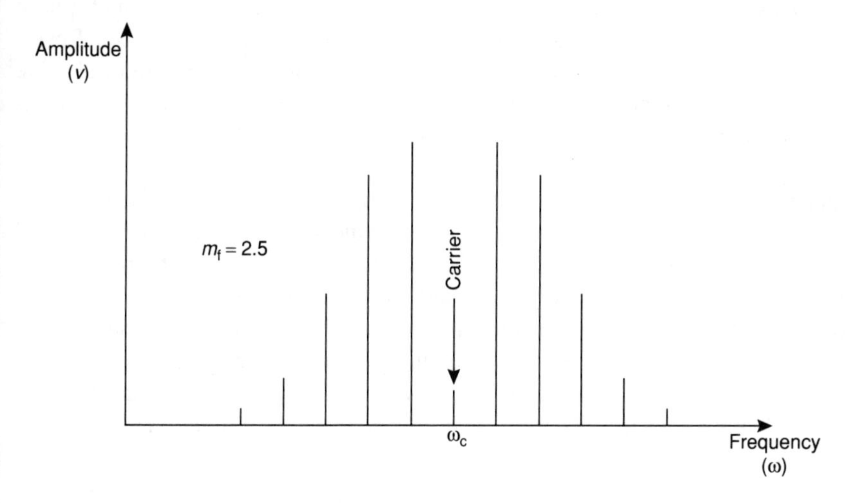

Figure 8.8

For a single-frequency modulation signal the number of side bands of significant amplitude increases with m_f. For example, if $m_f = 2$, there are four side bands; if $m_f = 2.5$, there are five (Fig. 8.8); if $m_f = 5$, there are eight. These side bands broaden out from single spectral lines into broader bands when the modulating signal contains mixed frequencies. Since m_f depends on the ratio of δ to f_m, m_f increases either with constant f_m and increasing δ, or with constant δ and decreasing f_m.

One of the advantages of *broad band* FM is that it is much less subject to noise than AM. Amplitude limiting in a receiver can be used to limit the noise on a received signal without affecting the instantaneous frequency. A disadvantage is the greater bandwidth required. For this reason FM transmission is restricted to the VHF and UHF bands (30 MHz to 3 GHz) where more bandwidth is available. This brings the disadvantage that reception is limited to line-of-sight.

8.1.3 Phase modulation

The third parameter that can be varied is the phase. Modulating the phase has an effect similar to that of modulating the frequency in that they both affect the phase angle (Fig. 8.6). The equation for a PM signal is:

$$v_c = V_c \sin(\omega_c t + m_p \sin \omega_m t)$$

This has the same form as that for FM, except that m_f is replaced by m_p, the phase modulation index, which is the maximum amount by which phase is modulated as $\sin \omega_m t$ varies between $+1$ and -1. Again we need Bessel functions to calculate the frequency spectrum and again there are multiple side bands. The important difference is that while m_f depends on δ and f_m, m_p is independent of f_m. For modulation at a constant single frequency, the results of FM and PM are identical. It is only as the modulating frequency changes that they differ in their outcomes. If f_m changes, there is no effect on m_p, but m_f varies inversely with f_m. If the frequency of the modulating signal falls, m_f increases. Thus, with a PM transmitter and receiver, the modulation index is unaffected by frequency and notes of all frequencies are reproduced with their original relative amplitudes. If the same PM transmission is received by an FM receiver, in which low-frequency notes would normally produce an increase in the modulation index, the constancy of the index in the PM transmitter results in a reduction of signal amplitude. There is bass cut (and treble boost). Conversely, a low-frequency note transmitted by FM and heard on a PM receiver has increased amplitude. There is bass boost and treble cut.

8.1.4 Pulse modulation

Pulses are primarily associated with transmission of binary data, but pulse modulation can also be used for analog signals. Many systems of pulse modulation have been developed, of which the more important ones are described below.

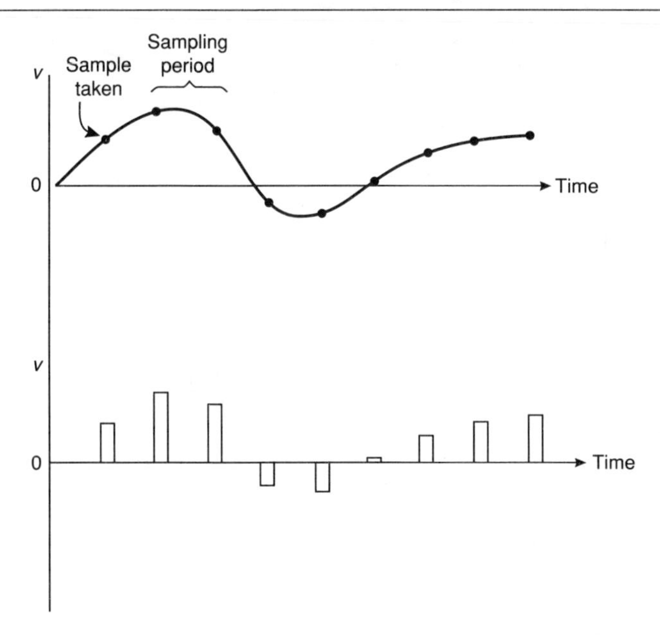

Figure 8.9

8.1.4.1 Pulse amplitude modulation

An analog signal (upper curve, Fig. 8.9) is sampled at regular intervals and its instantaneous amplitudes are used to modulate the amplitude of successive pulses. Provided that the frequency of sampling is at least twice that of the highest frequency present in the analog signal, the original analog signal can be recovered. The unmodulated pulse signal creates a Fourier series of sinusoids (2.6) at $f_s, 2f_s, 3f_s, \ldots$, where f_s is the sampling frequency. The spectrum of a modulated signal consists of these lines, each with side bands representing the modulating signal. In addition there is a line at f_m, which represents the original analog signal (Fig. 8.10). This can be recovered by passing the composite signal through a low-pass filter. PAM, like continuous amplitude modulation (8.1.1), is particularly subject to noise. A condition for successful operation is that the transmitter has to be able to generate enough power to produce a pulse of maximum amplitude, yet most of the time it is transmitting smaller pulses, which is wasteful. In many PAM systems the pulses are carried by frequency modulation, with constant amplitude.

Pulse modulation systems allow for *time division multiplexing* (TDM). Instead of generating single pulses with relatively large gaps between them, as in Fig. 8.9, we fill the gaps with a series of pulses representing concurrent signals (Fig. 8.11). Of the series of pulses, every eighth (in this example)

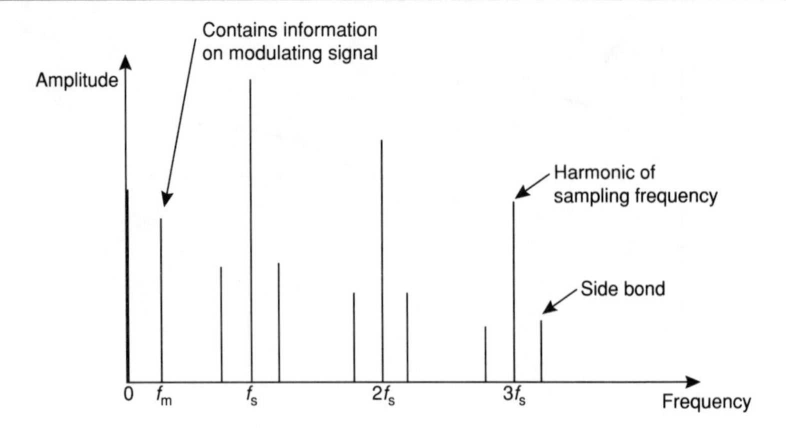

Figure 8.10

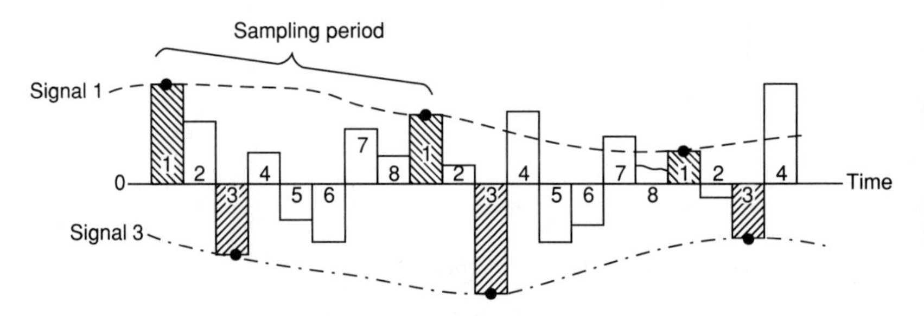

Figure 8.11

relates to a given signal. The pulses between relate to the seven other signals. Thus it is possible to send eight signals along the same line simultaneously. Special circuits are required to deliver the samples to the multiplexer at the right times, and to separate them out (de-multiplex them) at the other end of the line. In practice, many more than eight signals can be multiplexed. One standard system multiplexes 25 signals at 8000 pulses per second. This transmits 24 concurrent signals, the 25th pulse being used for synchronization.

8.1.4.2 Pulse width modulation

PWM is also known as *pulse duration modulation* (PDM). The signal consists of a chain of pulses at regular intervals and of fixed amplitude, their widths being proportional to the instantaneous amplitude of the modulating signal (Fig. 8.12). The leading edges of the pulses occur at equal intervals. The trailing edges occur at times dependent on their width. The frequency spectrum is similar to Fig. 8.10 and, as with PAM, the signal is recovered by low-pass

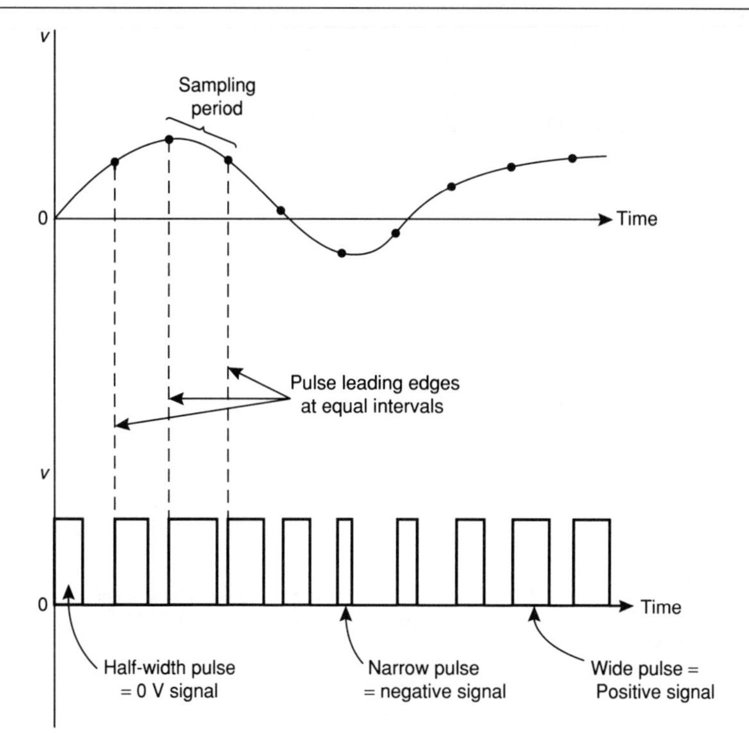

Figure 8.12

filtering. This system is less subject to noise than PAM for noise has less effect on pulse width than on amplitude. With varying pulse width, the transmitter has to be able to generate enough power for a full-width pulse though, on average, it is producing only half-length pulses, which is uneconomic.

8.1.4.3 Pulse position modulation

In PPM the pulses are all of the same length and amplitude (therefore full power is always being used, which is advantageous) and are advanced or retarded in time according to the instantaneous amplitude of the modulating signal. If this system is to work there must be synchronization between the transmitter and receiver (not required for PAM and PWM).

In Fig. 8.13 the signal is first pulse-width modulated, as in Fig. 8.12, to produce a series of pulses each beginning at regular intervals. The ending of each pulse depends on the modulating signal. Next the pulses are differentiated (5.2.2), to produce a positive spike at their leading edges and a negative spike at their trailing edges. Finally, the positive spikes are removed with a diode rectifier.

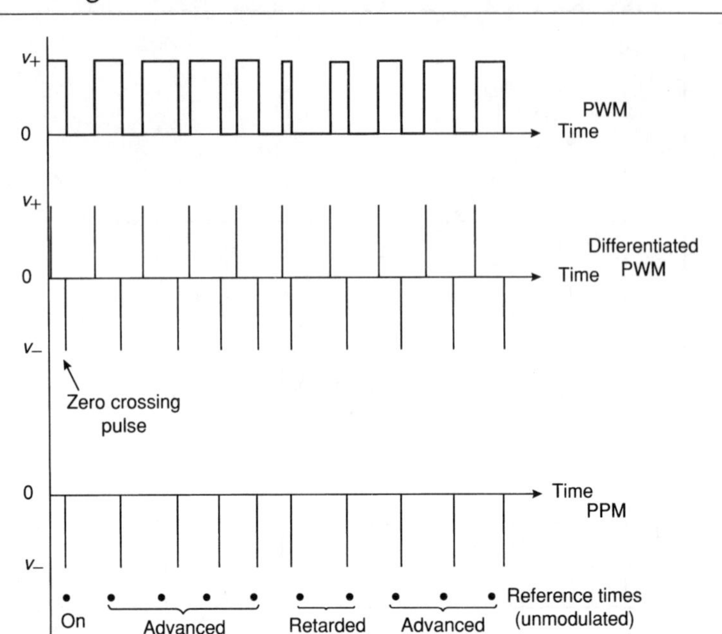

Figure 8.13

The pulse corresponding to a zero-crossing of the modulating signal is half the maximum length. Such a pulse is marked in Fig. 8.13, and it is taken to have zero displacement. The zero-displacement times are marked as dots below the PPM signal. If the signal were unmodulated, pulses would occur at these times. With a modulated signal, it can be seen that, when the modulation signal is positive (see Fig. 8.12), the pulses are retarded. When the signal is negative, the pulses are advanced.

At the receiving end the pulses are converted back to PWM by a flip-flop (or bistable circuit). This is turned on at regular intervals, as determined by the synchronizing circuit and is turned off by a received pulse. The length of time the flip-flop is on determines the length of the pulse and is the same as the original PWM pulse length. This action recovers the PWM signal which is then converted to the analog signal by low-pass filtering.

8.1.4.4 Pulse code modulation

This method differs from those described above in that it relies on digital techniques. The analog signal is sampled at regular intervals and each sample is converted into a binary code, using an analog-to-digital converter (9.2). The codes are processed in ways that are outside the scope of this book, to maintain checks for errors in coding and faults in transmission. The samples are transmitted as a series of pulses, in essence a sequence of high and low pulses representing the binary digits 1 and 0. On reception, the code groups are processed to correct for errors and to restore them to their original form. Finally they are converted to the analog signal by a digital-to-analog converter (9.3).

One of the advantages of PCM is that highs and lows, or ones and zeros, are distinct and distinguishable, even against a very noisy background. This makes transmission much less subject to error and noise, and allows a weak and noisy signal to be reconstituted exactly to its original form by repeater stations on a long transmission route.

PCM has become very widely used for transmitting analog signals, in spite of its greater bandwidth requirements and the complexity of its circuits. It is highly immune to noise and is suited for long-distance communications, especially by optical fibre. Its immunity to noise has made it ideal for space communications and remote telemetry. Its precision and reliability have made it the choice for compact disc technology.

8.1.4.5 Delta modulation

This is a version of PCM which has the merit of being very simple, both in principle and in execution. It contrasts with PCM itself, which requires very complicated circuits. Essentially each sample is sent as one bit (0 or 1), depending on whether it is bigger or smaller than the previous sample. In other words, it transmits differences. The system is described in 9.2.3.

8.2 Transmission lines

Any wire or strip of metal can be said to be a transmission line, but here we are concerned with the characteristics that make for transfer of a signal from one place to another with minimum loss of power, minimum distortion, and minimum added noise.

When the signal to be transmitted is a slowly changing current or voltage, the two main considerations are loss of power due to the resistance of the wires and the acquisition of interference. Power loss can be reduced by using thicker conductors, though cost may limit this for long distances. Interference may be countered by measures described in 7.1. One of many possible techniques for minimizing interference is illustrated in Fig. 8.14. This is a balanced cable suitable for connecting a microphone to an amplifier. The cable consists of twin screened wires (see Fig. 8.15a), the screens being connected to ground and the centre-tap of the microphone transformer. Note that the screening is not connected at the amplifier end as the loops so formed would be liable to

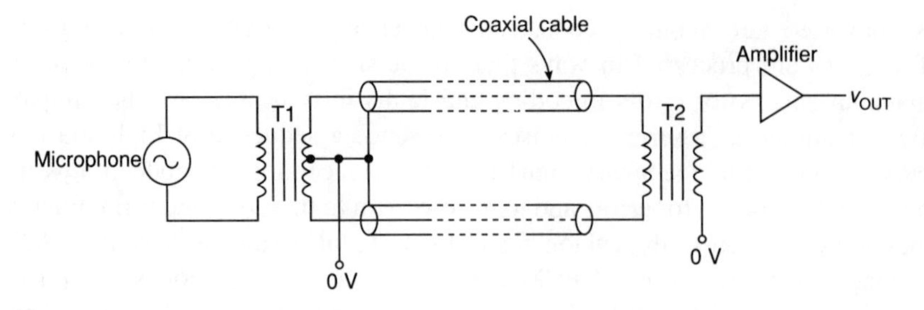

Figure 8.14

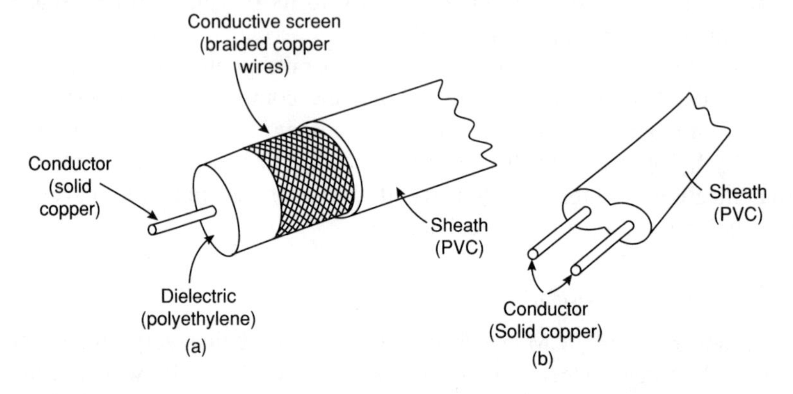

Figure 8.15

induced signals. The signals are balanced relative to screening and ground. The amplifier amplifies only the difference between the signals (that is, the output signal from the microphone). Common signals, such as mains hum and radio-frequency interference, occur on both wires equally, do not appear across the terminals of the amplifier transformer, and are ignored.

Other signals, such as Johnson noise (7.2.1) may be important if signal levels are low. Temperature may alter resistance and hence signal level (3.3.1). Each circuit has its own problems with its individual solutions.

For transmission of alternating (usually modulated) signals there are additional factors to be taken into account, particularly with RF signals. Two types of cable are used as transmission lines, either a coaxial cable or parallel leads (Fig. 8.15). Other forms of transmission line are used for microwave transmissions in the UHF bands (8.2.2).

8.2.1 RF transmission lines

A transmission line, whether it is a length of *coaxial cable* or a pair of *parallel leads*, may be considered as an electronic circuit in its own right, with input and output terminals. The goal is to feed the signal into it at one end and to recover it at the other end with minimum loss or distortion. The line is equivalent to the circuit of Fig. 8.16. The circuit elements, resistance, capacitance, conductance and inductance actually exist continuously along the line but here we show them 'lumped' as individual resistors, capacitors, conductors and inductors, so that we may understand their action more easily. The resistance and inductance in the line are the resistance and inductance of its two conductors. The capacitance is the capacitance between the two conductors with the insulating dielectric between them (some types of transmission line have air between the conductors). The conductances represent leakage across the dielectric. With a high-frequency signal, the effects of the reactive elements predominates. Resistance and conductance can be ignored and the equivalent circuit simplifies to Fig. 8.17.

The *characteristic impedance* Z_0 of a line is defined as the ratio between voltage and current at the input of a line (Fig. 8.18a), which is of infinite length. The way in which the far end of the line is terminated, whether by open circuit, short circuit, or by a resistor, is of no consequence if the line is infinitely long.

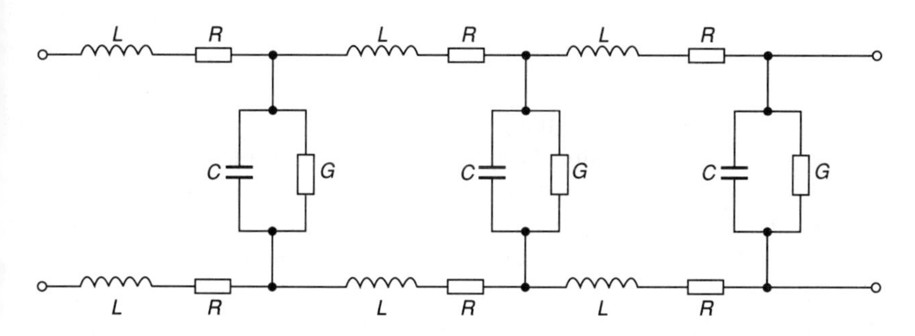

Figure 8.16

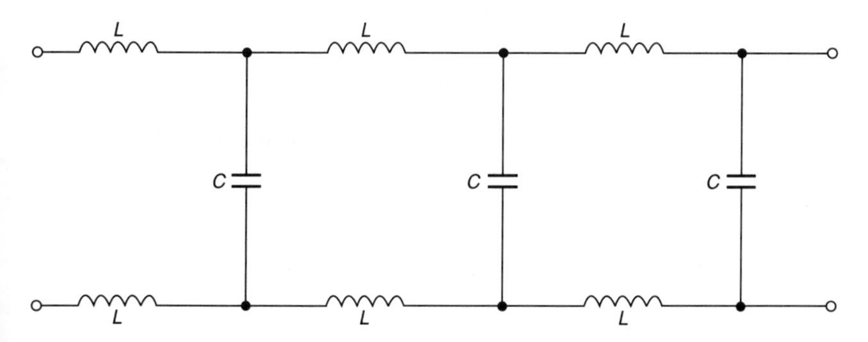

Figure 8.17

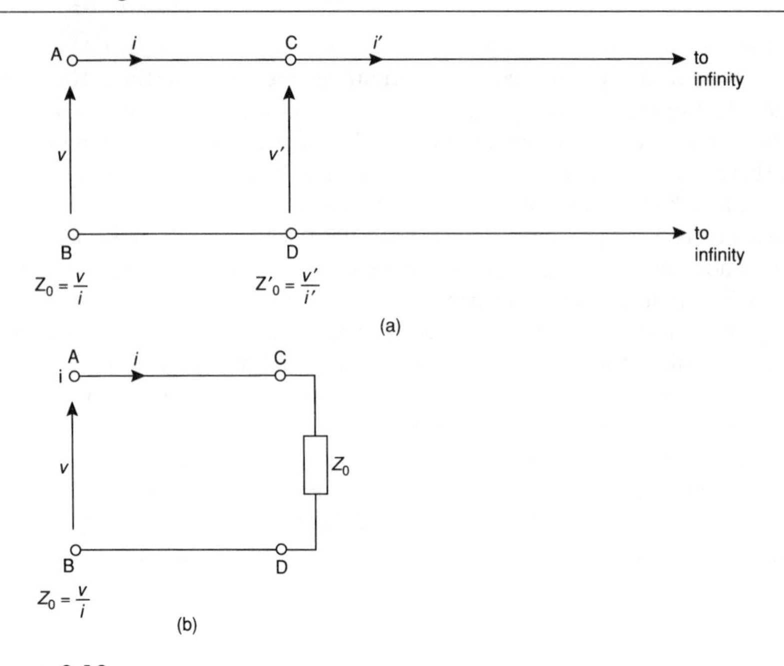

Figure 8.18

If the line is infinitely long, the power fed into it is completely absorbed and is conducted away to infinity. Proceeding further from the input, we expect that the voltage and current will decrease, owing to the effects of inductance and capacitance. Between points C and D, for example, the voltage will be $v'(v' < v)$ and the current will be $i'(i' < i)$. But, since the line is infinitely long, CD is no nearer to the end of the line than AB. The impedance looking into CD is exactly the same as that looking into AB, in other words it is Z_0. This being so, we can replace the line beyond C and D with an impedance equal to Z_0 (Fig. 8.18b). The distance between AB and CD is immaterial to this discussion so we can say that the input impedance Z_{in} equals Z_0, when the line is terminated in an impedance equal to Z_0. Given a signal source with Z_{out} equal to Z_0 there is complete transfer of power to the line and, given a termination with impedance Z_0, there is complete transfer of power out of the line at the far end. Theoretically there is no loss of power along the line, which is to be expected since, at high frequency, the characteristics of the line are dominated by capacitance and inductance, which may store and release energy but not absorb it.

With reference to Fig. 8.16, it can be shown that:

$$Z_0 = \sqrt{\frac{R + j\omega L}{G + j\omega C}}$$

where R, L, G and C are given for unit length of line. For radio frequencies the line simplifies to Fig. 8.17 and:

$$Z_0 = \sqrt{\frac{j\omega L}{j\omega C}} = \sqrt{\frac{L}{C}}$$

Note that frequency (ω) is eliminated from the equation, so the characteristic impedance at RF is independent of frequency. The units of Z_0 are found by considering the units of the quantities in the expression. The emf induced in a coil is $V = L\,\mathrm{d}I/\mathrm{d}T$ (ignoring sign), so L has the units VT/I. Since, for a capacitor, $C = Q/V$ (where Q is the charge) and $Q = IT$, then $C = IT/V$. Therefore the units of Z_0 are:

$$\sqrt{\frac{L}{C}} = \sqrt{\left(\frac{VT}{I} \times \frac{V}{IT}\right)} = \sqrt{\frac{(V^2)}{I^2}} = \frac{V}{I} = R$$

The impedance is a resistance, with ohms as the unit. Although the line is considered to have capacitance and inductance, its characteristic impedance is shown to be pure resistance.

Z_0 for a coaxial line is given by:

$$Z_0 = (138/\sqrt{k}) \log(D/d)\,\Omega$$

where k is the dielectric constant of the insulation, D is the diameter of the screen and d is the diameter of the conductor, dimensions in millimetres.

Z_0 for a parallel lead line is given by:

$$Z_0 = 276 \log(2s/d)\,\Omega$$

where s is the separation between the two conductors (centre-to-centre) and d is their diameter, both dimensions in millimetres.

Depending on the diameter of the conductor and screen, or on the spacing between parallel wires, and also on the nature of the dielectric, cables are made with standard values for Z_0, such as $50\,\Omega$ and $300\,\Omega$. These values apply whatever the frequency of the signal (in the RF range) and whatever the length of the cable. The fact that the separation of parallel leads cannot be less than d, or that the diameter of the conductor must be appreciably less than that of the screen, restricts the possible range of Z_0 for both types of cable.

In theory, we refer to *lossless lines* but, in practice, all lines show losses of energy. Part of the energy may be radiated as electromagnetic radiation, the line acting as an antenna. This is less serious in coaxial lines because of the screening, but occurs in parallel lead lines, depending on the spacing between the leads. Conductor heating is a well-known means of energy loss, occurring also in power distribution lines. This is simply due to dissipation of energy owing to the resistance of the conductors. Heating of the dielectric also accounts for energy loss and is proportional to the voltage between the conductors. It is very much lower when the dielectric is air.

A major contrast between a transmission line and ordinary circuit connections is that pulses, waves and other signals take a finite time to pass from one end to the other. When a pulse is fed into one end of a line, it travels at a speed ranging between 65% and 97% of the speed of light. A pulse takes between 1.03 μs and 1.54 μs to travel 100 m. This is a very short time but is comparable with the length of a pulse, or the period of a signal of frequency 1 MHz or over. With periodic signals of several MHz or GHz, there is a succession of voltage and current waves passing along the line, transferring energy from source to load.

When a line is correctly terminated, with a load equal to Z_0, all the energy fed into it from the source is transmitted to the other end and transferred to the load (ignoring losses mentioned above). When the line is not correctly terminated, we say it is *mismatched*. The energy is not all transferred to the load. Some or all of the energy is reflected back, causing *standing waves* to form. There are four cases:

A Line terminated by a short circuit:

The source sets up a series of voltage and current waves in the line, travelling from the source to the far end. At the far end, these waves are reflected back toward the source, and, as long as frequency is constant, set up a pattern of standing waves. This is just the same as the acoustic case, when we blow across the end of a pipe which is closed at the far end (Fig. 8.19a). Standing waves are set up and there is zero motion of the air molecules (a node) at the closed end because of the physical barrier there. Note that the figure represents the motion as transverse waves, for clarity, though in fact they are longitudinal waves, causing compression and rarefaction of the air. There are *nodes* equally spaced half a wavelength ($\lambda/2$) apart along the length of the tube. If the tube is a quarter-wave plus an integral number of half-wavelengths long, the air resonates and a note is heard.

Similarly, there can be no voltage changes at the short-circuited end. The reflected wave is 180° out of phase with the incident wave, so their voltages cancel out. This situation recurs at a distance of $\lambda/2$ (half a wavelength) from the end and is repeated at $\lambda/2$ intervals all the way back to the source. In between are points where the incident and reflected voltages reinforce each other, producing maximum voltage changes (antinodes). The outcome is that standing voltage waves are set up in the cable, with nodes at the end of the line and at points at $\lambda/2$ intervals along the line (Fig. 8.19b). In the figure, the voltage waves are drawn alongside the cable for clarity, but in fact they are within it.

In the air-filled pipe, the points of zero motion are also points at which change of pressure is maximal. Figure 8.19c depicts the standing pressure waves. There is maximum change of pressure (an *antinode*) at the closed end, with a node (no change of pressure) $\lambda/4$ away, the pattern repeating at intervals of $\lambda/2$.

In the short-circuited transmission line there is a comparable set of standing current waves. The current wave is reflected with no phase change, so there is a maximum (antinode) at the short-circuited end where there is no resistance to current flow. The first current node is $\lambda/4$ from the end and nodes recur at $\lambda/2$ intervals along the line (Fig. 8.19d).

B Line terminated by an open circuit:

In an air-filled pipe which is open at the far end, there is a motion antinode at the open end, and the first motion node is $\lambda/4$ away (Fig. 8.20a). Correspondingly, if the transmission line ends in an open circuit, the voltage wave is reflected without phase change. The result is the set of standing waves shown in Fig. 8.20b. Conversely, there can be no current flow with an open circuit so the current wave is reflected as a wave 180° out of phase. It has a node at the open-circuited end, with nodes spaced $\lambda/2$ apart along the line.

C Line terminated by an impedance not equal to Z_0:

This is an intermediate situation in which part of the energy of the incident wave is transferred to the load. The reflected wave has smaller amplitude than the incident wave, so the standing waves have reduced amplitude. The amplitude of the standing waves is found by calculating two quantities:
Voltage standing wave ratio, VSWR, is the ratio between the maximum and minimum voltages on the line:

$$\text{VSWR} = \frac{V_{\text{max}}}{V_{\text{min}}} = \frac{V_{\text{i}} + V_{\text{r}}}{V_{\text{i}} - V_{\text{r}}}$$

where V_{i} and V_{r} are the amplitudes of the incident and reflected waves. The VSWR is thus the ratio between the voltage where the incident and reflected waves reinforce each other (antinodes) and where they neutralize each other (nodes).
Reflection coefficient, ρ, is a number between 0 and 1, expressing the proportion of the energy reflected back at the far end of the line:

$$\rho = \frac{|Z_{\text{L}} - Z_0|}{Z_{\text{L}} + Z_0}$$

where Z_{L} is the load impedance. With a correctly terminated line, $Z_{\text{L}} = Z_0$ and so $\rho = 0$. There is no reflection.

If, for example if $Z_{\text{L}} = 30\,\Omega$ and $Z_0 = 50\,\Omega$, then:

$$\rho = \frac{|30 - 50|}{30 + 50} = \frac{20}{80} = 0.25$$

25% of the energy is reflected. As Z_L decreases, ρ increases; more and more energy is reflected until, when Z_L becomes a short circuit, $\rho = 1$ and there is total reflection. Because the limit is a short circuit, the standing waves are located as in Fig. 8.19b whenever $Z_L < Z_0$.

If $Z_L > Z_0$, for example if $Z_L = 200\,\Omega$ and $Z_0 = 50\,\Omega$, then:

$$\rho = \frac{|200 - 50|}{200 + 50} = \frac{150}{250} = 0.6$$

60% of the energy is reflected. As Z_L increases, ρ increases; more and more energy is reflected until, when Z_L becomes an open circuit, $\rho = 1$ and there is total reflection. Because the limit is an open circuit, the standing waves are located as in Fig. 8.20b whenever $Z_L > Z_0$.

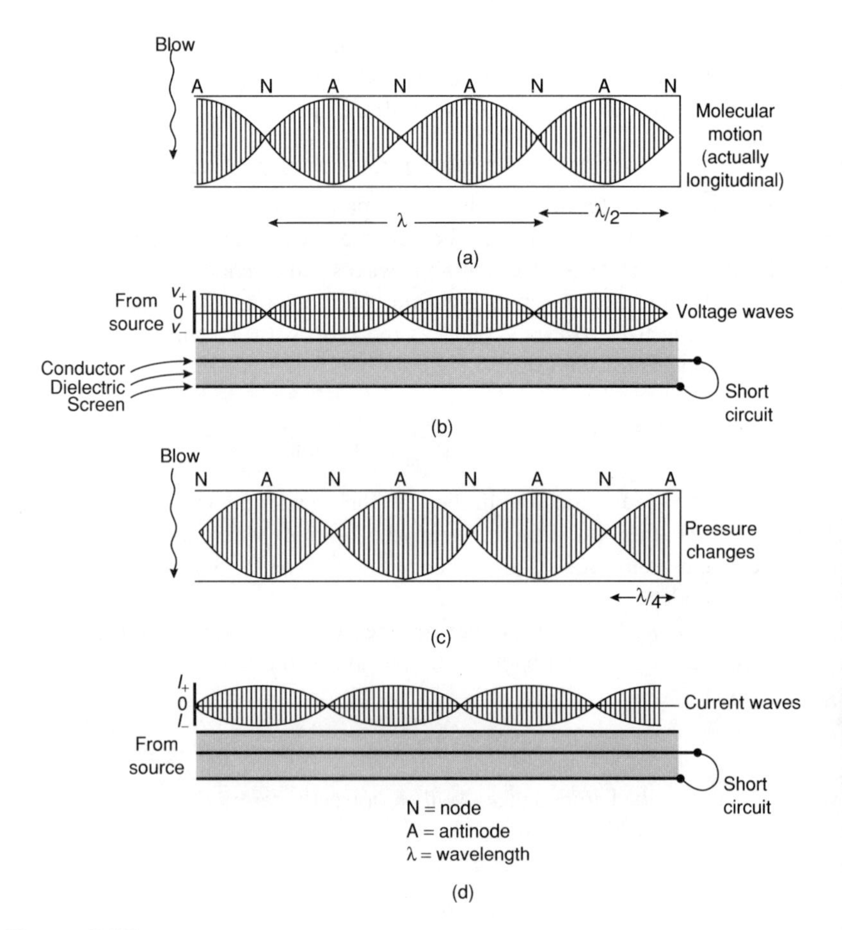

Figure 8.19

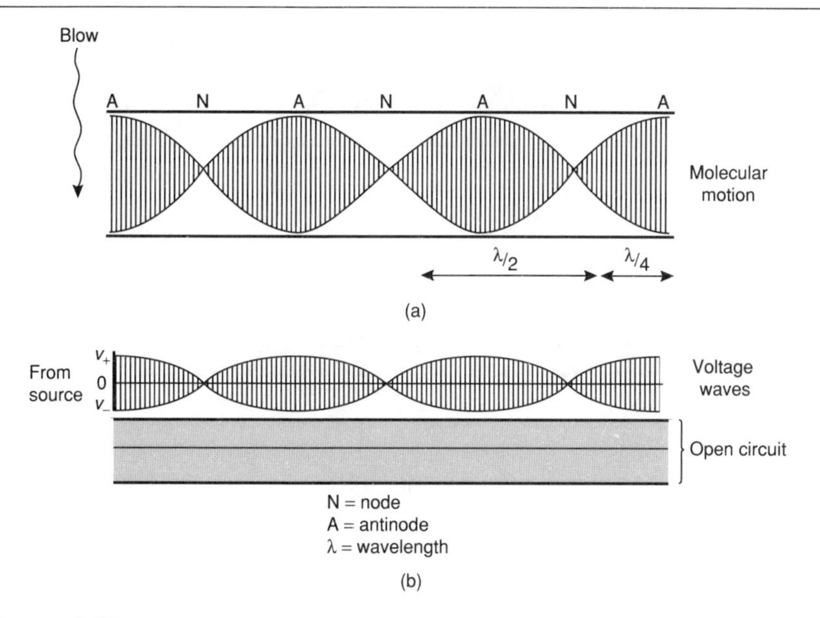

Figure 8.20

D Line terminated by an impedance:

The reactance, inductive or capacitative, causes the waves to be reflected with change of phase depending on the value of the reactance and the frequency of the signal, as in a filter (6.1.1). This produces standing waves with the first node of the voltage wave between λ/2 and λ/4 from the load in the case of an inductance, or less than λ/4 from the load in the case of a capacitance. Because inductors and capacitors do not absorb energy, all the power is reflected.

8.2.2 Transmission lines as components

The formation of standing waves on a mismatched line suggests that the length of the line in terms of wavelength has a significant effect on transmission. It can be shown that a line λ/4 long (Fig. 8.21a) and terminated by a load Z_L has an impedance, looking into the line, of $Z_{in} = Z_0^2/Z_L$. If the line is correctly terminated, with $Z_L = Z_0$, then $Z_{in} = Z_0$, as required. The same expression applies to all lines that are an integral number of quarter-wavelengths long.

This property of a *quarter-wave line* is used for matching the impedances of lines and loads when they are unequal. The equation above when rearranged gives:

$$Z_0 = \sqrt{(Z_{in}Z_L)}$$

In Fig. 8.21b a quarter-wave line, impedance Z_0, is being used to match a load Z_L to a main line, impedance Z_m. To match the quarter-wave line with

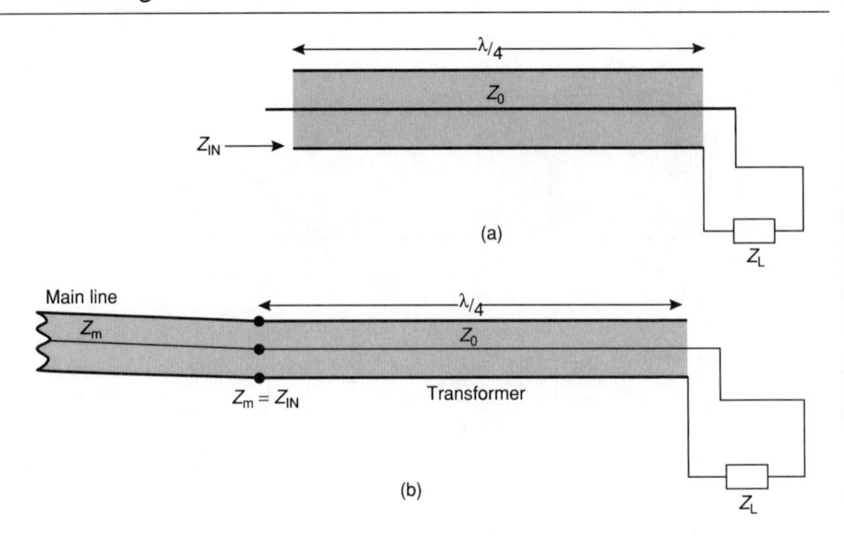

Figure 8.21

attached load to the main line, we need to make $Z_{in} = Z_m$. For example, if the main line impedance is 50 Ω and the load impedance is 60 Ω, the impedance of the quarter-wave line should be:

$$Z_0 = \sqrt{(50 \times 60)} = 54.8\,\Omega$$

A length of transmission line used in this way is a circuit component, in this case a *transformer*.

In connection with Fig. 8.18a, it was mentioned that the air in the tube resonates at a fixed frequency, in this example when the tube is 1.75 wavelengths long. This is the principle on which wind instruments such as the flute and the pipe organ operate. Similarly electrical oscillations can be set up in a short length of transmission line (either open-circuit or short-circuited at the far end) when a signal of the correct frequency is applied to it. In other words, the line acts as a resonant or tuned circuit, equivalent to the capacitor–inductor circuit in Fig. 2.19, for example. Since the exact number of quarter-wavelengths must fit into the length of line, resonance occurs only when the frequency is very close to the correct value. It may also occur at harmonics, but circuit design can ensure that the harmonics do not occur. Such lengths of line, known as *stubs*, may be connected in parallel or series with a transmission line to match it to an inductive load. Stubs have the advantage that they are able to resonate at frequencies higher than it is practicable to obtain with a capacitor–inductor circuit. More illustrations of the use of transmission lines as components are given in the next section.

8.2.3 Microwave transmission lines

At high frequencies, the flow of current through a metallic conductor is confined to the surface layer. This is known as the *skin effect* and the skin depth is defined as:

$$\delta = \sqrt{\frac{2}{\omega\mu\sigma}}$$

where δ is the depth at which the current is $1/e(= 1/2.718 \approx 37\%)$ of its surface value. In this expression, μ is the permeability in henrys/metre and σ is the conductance in siemens/metre.

Example: For copper sheet at 10 GHz, $d = 0.00066$ mm. Typically the copper on a pcb is 0.035 mm thick, so only the outer 1.9% is being used. The skin effect substantially increases the resistance of the conductor to microwave signals. Consequently, although coaxial lines may be used for the lower-frequency microwave bands, UHF transmission lines take the form of *waveguides*, *stripline* and *microstrip*. But, whatever form the transmission line takes, characteristic impedance, standing waves and related concepts still apply.

Wavelengths are very short at microwave frequencies.

Example: For a 10 GHz signal the wavelength is only 30 mm. The wavelength is commensurate with the lengths of the conductors in the transmitter or receiver. Almost every connection in the circuit is a transmission line.

A waveguide is a hollow metal tube, usually rectangular in section, which is used to transport a microwave signal. It can be thought of as a pair of conductors connected together with quarter-wave stubs (Fig. 8.22). The stub is a resonant circuit (8.2.2) with high input impedance, which means that they will not short-circuit the conductors. If they were joined together at their ends by wires the wires would act inductively, with loss of power. But if they are joined by a metal plate, the lines of force are less able to encircle this so the inductive effect is minimized. The final stage is to merge an infinite number of stubs and their attached plates to form a tube of rectangular section, the waveguide. Since the stubs are $\lambda/4$ long, the sides of the waveguide are just over $\lambda/2$ high.

A waveguide is very simple and inexpensive to construct, which makes it preferable to coaxial cable. The walls of the waveguide do not conduct current in the normal sense, that is, from source to load. We must think in terms of electrical and magnetic fields rather than potential differences and currents. In a waveguide the signal is carried along as an electromagnetic plane

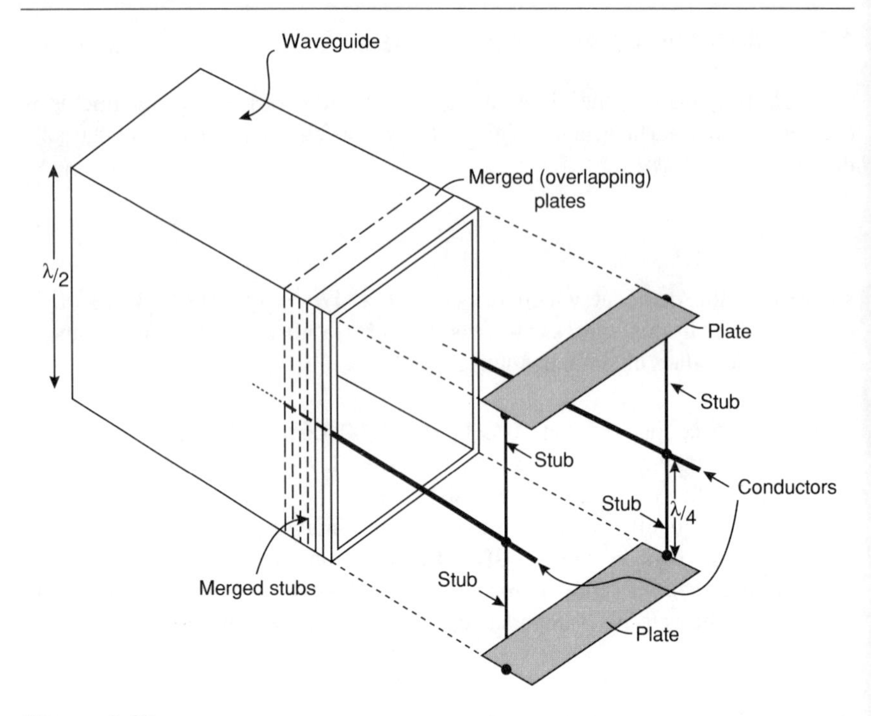

Figure 8.22

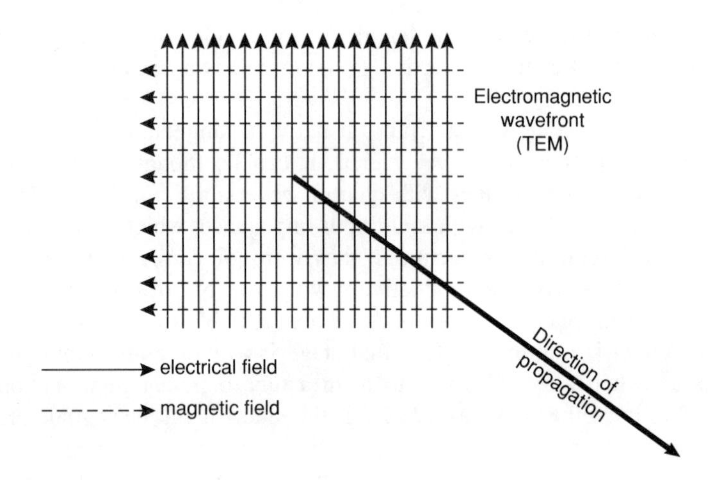

Figure 8.23

wave, reflected along inside the guide in the dielectric, which is usually air. A *transverse electromagnetic* (TEM) wave consists of an electrical field and magnetic field which are perpendicular to each other, both being perpendicular to the direction of propagation (Fig. 8.23). The waves must never have their electric field parallel and close to the walls of the tube, for this would result in the field being short-circuited by the wall. To prevent this from happening, the waves are made to zigzag along inside the waveguide, bouncing off the walls and causing the electric field to be strongest near the centre of the waveguide and weakest near the walls. Then the walls cannot short-circuit the electric field and energy is not lost. The direction of propagation is along the axis of the guide and one of the fields (electrical or magnetic) has a component in the

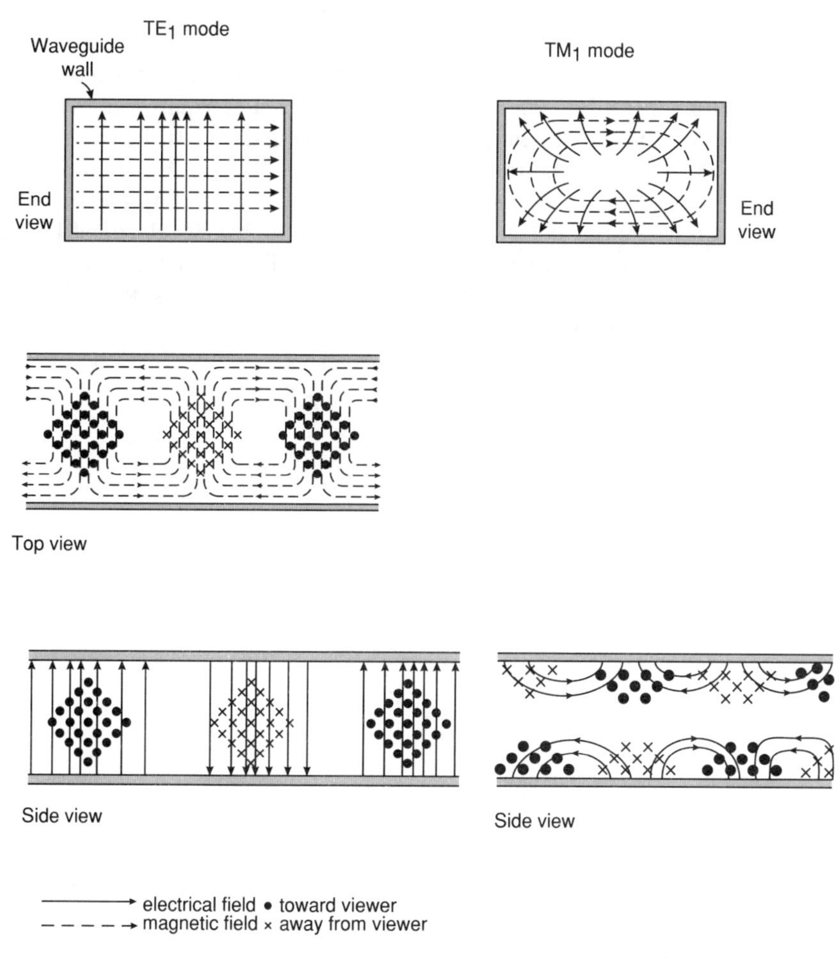

Figure 8.24

direction of propagation. So the wave is no longer a TEM wave. A number of different field patterns or *modes* are possible, depending on whether it is the electrical field or the magnetic field that is transverse. Figure 8.24 illustrates two of the possible modes seen from different viewpoints. Think of these patterns passing along the waveguide from the source to the load. The TE mode (transverse electrical) is the most commonly employed. Other TE modes are available but $TE_{1,0}$ is the dominant mode, in which the waveguide is $\lambda/2$ wide. It requires less power than other modes, and the waveguides are smaller, lighter and cheaper to make. The second example in Fig. 8.24 illustrates one of the TM (transverse magnetic) modes. One of the features of waveguides is that different signals propagated in different modes may pass along the waveguide simultaneously. It is not necessary for them to have different frequencies as in coaxial lines.

Signals are introduced into a waveguide by an antenna inserted through a hole in the wall (Fig. 8.25). The location of the antenna determines the mode of propagation in the guide. Two joined antennae are required for certain modes. Conversely, a probe may be used to extract signals. If the end of the waveguide is left open, a narrow microwave beam emerges. A slot cut in the wall allows either an escape of energy or the creation of a magnetic field, depending on where the slot is cut. What happens is that the electrical fields in the guide cause local currents to flow in the walls. The slot interrupts the flow of these currents. If two waveguides are placed side by side with slots coincident, energy can pass from one guide to the other.

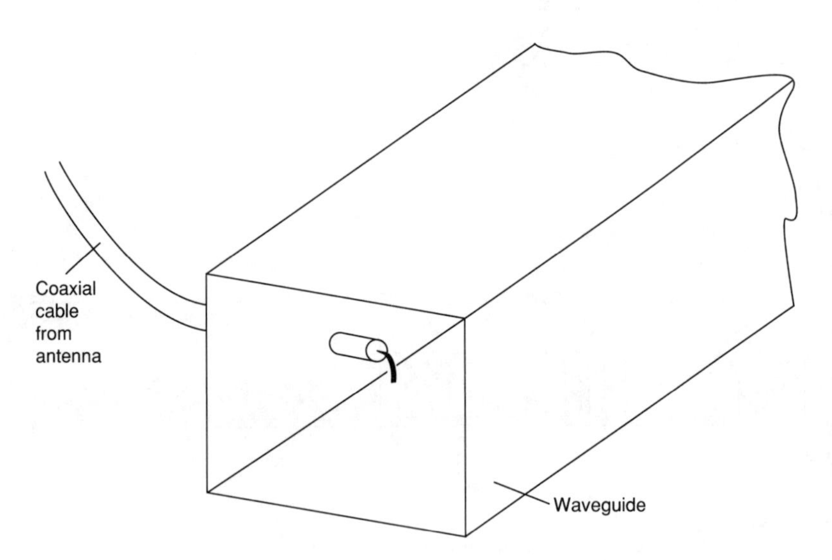

Figure 8.25

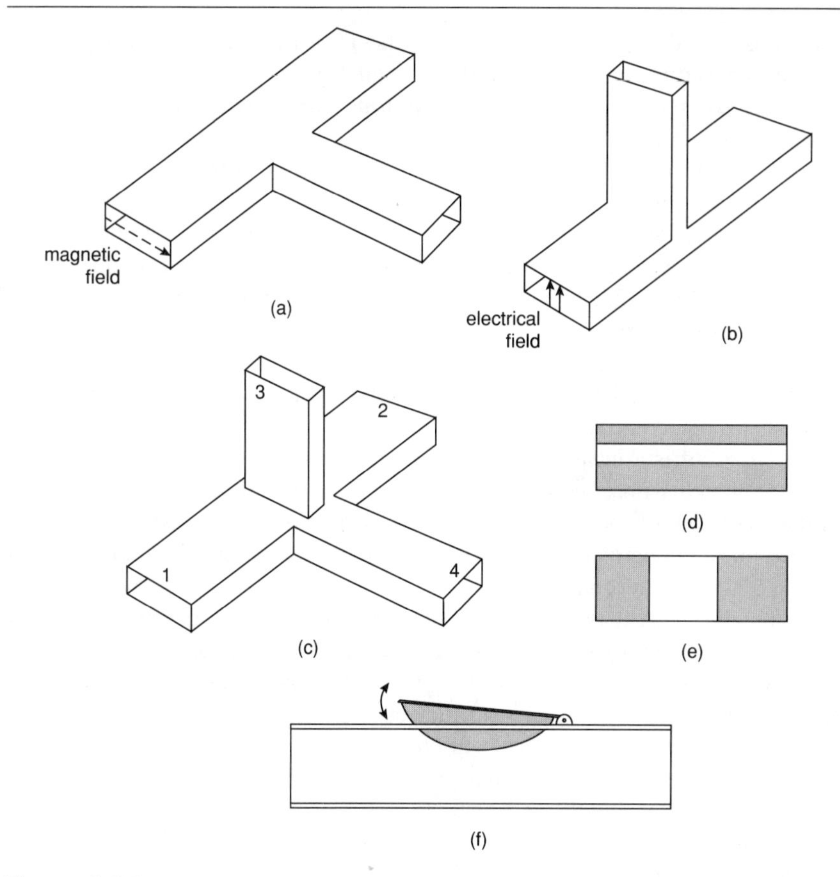

Figure 8.26

As with other forms of transmission line, waveguides can be designed to behave as circuit components. Even more device types can be realized by including diaphragms and pins in the waveguide. Figure 8.26 depicts some waveguide components. Figure 8.26a is an *H-plane tee-junction*. H-plane means that it is in the plane of the magnetic field (in $TE_{1,0}$ mode). It is used for splitting a signal that is fed into the central arm into two signals of equal magnitude, or for combining two signals into one. The *E-plane tee-junction* (Fig. 8.26b) in the plane of the electrical field is also used for signal splitting or combining. It can also be used as a stub for tuning. In this application a plunger in the central arm is moved up or down to trim the arm to the correct length.

The two tee-junctions are combined in Fig. 8.26c to make a *hybrid-tee*, sometimes known as a *magic-tee*. Signals can pass between arm 3 (the E arm) and arms 1 and 2, or between arm 4 (the H arm) and arms 1 and 2, but not between arms 3 and 4. This allows signal generators coupled to arms 3 and 4 to send their combined signals to arms 1 or 2, without being coupled to each other. This junction has many other applications of a similar nature. Figures 8.26d and 8.26e illustrate the use of diaphragms (*irises*) in wave-guides. A metal diaphragm with a horizontal slit placed across the waveguide acts as a capacitor. This is because the pd between the top and bottom walls of the guide now exists across the narrow slit, increasing capacitance. The diaphragm in Fig. 8.26e conducts current between the top and bottom surfaces, causing additional energy to be stored in the magnetic field. This has the action of an inductor. By making a narrow horizontal slit of the correct width it is possible to combine both capacitance and inductance in the same diaphragm. If these are made equal it produces a resonance.

Figure 8.26f is a *vane attenuator*. The vane is made from a dielectric such as glass coated with conductive material such as a carbon film. The vane is moved in or out of the guide to adjust the degree of attenuation.

Stripline and microstrip are two types of microwave transmission line which are in general more convenient to use and less expensive than waveguides. Stripline (Fig. 8.27a) is the equivalent of a coaxial cable squashed flat. It is made from a sheet of dielectric such as Teflon with a copper layer on both sides. One layer is etched to produce the conductive track. The other is left as a ground plane. Then a second dielectric sheet, with copper on one side, is placed dielectric side down on the first sheet and bonded to it by heating. Usually the edges of the ground planes are bonded together by conductive tape.

Microstrip is a simpler and even less expensive transmission line, made from a sheet of dielectric covered on both sides with copper, one side being etched to form the microstrip, the other side being left intact as a ground plane. The advantage of microstrip is that it is much easier to solder components such as transistors directly to the strip. It is often used for miniaturized circuits.

The wavelength λ of the microwave signal in stripline is:

$$\lambda = \lambda_0/\sqrt{\varepsilon}$$

where λ_0 is the wavelength in free space and ε is the dielectric constant.

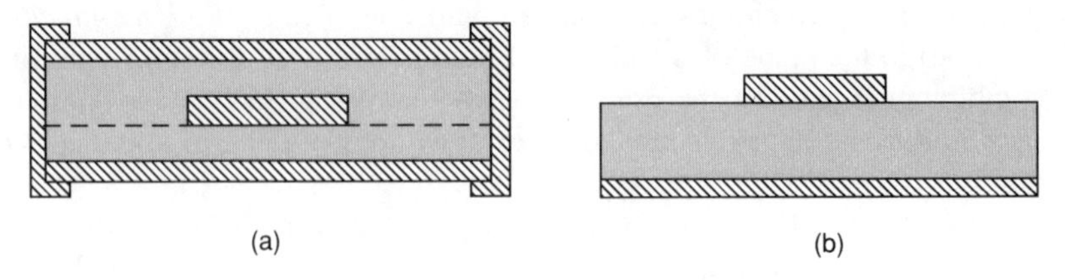

Figure 8.27

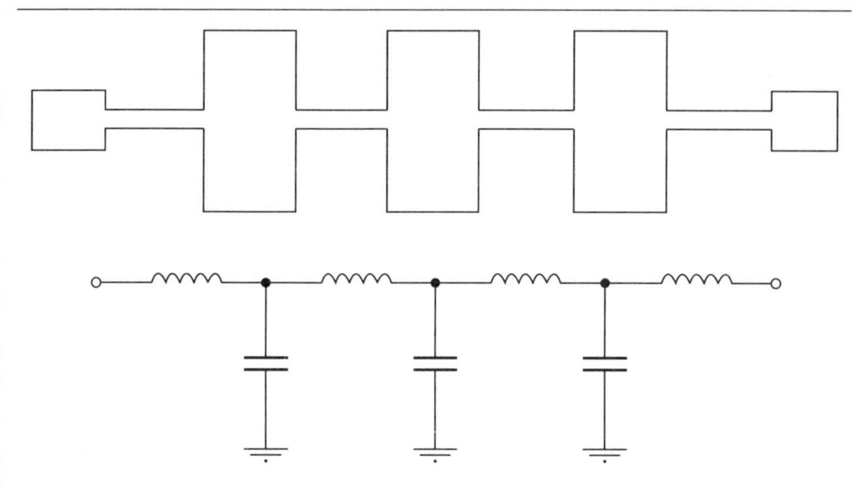

Figure 8.28

Example: With Teflon or alumina, for which $e \approx 10$ and a frequency of 10 GHz ($\lambda_0 = 30$ mm), we find $\lambda = 9$ mm. This means that a $\lambda/4$ stub is only 2.25 mm long. For precise tuning, the stripline must be precisely etched, so there is an upper limit to the frequency at which stripline can be used. Using a dielectric with lower ε can extend the range.

The characteristic impedance of stripline depends on the dielectric constant, the width and thickness of the conductor and the thickness of the dielectric. The relationships are complicated and it is usual to employ prepared charts when designing.

A wide range of microwave components can be made from stripline or microstrip. In general they lack the facility for mechanical adjustments provided by the vanes, pegs and irises found in waveguides, but they are so much simpler to manufacture. As an example, Fig. 8.28 shows a low-pass filter realized in microstrip, together with its equivalent in conventional components. The mathematics is complicated but it can be seen that a narrow stripline acts as an inductor, and a broad stripline acts as a capacitor shunted to ground. High-pass and band-pass filters can be implemented in a variety of ways.

8.3 Modulators

Figure 8.29 is the circuit of one of the simpler types of modulator used in AM radio transmitters, a *diode balanced modulator*. The audio (or other low-frequency signal) is amplified by a JFET CS amplifier then fed to one end of the primary of the output transformer. The radio-frequency carrier (usually

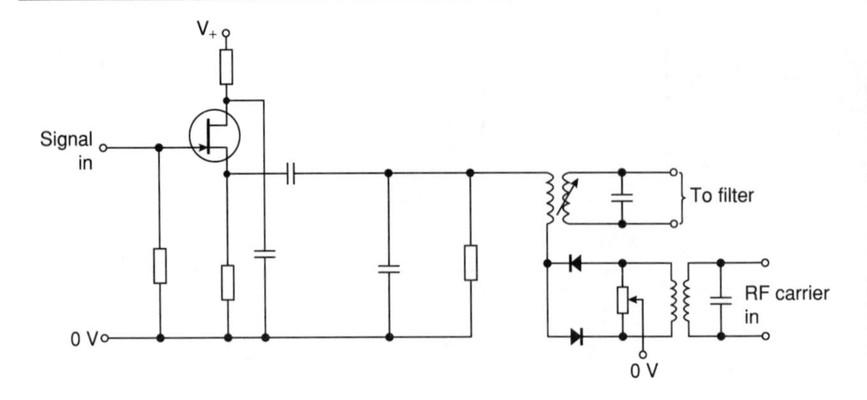

Figure 8.29

generated by a crystal-controlled oscillator) is coupled through a transformer to a diode circuit. The diodes are matched and, as long as no signal is present, prevent the carrier from getting through to the primary coil. If a signal is present, the positive and negative swings unbalance the diode circuit, allowing the RF to appear on the primary coil. But voltage levels are such that the signal and carrier produce a modulated waveform. This is picked up by the secondary coil and fed to the next stage, usually a filter.

This modulator produces an output only when the low-frequency signal is present, which means that the carrier alone does not appear at the output. Only the sum and difference signals (Fig. 8.2), the upper and lower side bands, appear. The pure carrier is suppressed. Depending on the true balancing of this simple circuit, it may not be completely suppressed but it is at least considerably (up to 67% of the power) reduced, so reducing power wastage in later stages. The two side band signals are then sent to a filter or equivalent circuit to remove one of them, resulting in a single-side-band (SSB) signal. This is then fed to RF amplifiers and possibly other stages before being sent to the transmitting antenna. These aspects of radio transmitter circuitry and the topic of the propagation of radio waves are outside the scope of this book.

Frequency modulation requires the frequency to be varied while amplitude remains constant. The principle of the direct method of FM is to use a reactance modulator to vary the reactance of the tuning circuit of an RF oscillator. Reactance modulators are based on BJTs, FETs or on varactor diodes (B.3). Figure 8.30 is a simple varactor-based reactance modulator. The varactor is reverse-biased so it functions as a capacitor, the capacitance varying with the audio signal. The varactor is coupled to the resonant circuit of an RF oscillator, thus increasing or decreasing its capacitance in it in accordance with the audio signal and varying its frequency as a result.

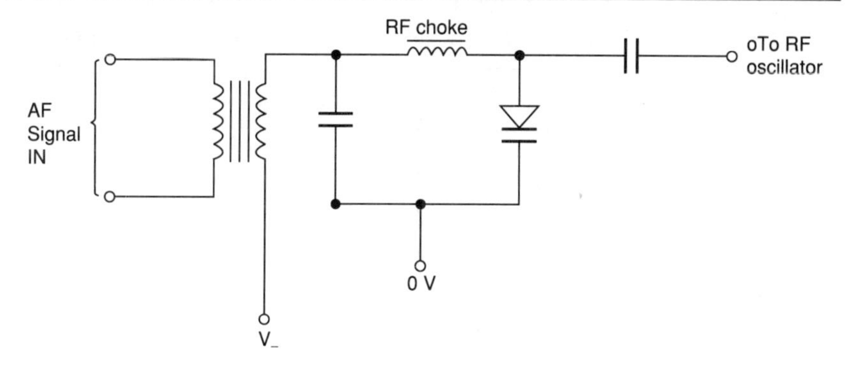

Figure 8.30

8.4 Demodulators

Demodulating an AM signal to recover the original modulating signal involves rectifying the signal and removing the RF component from it to leave the modulating signal. The most commonly used rectifier is the diode, the descendant of the original cat's-whisker. Figure 8.31 is a diode rectifier or detector circuit. In a simple radio receiver the RF signal comes directly from the antenna. Next is the tuner, consisting of the secondary winding of the input transformer and the variable capacitor. The resonant frequency of this circuit is tuned to that of the transmitter. The diode rectifies the signal which appears as an alternating pd across the resistor. The RF component is filtered off by the low-pass filter, consisting of a capacitor and inductor, leaving the AF signal, which is then amplified. This is an extremely simple rectifying circuit, one of its drawbacks being that the signal may be too small to pass through the diode. More often a radio receiver has several stages of RF amplification, using tuned amplifiers, before rectification, so as to minimize the effect of diode drop.

There are several ways of demodulating FM, of which one of the more commonly used ways is the Foster-Seeley *frequency discriminator* pictured

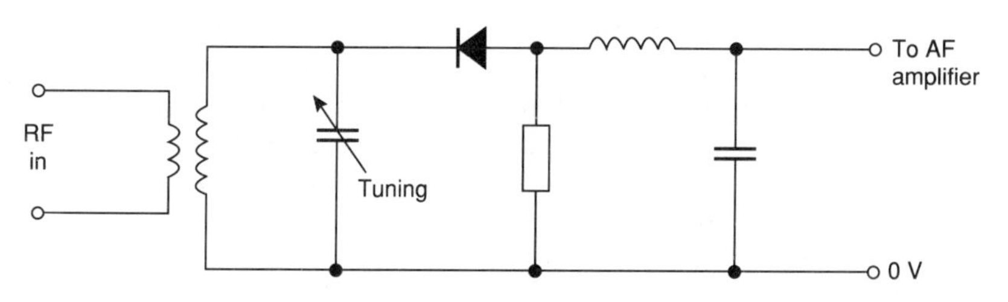

Figure 8.31

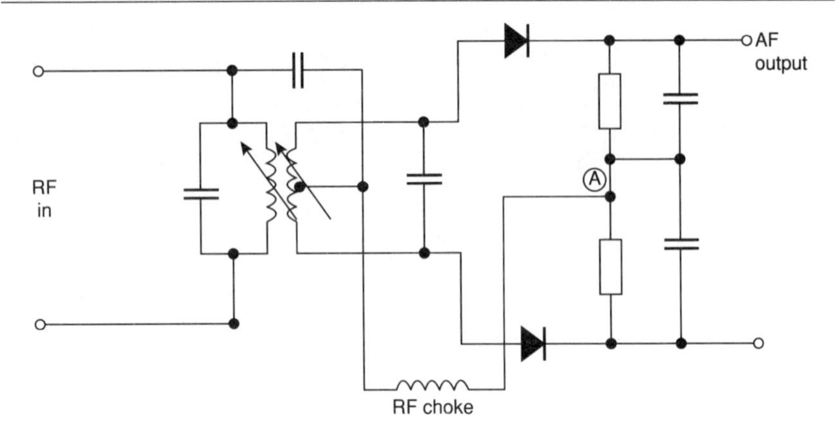

Figure 8.32

in Fig. 8.32. The RF signal is fed to the primary of the transformer inducing a signal in the secondary that is 90° out of phase with it. At the same time, the signal fed to the primary is also fed through a capacitor (to eliminate different DC levels) to the centre tap in the secondary. On one side of the tap, the combined primary and secondary signals lead the primary signal and, on the other side of the tap, it lags it by an equal amount. These signals when rectified by the diodes are equal in size but opposite in polarity with respect to the tap and to point A. They cancel out and no signal appears at the audio output. This applies only if the signal is at its central (unmodulated) frequency. If the frequency changes there is a change of phase on one side of the tap, accompanied by an opposite change of phase on the other side. This produces an increase of voltage on one side and a decrease on the other, resulting in a pd across the output resistor. This is a variable resistor allowing the amplitude of the audio output to be adjusted.

8.5 Gunn and IMPATT diodes

A Gunn diode is not a rectifying diode. It is called a diode simply because it has two terminals. The diode consists of a layer of n-type gallium arsenide sandwiched between two layers of low-resistance gallium arsenide. If the pd across such a diode is gradually increased from 0 V, at first the current increases until it reaches several tens of milliamps, as might be expected. When the pd across the diode exceeds a certain value, usually in the region of 3 V, depending on the thickness of the n-type layer, current unexpectedly begins to decrease. There is negative resistance. It is found that negative resistance begins when the voltage gradient across the n-type layer exceeds 330 V/m. Current does not flow smoothly under these conditions but pulsates. A bunching of electrons occurs, forming what is called a *domain* of electrons and this passes across the n-type layer. When it reaches the other side, a new domain forms. Thus a sequence of domains crosses the n-type layer, and the current through the device pulsates accordingly. The period of the pulsation is regular and equals the transit time of the electrons through the layer. With a layer 10 μm thick, for example, the frequency of pulsation is 10 GHz. The *Gunn effect*, as it is called, provides the basis for UHF oscillators, particularly for generating microwaves. The diode is coupled to a resonant circuit and keeps it in oscillation. The Gunn diode is able to operate at powers up to 1 W and it has low noise.

The negative resistance of gallium arsenide is thought to be due to the existence of an energy level for GaAs that does not occur generally in semiconductors. Above the usual conduction band (see Fig. A.6), which is partly filled, there is (in GaAs) a narrow forbidden energy gap with a usually empty conduction gap above it. Under a high voltage gradient, some electrons acquire sufficient energy to take them into this higher energy band. But in this band they are less mobile and the current is reduced.

Gunn diodes are primarily used in microwave oscillators but have applications in parametric amplifiers (5.4), security circuits, speed sensors, and related fields.

The *impact avalanche and transit time* (IMPATT) diode is another negative resistance device but this is a true diode with a pn junction. Its action depends on generating an electron avalanche in the n-type region. Adjacent to the n-type is a wider drift region of n-type material into which the shower of electrons can drift. The quiescent voltage across the junction is carefully held at the level at which the avalanche does not occur. Any increase in voltage due to an RF signal increasing from zero in the positive direction triggers off an avalanche. This builds up during the whole positive phase to produce a pulse of current just as the phase ends and the voltage crosses through zero in the negative direction. During the first half of the negative phase, the pulse drifts through the drift region, reaching the cathode just as the voltage reaches its most negative value. Thus the maximum current occurs when the voltage is a minimum; voltage and current are 180° out of phase, which is equivalent to negative resistance. This action is used to drive oscillators and amplifiers in the 50 GHz band upward. They are faster acting than Gunn diodes, but have the disadvantage of being noisy. The oscillating frequency depends very much on the physical dimensions of the diode, particularly the width of the drift region so the bandwidth of an IMPATT oscillator is narrow.

8.7 Optical fibre communications

The principle of optical fibre technology is that, instead of communicating by sending electrical signals along a copper cable, we send light signals along a transparent fibre of pure quartz. The light is reflected internally from the sides of the fibre at the boundary between the *core* and a transparent *cladding* material (quartz that is less pure and has a lower refractive index). The lower refractive index of the cladding causes total internal reflection of light as it strikes the core–cladding boundary. Since the core is of such high purity, and total reflection is so efficient, there is little absorption or loss of light energy. Taking into account absorption and reflection losses, together with losses due to microscopic irregularities in the surface of the core, the loss of energy can be as little as 0.21 dB/km in high-quality optical fibre. Repeaters spaced 60 km apart are usually adequate (compared with every 1–2 km for copper cables). The optical fibre, even when sheathed in a strengthening and protective jacket, is lighter than copper cable. Optical fibre, unlike copper cable, is not subject to electro-magnetic interference. Conversely, it is secure in the sense that the passage of signals cannot be detected from outside the cable. If several fibres are enclosed in the same sheath, there is no transfer of signal from one fibre to its neighbours, as there is with copper wires. Unlike copper cable, optical fibre does not corrode.

Although optical fibre can operate in up to 100 000 modes, it is more usual in long-distance transmission for the fibre to be a single strand with only one transmission passing along it. Signals are pulse modulated. The available bandwidth is so large that there is room for up to 100 000 voice channels to be carried simultaneously at different frequencies.

The source of light is usually a light-emitting diode, though lasers and laser diodes are also used, especially for the higher-frequency systems. They are driven by high-frequency amplifiers, operating at 40 GHz, often more. The light is detected by a photodiode, usually a PIN photodiode (B.3), which is better able to respond at high frequency. An *avalanche photodiode* (APD) is superior to the PIN diode in that, when a photon has created an electron-hole pair, an avalanche action takes place, increasing the number of electron-hole pairs generated. This gives the APD much higher sensitivity that the PIN photodiode. Their response time is less than half that of the PIN photodiode, which makes them preferred for high-frequency systems. Multiplexing signals at the transmitting end and demultiplexing them at the receiving end is performed by digital circuits. In addition, optical devices are used to multiplex and demultiplex signals transmitted by light of different wavelengths.

With their many advantages over copper-wire systems, it is not surprising that there is a dramatic and continued increase in the use of optical fibre for national and international telecommunications.

9 Storing analogs

Compared with the bewildering variety of sensors and other devices for capturing analogs (Chapter 1) there are few really different ways in which analog circuits influence the outside world. Their main actions rely on the conversion of electrical energy into:

- light–lamps, LEDs, lasers, photographic recorders (including cinema soundtracks), TV and oscilloscope tubes;
- magnetic field–motors, relays, mechanical latches, meters (moving-coil and moving-iron), chart recorders, tape recorders (audio, video), loud-speakers, magnetically actuated machinery (including cutters for shellac and vinyl recording discs);
- piezo-electric force–sounders, crystal headsets.

Heat, although a common by-product of analog circuits, rarely features as a useful output. Several of the examples quoted above can be used to produce a permanent record of the analog. Some, such as the chart recorder, produce a pen-and-ink record that can be studied, but it cannot be used to recreate the original electronic analog. Others, such as the cinema soundtrack, the vinyl disc, and the tape recording can be used to reproduce not only the original sounds but also the analogs derived from them. But the analogs are stored in analog form, as varying image density or as a wavy groove. These methods have been or are being replaced by digital storage, for example the compact disc, which is used not only for storage of video and audio but also for the storage of data of all kinds. This is paralleled by the storage on magnetic and other forms of computer disc and in the memory of computers. Digital storage has many advantages but is not a topic for study in this book. Here we are concerned only with ways of signal conversion, analog to digital and digital to analog. Many techniques have been invented but we deal only with those most commonly in use.

9.1 Temporary storage

A current can be stored for several years in a loop of superconductor, but this is not routinely done. Voltage is easier to store, using a capacitor. The *sample-and-hold* circuit of Fig. 9.1 is a useful temporary device. It can be used to sample a rapidly changing voltage to provide an interval long enough for its value to be read on a meter. Similarly, it can hold a voltage constant while being converted into digital form.

The op amps, which have FET inputs, are wired as voltage followers (D.2.2), and act as buffers. The MOSFET switch is opened or closed by a high or low control voltage. It is normally closed → v_{OUT} follows v_{IN}. Because of the high Z_{IN} and low Z_{OUT} of OA1, there is an ample supply of current to keep the capacitor charged to the v_{IN} level. To sample v_{IN} the switch is opened. Both the MOSFET switch and the (+) input of OA2 offer high resistance → the charge on the capacitor remains constant → v_{OUT} is held at the value v_{IN} had when the switch was opened. v_{OUT} remains constant until the switch is closed again, when it again follows v_{IN}.

Errors in this circuit are principally the result of current leakage through the switch and OA2. Leakage through the switch is a nanoamp or less; leakage through the input of OA2 is a few picoamps. There is also leakage of a few tens of picoamps through the capacitor itself. For example, if the switch leakage is 500 pA, the op amp leakage is 5 pA, and the capacitor leakage is 50 pA, the total leakage is approximately 600 pA. Charge leaks at 600 pC/s, and with a 100 nF capacitor, voltage falls at 600 pC/100 nF = 6 mV/s. With hold times less than a second, leakage is generally negligible but, if the sample is held for a minute or more, droop (Fig. 5.25) causes serious error. Leakage can be reduced by using a low-leakage capacitor (polypropylene has resistance of over 10^{11} Ω). The effects of leakage can be reduced by increasing capacitance, but this may mean that the capacitor does not have time to charge or discharge

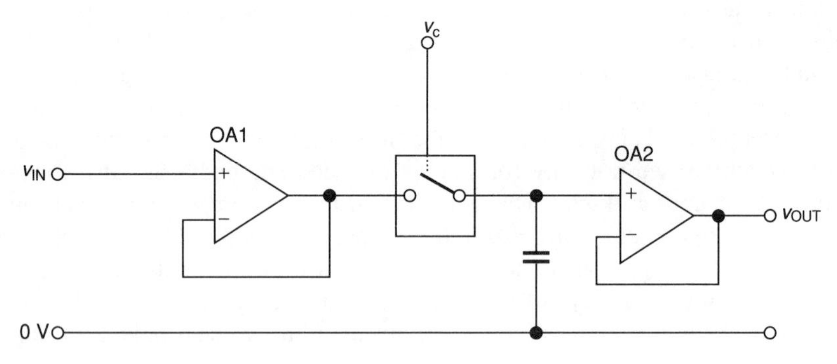

Figure 9.1

fully when sampling rate is high. Other errors may arise from the time taken for the op amps to respond to voltage changes. Op amps with a high slew rate should be used.

Charging and discharging of a capacitor is exponential, so that theoretically a capacitor never charges completely to v_{IN}. However, starting from no charge, a capacitor reaches 99% of full charge in 5RC seconds. Given a 100 nF capacitor and a switch resistance of 50 Ω, the time taken to reach 99% charge is $5 \times 50 \times 100 \times 10^{-9} = 25\,\mu s$. This is the minimum time for which the switch should be opened between samples, unless it is known that only small alterations in v_{IN} occur between samples.

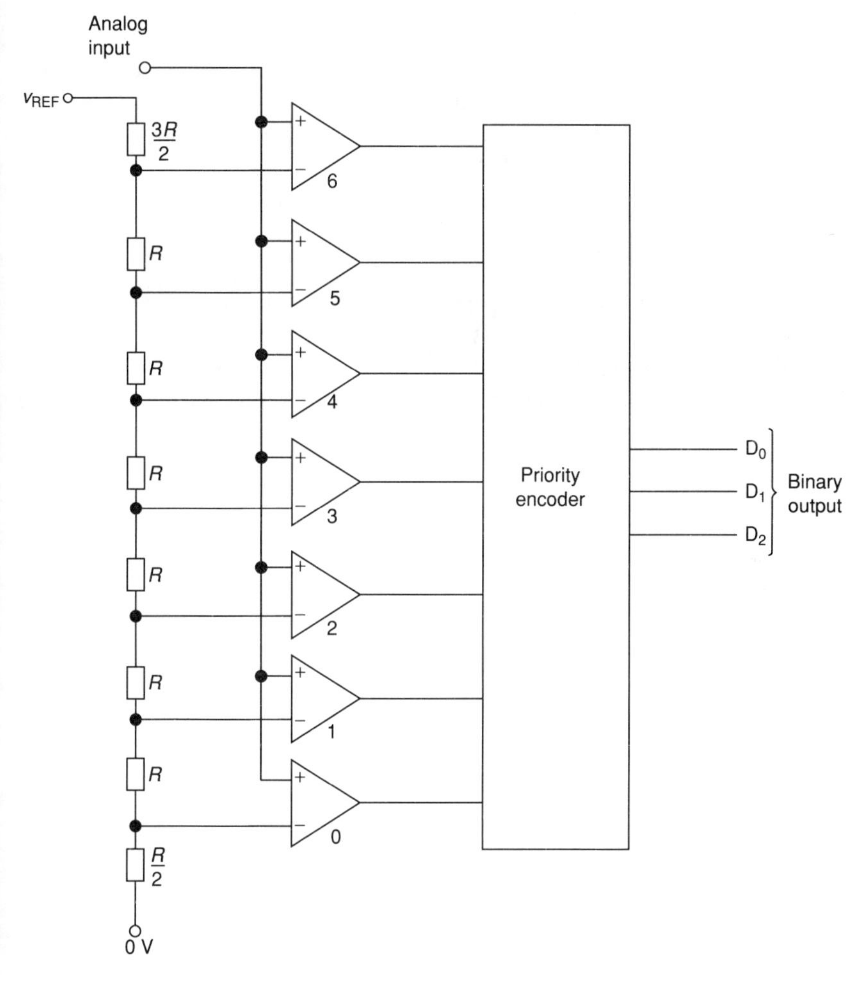

Figure 9.2

9.2 Analog to digital conversion

Analog signals are converted into digital form for:

- storage;
- processing, for example for digital filtering;
- to allow the use of a digital display, for example in a digital panel meter or testmeter.

The topic of the noise generated by conversion is dealt with in 7.2.4.

9.2.1 Flash converter

This is the fastest converter so that it does not need a sample-and-hold stage before it. It consists of a resistor chain and an array of comparators (Fig. 9.2). The resistors act as a potential divider. Given the values shown in the figure, the voltages at the (−) inputs of the comparators are:

$$V_{REF} \times \frac{nR + R/2}{6R + R/2 + 3R/2} = V_{REF} \times \frac{2n+1}{16}$$

where n ranges from 0 (at comparator 0) to 6 (at comparator 6). Thus the voltages run from $V_{REF}/16$ to $13V_{REF}/16$, in steps of $2V_{REF}/16 = V_{REF}/8$. Note that the value of R does not affect the voltages. All that is necessary is that the resistances shall be strictly in the ratio shown in the figure. The

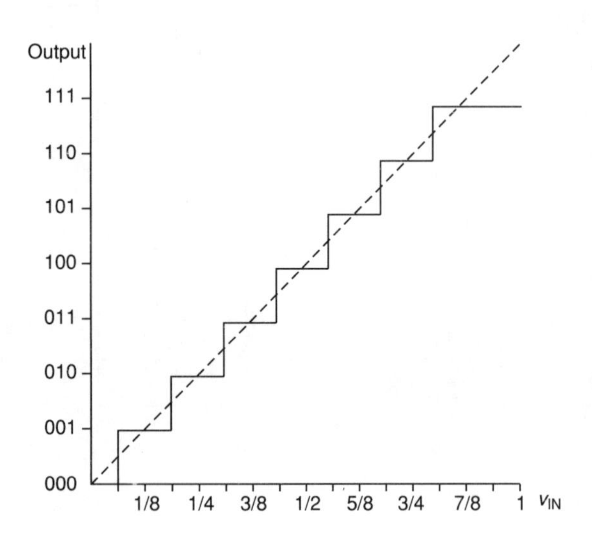

Figure 9.3

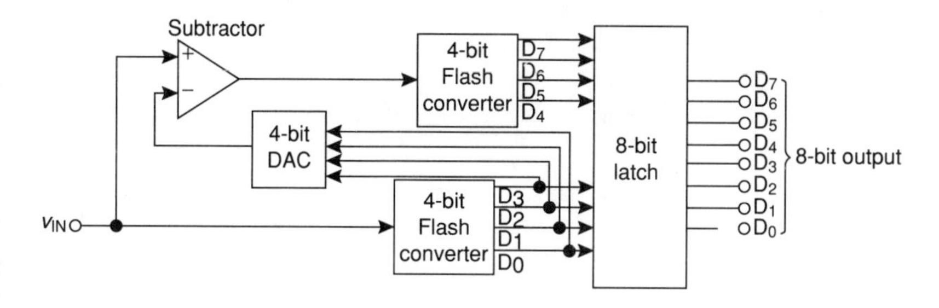

Figure 9.4

converter is fabricated on a single chip so it is easy to ensure that the ratios are precise.

The analog voltage is applied simultaneously to the (+) inputs of all comparators, each of which compares this voltage with a voltage on the chain. When v_{IN} is 0 V, the (−) input of every comparator is at a higher voltage than the (+) input. The output of every comparator is logical low. As v_{IN} rises from 0 V to V_{REF}, first comparator 0, then comparator 1, then comparator 2 have a logical high output, and so on up the chain. Eventually, when v_{IN} exceeds $13V_{REF}/16$, all comparators have a high output. The priority encoder is a logical circuit which determines which is the highest comparator in the chain to have a high output. It produces a 3-bit output accordingly. The digits are numbered D0 to D2, with D0 being the least significant digit.

Three digits allow for eight voltage ranges to be expressed:

Output	Voltage range	
	≥	<
000	0	1/16
001	1/16	3/16
010	3/16	5/16
.	.	.
111	13/16	1

The voltage values are given as fractions of V_{REF}. Figure 9.3 plots the relationship between input and output. The ideal conversion curve is shown as a dotted line. The stepped curve is the digital output, which has a resolution of 1/8 of v_{IN}. There is a maximum offset error of half a bit, that is 1/16 of v_{IN}. Maximum output is reached when v_{IN} exceeds $13V_{REF}/16$, since there are only three digits in the output.

To make a conversion, the flash converter requires only the comparator settling time plus the transit time of the encoder. Conversion times are in the region of 0.01 μs. Converters are able to sample at 25 MHz, which means they can sample signals up to 12.5 MHz (8.1.4.1).

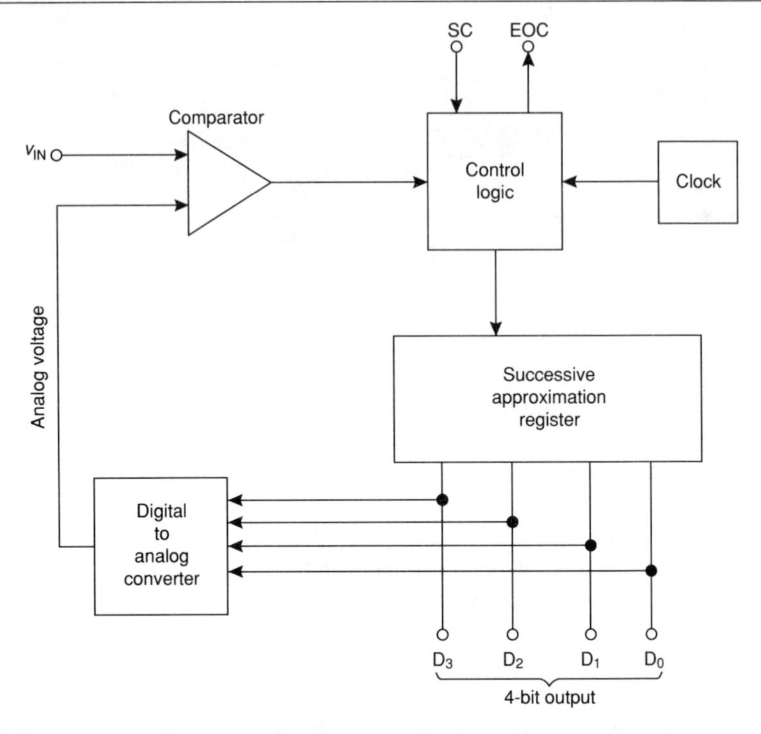

Figure 9.5

An n-bit flash converter requires 2^{n-1} comparators, which makes even 8-bit flash converters expensive because they need 255 comparators. For numbers with 8 or more bits, a compromise is the half-flash converter (Fig. 9.5), which converts the analog voltage in two halves. The first stage is a 4-bit flash converter, like that in Fig. 9.2, but with 15 comparators. The output of this goes to the four least significant bits in an 8-bit latch. It also goes to a digital-to-analog converter (9.3) which turns it back into an analog voltage again. This voltage is subtracted from the analog input voltage and the difference is converted by another 4-bit flash converter which provides the four most significant bits of the conversion. Thus a total of only 30 comparators is required. Conversion takes around 2 µs which is slower than the full flash converter, but is still faster than most other types. A half-flash converter is relatively inexpensive.

9.2.2 Successive approximation converter

Because this converter is slower than the flash converter (typically the conversion time is between 10 µs and 100 µs) the analog signal is usually held in a sample-and-hold circuit (9.1) while it is being converted to digital form. Some converters have a sample-and-hold circuit on the same chip, driven from the same clock.

The centre of the converter is the successive approximation register (SAR) which has between 8 and 16 bits. It is operated by the control logic which is clock driven to perform a cycle of operations for each bit. At the beginning of a conversion the most significant bit is set to 1 and all others are set to zero. If we assume that we are describing an 8-bit converter for which the maximum input voltage is 16 V, a setting of 1000 0000 corresponds to 8 V, half the maximum value. The output from the SAR goes to a digital to analog converter (9.3), the output of which is compared with v_{IN}. If the output of the DAC is less than or equal to v_{IN}, the 1 is left unchanged. If the output of the DAC is more than v_{IN}, the 1 is changed to 0. At the next clock cycle the next bit is made 1, the DAC output is compared again with v_{IN} and this bit is either left as 1 or changed to 0. This operation is repeated for each bit in succession, so conversion takes 8 clock cycles.

As an example, this is the sequence of values in the SAR when v_{IN} is 3.265 V:

Clock cycle	SAR is	Output of DAC (V)	More or less than v_{IN}	SAR becomes:	Value in SAR (V)
1	1000 0000	8	more	0000 0000	0
2	0100 0000	4	more	0000 0000	0
3	0010 0000	2	less	0010 0000	2
4	0011 0000	3	less	0011 0000	3
5	0011 1000	3.5	more	0011 0000	3
6	0011 0100	3.25	less	0011 0100	3.25
7	0011 0110	3.325	more	0011 0100	3.25
8	0011 0101	3.3125	more	0011 0100	3.25

The SAR holds 3.25 V for the last three cycles because this is the nearest value to 3.265 V within the 8-bit resolution of the converter. Also shown in Fig. 9.5 are a logic input SC to instruct the control logic to start conversion and a logical output EOC to indicate the end of conversion.

9.2.3 Dual slope integration

This technique employs an integrator (5.2.1). In the first stage of conversion, the control logic switches the integrator to v_{IN} (Fig. 9.6). Assuming that v_{IN} is negative, the output of the integrator ramps up at a rate dependent upon v_{IN}. The control logic allows the voltage to ramp up for a fixed period of time t_1 (Fig. 9.7). The voltage V_R reached in that time is proportional to v_{IN}. In the second stage, the counter is first reset to zero, then the input of the integrator is switched to V_{REF}, which is a fixed positive voltage. This makes the output

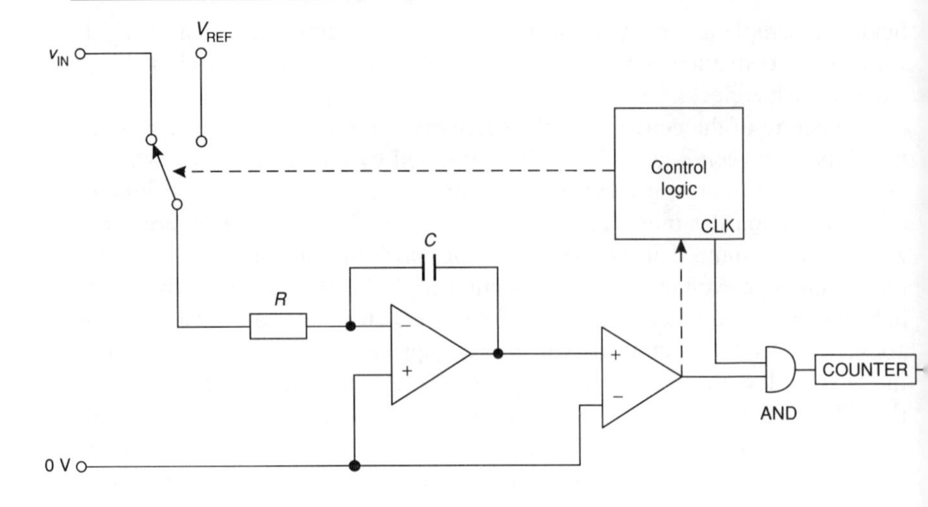

Figure 9.6

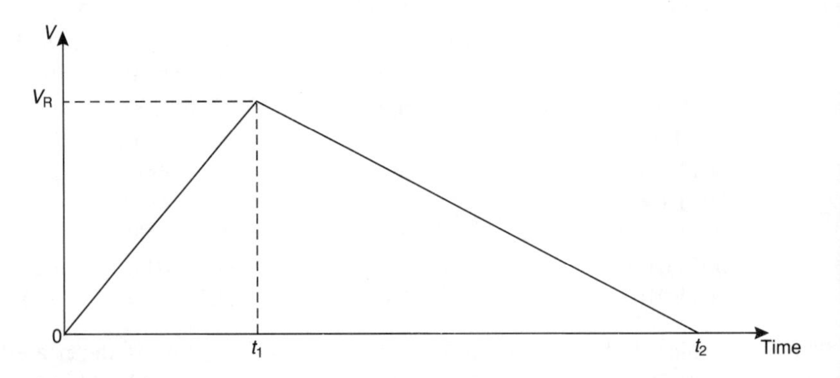

Figure 9.7

ramp down but, since V_{REF} is constant, the slope of the downward ramp is fixed. As long as the integrator output is positive, it allows clock pulses to pass through the AND gate, to be counted by the counter. When the ramp reaches zero, at t_2, pulses no longer pass through the AND gate and counting stops. The length of time that has been taken to ramp down at fixed rate from V_R to zero is proportional to V_R, which is proportional to v_{IN}. Thus the count is the digital equivalent of v_{IN}.

Both rates of ramping depend not only on the voltages but on the values of R and C. However, the values of R and C affect both upward and downward ramps, so their exact values do not matter and neither do changes in their values due to ageing. In addition, the precision and stability of the clock is unimportant since if, for example, the clock runs slow → the upward ramp lasts longer → V_R is greater → the downward ramp lasts longer, but is being timed by a slower clock → the effects of slow clock rate cancel out. Another advantage of this converter is that an integrator, having the properties of a low-pass filter, is insensitive to noise. As long as the V_{REF} is precise, results of high accuracy are guaranteed. The main disadvantage is that this is a slow ($\approx 10\,\mu s$) method of conversion, which limits its usefulness. It is often used in the measuring circuits of high-precision multimeters, or in other measuring applications where only a low refresh rate is needed.

9.2.4 Sigma delta converter

Instead of sampling at double the signal frequency rate, and converting the analog into 8 or 12 bits, the sigma delta converter samples at a much higher rate (oversampling) but converts into only one bit. The output of the converter at any instant is either logical high or logical low. In this way it produces a rapid succession of highs and lows, known as a bit stream. It is the varying proportion of highs to lows that corresponds with the varying value of v_{IN}. Naturally, there are problems with operating at very high frequencies, but the circuits required are simpler than in multi-bit converters.

Figure 9.8 is a block diagram of one type of sigma delta converter. Imagine it operating without the feedback loop, in which case the output of the summer block takes the present value of v_{IN}. At the integration block this is added to previous samples. If v_{IN} is positive, the output of the integrator ramps steadily up. The output of the one-bit quantizer (Q) is high (= 1) if its input is zero or positive, or it is low (= 0) if its input is negative. If the output of the integrator is positive and ramping upward, the output of the quantizer is a series of high bits. A continuous series of high bits indicates that v_{IN} is positive. If v_{IN} then becomes negative, the integrator ramps down, but the output of quantizer stays high for some time, until the integrator output has ramped down below zero.

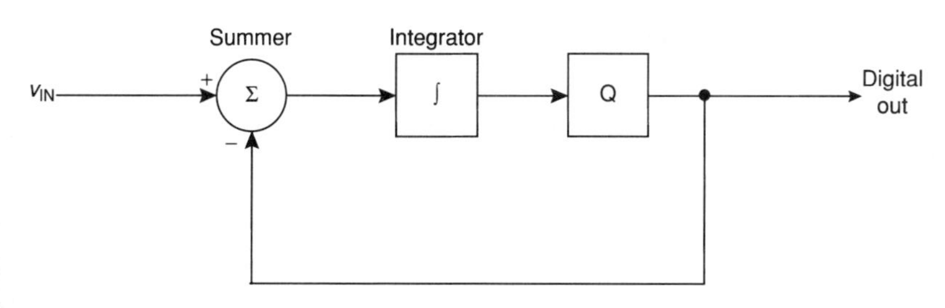

Figure 9.8

The time taken for this depends on for how long v_{IN} was previously high. To remove this effect of previous levels of v_{IN}, we introduce feedback which continually compensates by subtracting the present output from v_{IN} before it is integrated. The output of the circuit is a continuous series of highs if v_{IN} is positive and rising, a continuous series of lows if v_{IN} is negative and falling, or a mixture of highs and lows if v_{IN} is changing from rising to falling, or from falling to rising. In other words, the sigma delta DAC output represents the differences between successive samples, not their absolute values. The output signal could also be a mixture of highs and lows if v_{IN} was rapidly alternating between rising and falling, but the sampling frequency is well above the highest frequency of interest in v_{IN} so this is not of importance. The only very high frequency likely to be present in v_{IN} is noise, but the integrator acts as a low-pass filter and eliminates this.

The next stage of processing is usually a *digital decimation filter*. This converts the high-frequency one-bit signal into a multibit signal at lower frequency. It averages the bits by taking them in groups. For example, if the output from the quantizer is:

0 0 1 1 0 0 1 0 0 0 1 1 1 0 1 1 0 1 0 1

Group the bits in fives:

00110   01000   11101   10101

Select the majority bit:

0   0   1   1

Output from the decimation filter is a 4-bit value:

0011

Twenty successive bits at high frequency have been converted to a single 4-bit value at 1/20 of the frequency. This is still a high-frequency sample, since the original signal was oversampled, so it is suitable for further processing or storage.

9.3 Digital to analog conversion

Digital signals are converted to analog signals so that stored values or values calculated by a computer (for example) may be used to control devices such as lamps or motors. Motor speed can be varied under computer control or, as in a washing machine, not by a computer in the generally accepted sense of the word, but by an integrated circuit known as a microcontroller. Data stored on a compact disc may, after being read and processed digitally, be converted into analog signals to drive a loudspeaker.

The most commonly used DAC technique is based on a 'ladder' consisting of resistors of values R and $2R$ (Fig. 9.9). Commonly $R = 10\,\text{k}\Omega$ and $2R = 20\,\text{k}\Omega$, but the exact values do not matter as long as the proportions are precisely correct. This is relatively easy to achieve if the resistors are all formed on the same chip. For simplicity, Fig. 9.2 illustrates a 4-bit converter, but converters for up to 16 bits are made. The digital input connects the analog switches S_0 to S_3, either to the 0 V line (equivalent to the binary digit 0) or to the (−) input of the op amp (equivalent to the binary digit 1). The converter in Fig. 9.9 is set to a digital input of 0110.

The (−) input of the op amp is a virtual ground so all $2R$ resistors are in practice connected to 0 V, whatever the settings of the switches. The ladder part of the circuit can be re-drawn as in Fig. 9.10a, but with the parallel $2R$ resistors at node A replaced by their equivalent, a resistor R. We can then replace the two R resistors at node A with a $2R$ resistor from node B to ground. This gives two $2R$ resistors in parallel at node B, which can be replaced with a single R resistor (Fig. 9.10b). Going one stage further, replace the two R resistors at B with a $2R$ resistor from C to ground, then replace these two parallel resistors with an R resistor from C to ground (Fig. 9.10c). In Fig. 9.3c the voltage at node D is V_{REF}. The two R resistors act as a potential divider, so the voltage at C is $V_{\text{REF}}/2$. Returning to Fig. 9.10b, the voltage at C is $V_{\text{REF}}/2$, so the two R resistors are a potential divider and the voltage at B is $V_{\text{REF}}/4$. In Fig. 9.10a, the voltage at A is $V_{\text{REF}}/8$.

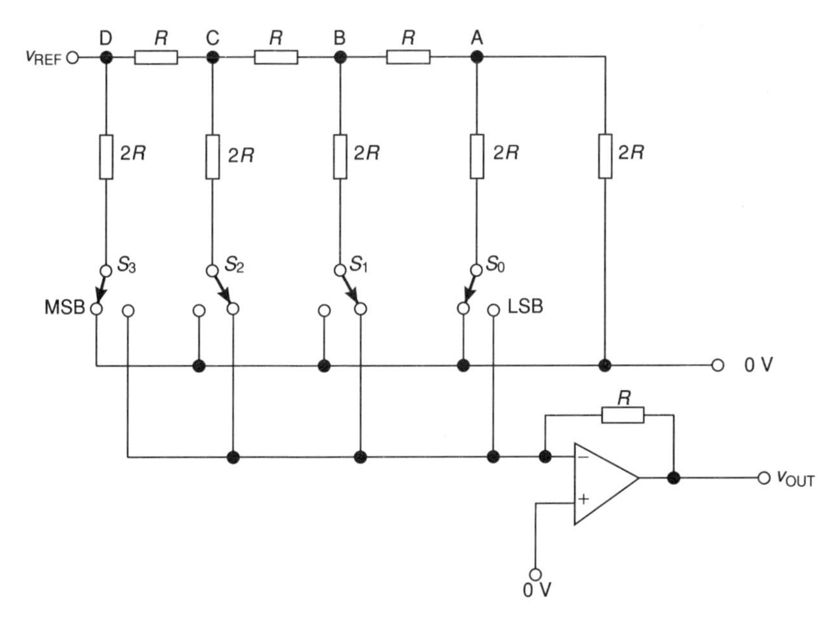

Figure 9.9

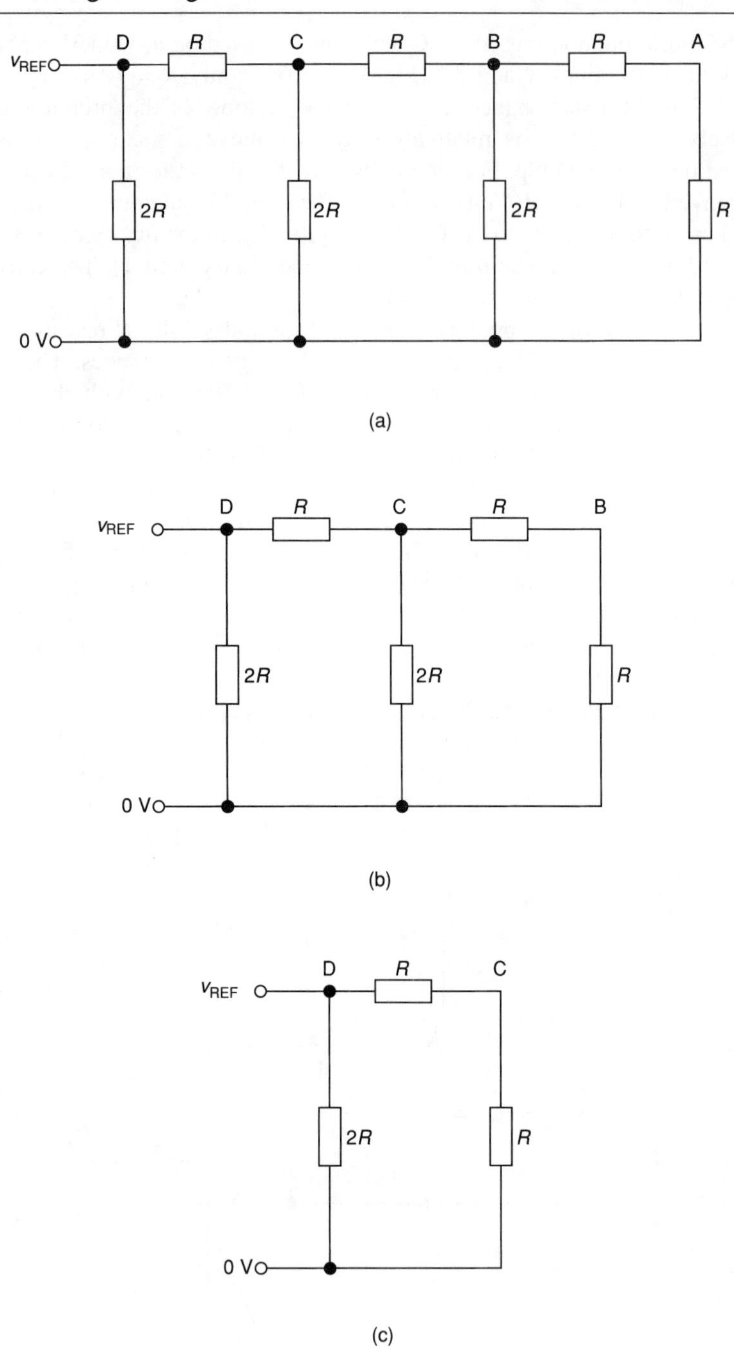
D
R
C
R
B
R
A
v_{REF}
2R
2R
2R
R
0 V
(a)
D
R
C
R
B
v_{REF}
2R
2R
R
0 V
(b)
D
R
C
v_{REF}
2R
R
0 V
(c)

Figure 9.10

When any one or more of the switches are connected to the op amp, the currents flowing through to the op amp are proportional to the voltages at the nodes A to D. For example, if the digital input is 0001, only S_0 is switched to the op amp. The voltage at A is $V_{REF}/8$, so the current is:

$$i_{0001} = \frac{V_{REF}}{8 \times 2R}$$

The current flows on through the feedback resistor, so that:

$$V_{OUT} = \frac{-V_{REF}}{16R} \times R = \frac{-V_{REF}}{16}$$

Note that the output does not depend on the value of R. The digital input has 16 values, 0000 to 1111 (0 to 15 in decimal) and each increment of 1 increases V_{OUT} by $V_{REF}/16$.

The superposition theorem (D.4) applies, so the effects of switching are additive. For example, with the switches set as in Fig. 9.10, the total current to the (−) input is:

$$V_{REF}/4R + V_{REF}/8R = 3V_{REF}/8R$$

This current flows through the R feedback resistor so:

$$v_{OUT} = \frac{-3V_{REF}}{8R} \times R = \frac{-3V_{REF}}{8}$$

The digital input in this example is 0110, which in decimal is 6 and v_{OUT} is −6/16, or −3/8 of V_{REF}. As the digital count runs from 0000 to 1111, v_{OUT} decreases from 0 to $-15V_{REF}/16$. An inverting op amp may be used to convert the output to positive voltages. Conversion times vary widely but are of the order of 1 μs.

This type of converter is one of a group known as *multiplying DACs*, so-called because v_{OUT} is the result of multiplying V_{REF} by the digital input. Certain multiplying DACs are designed to be used as attenuators. If V_{REF} is replaced by a varying analog signal v_{in}, this can be attenuated in 15 steps, with the 16th step (0000) turning off the signal altogether. In this way we have a logically controlled attenuator.

Appendix A Semiconduction

A.1 Charge carriers

Electric current: A mass flow of charged particles (*charge carriers*), moving under the influence of an electric field.
Negative charge carriers: Electrons, anions.
Positive charge carriers: Holes, cations.
Conduction in metals: By conduction electrons, which are valence electrons from outer atomic orbit that have gained extra energy and escaped.
Energy: Of an electron is measured in electron-volts (eV), the kinetic energy that it gains when it falls through a potential of 1 V in an electrical field. $1\,\text{eV} = 1.6 \times 10^{-19}\,\text{J}$.
Energy-band diagrams (Fig. A.1) show energy levels in a material and which levels contain carriers:

(1) Conductors: conduction and valence bands adjacent or overlapping, electrons leave and re-join atoms freely. Ample supply of conduction electrons.
(2) Non-conductors: large *forbidden energy gap* separates bands, valence electrons unable to obtain sufficient energy to become conduction electrons.
(3) Semiconductors: forbidden energy gap is relatively narrow in (e.g.) silicon and germanium, electrons able to escape at room temperature and thereabouts. Escaping electron leaves a *hole* (vacancy), filled by another electron within a few hundred nanoseconds. Only about 1 in 3×10^{10} atoms lose an electron, so conductivity of a pure (*intrinsic*) semiconductor is very low. The equilibrium between escape and recombination depends on temperature, approximately proportional to $T^{1.5}$ (T in kelvin).

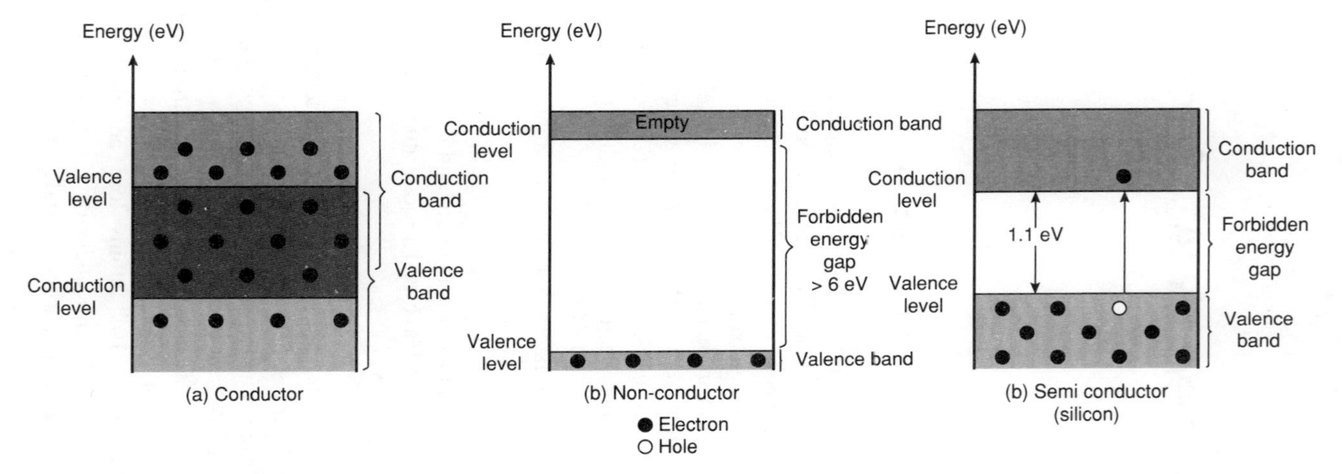

Energy (eV)
Valence level
Conduction level
Conduction band
Valence band
(a) Conductor
Energy (eV)
Empty
Conduction level
Valence level
Conduction band
Forbidden energy gap > 6 eV
Valence band
(b) Non-conductor
Electron
Hole
Energy (eV)
Conduction level
1.1 eV
Valence level
Conduction band
Forbidden energy gap
Valence band
(b) Semi conductor (silicon)

Figure A.1

A.2 Current

In silicon (Fig. A.2), eight electrons are shared between an atom and its four neighbours (*covalent* bonding). An electron escapes, leaving a hole, and moves randomly in the inter-atomic space, causing a *diffusion current* (Fig. A.3a). With an electric field present a *drift* toward the positive end of the gradient is superimposed (Fig. A.3b). The hole left by an electron is soon filled by an electron from an atom nearer to the negative end of the gradient (Fig. 2.4). The holes behave as if they were mobile positive charge carriers (*hole current*).

Current is conventionally taken to be a flow of positive charge from positive to negative. In metals and intrinsic semiconductors, it is a flow of negative charge from negative to positive.

A.3 Conductivity

Conductivity σ (Ω.cm) depends on:

- Electron density, n_e, number of free electrons per cm^3.
- Electron mobility, μ, speed of drift of electrons (cm/s) in a given conductor in an electrical field of unit intensity.

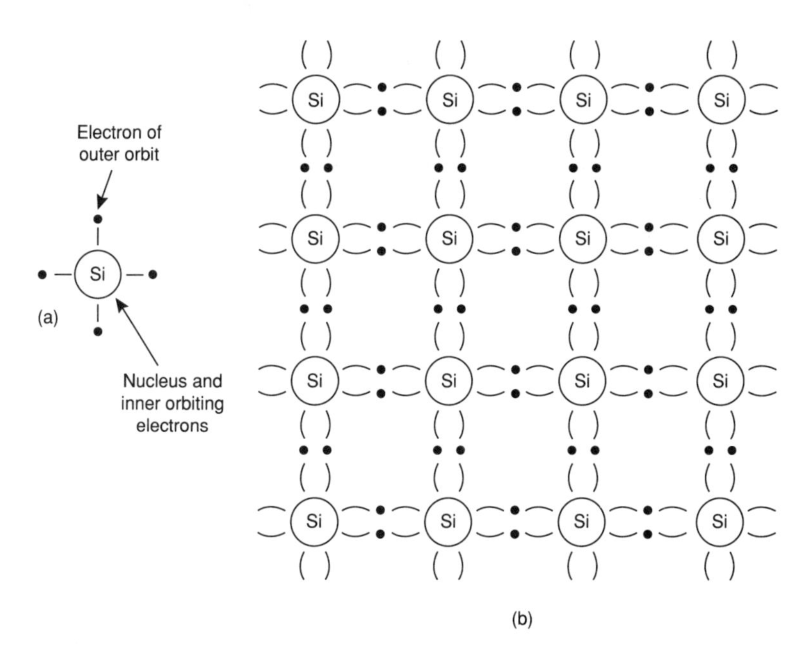

Figure A.2

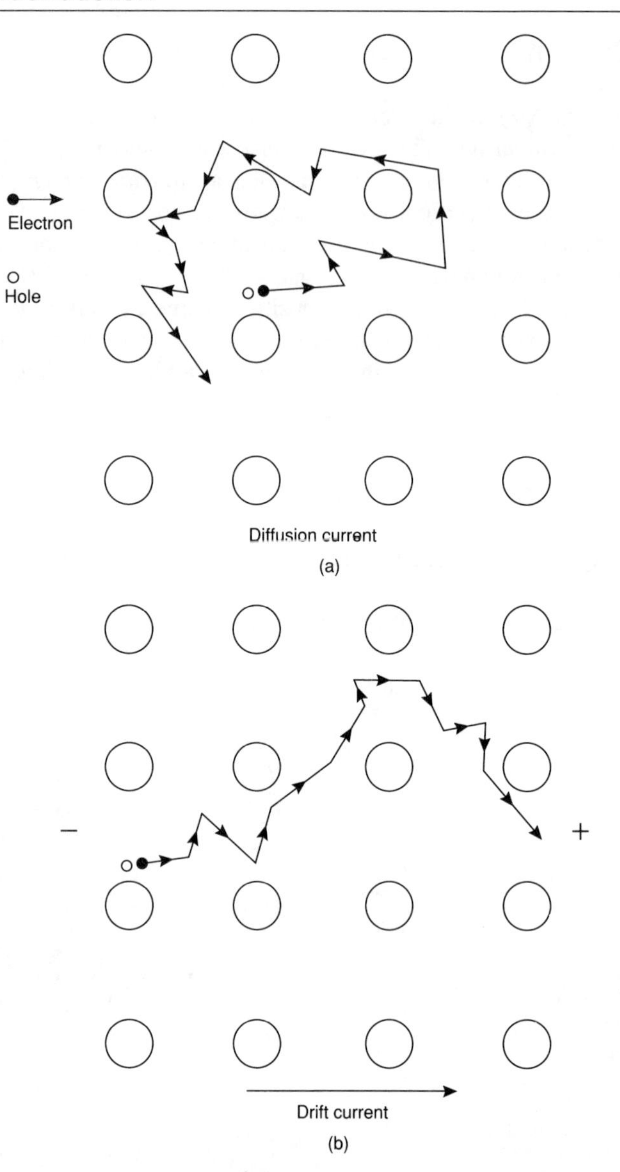

Figure A.3

- Charge, q, on an electron.

$$\sigma = n_e q \mu$$

In metals, increasing temperature → increasing agitation of atoms → increasing obstruction to drift of electrons → reduced mobility → reduced

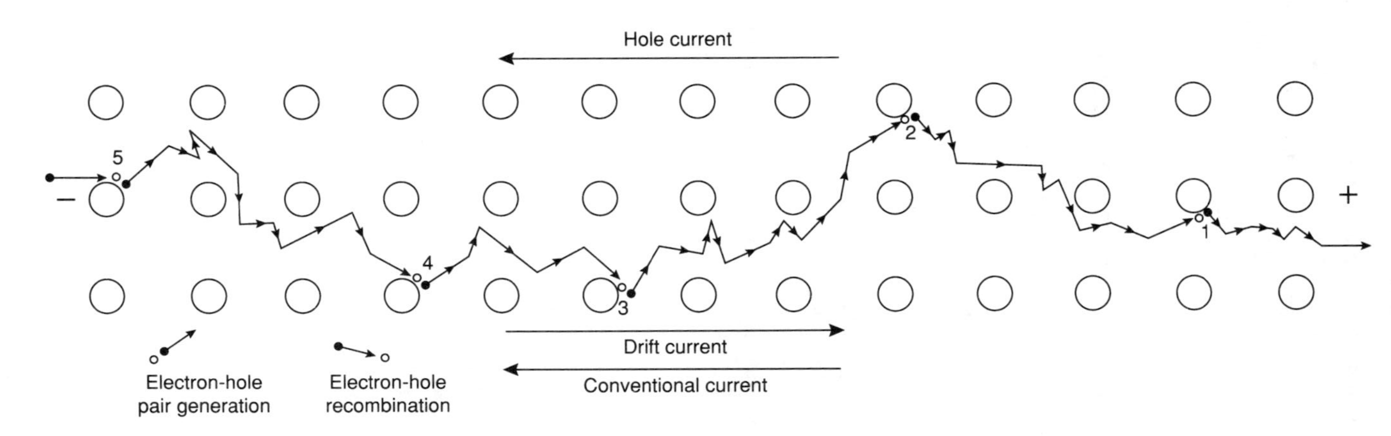
Hole current
5
4
3
2
1
−
+
Drift current
Conventional current
Electron-hole
pair generation
Electron-hole
recombination

Figure A.4

conductivity (or increased resistivity). In intrinsic semiconductors, increased temperature → more electron-hole pairs → increased electron density → increased conductivity (or reduced resistivity). Electron mobility is reduced, as in metals, but this is less significant than the effect of increasing electron density. Result: conductivity of semiconductors increases with increasing temperature.

A.4 Extrinsic semiconductors

Produced by *doping*, often only about one atom in a million replaced.

n-type silicon: Dopant is pentavalent (As, Sb, P, Fig. A.5) and called a *donor*. The energy level of the extra electron is only 0.05 eV below the conduction band, so it easily escapes to become a conduction electron (Fig. A.6) → permanent surplus of free electrons in the conduction band for which there are no corresponding holes. Electrons are *majority charge carriers*; holes are *minority carriers*. Each donor atom provides an electron, so resistivity of n-type is several thousand times less than intrinsic semiconductor. Note majority carriers from donor, minority carriers from silicon (from electron-hole pairs).

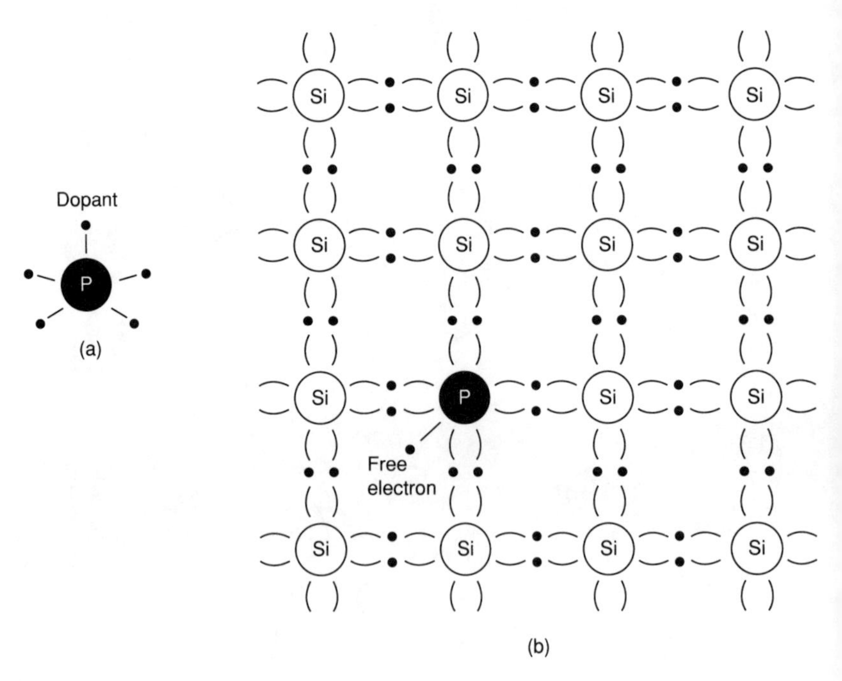

Figure A.5

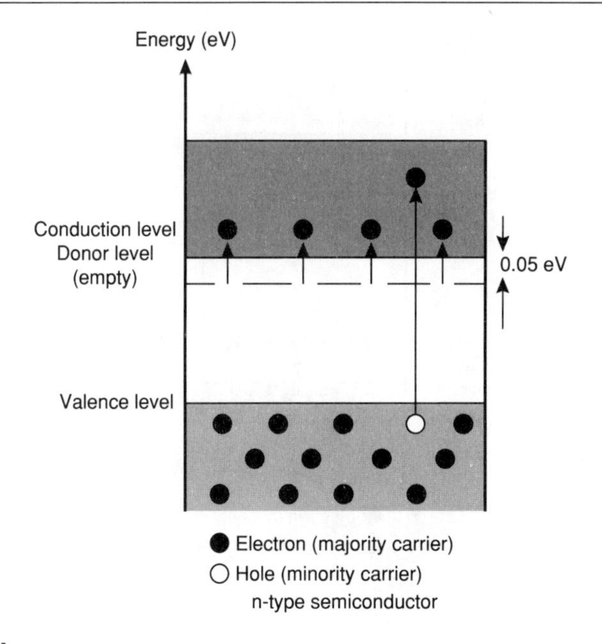

Figure A.6

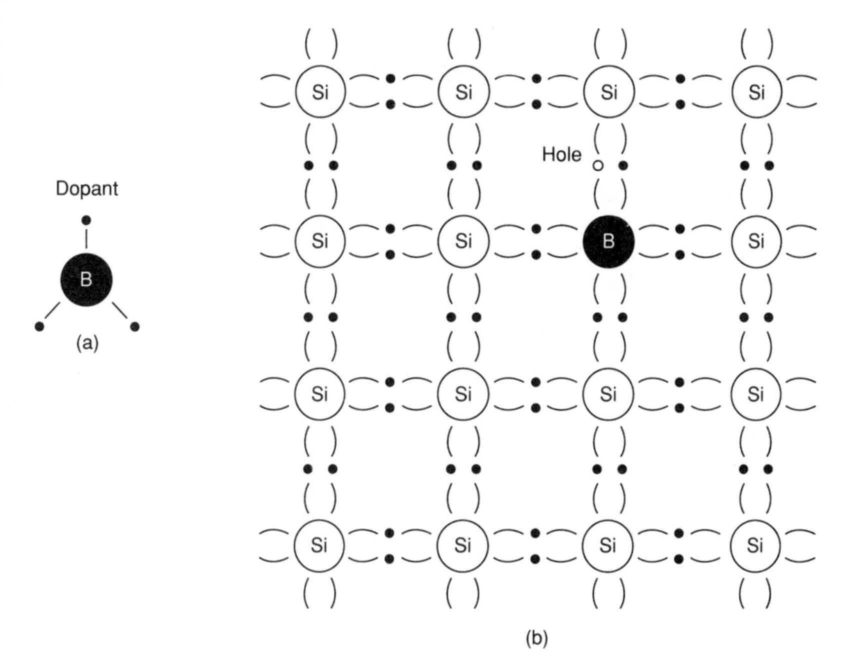

Figure A.7

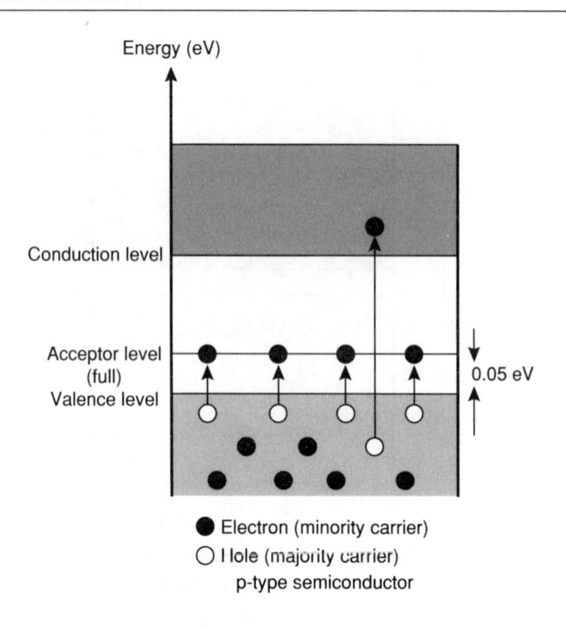

Figure A.8

p-type silicon: Dopant is trivalent (B, Ga, In, Fig. A.7) and is called an acceptor. It creates a hole. The energy level of the dopant is only 0.05 eV above the valence band. Valence electrons readily escape and are captured by the acceptor holes, leaving a surplus of holes in the valence band (Fig. A.8). Electrons are minority charge carriers; holes are majority carriers. The resistivity of p-type is lower than intrinsic semiconductor but holes lack the mobility of electrons so the resistivity is not as low as n-type.

Other semiconductors: Germanium doped as above, n-type or p-type.

Binary compounds: III–V compounds: gallium arsenide (light-emitting diodes) is a compound of Ga (tetravalent) and As (pentavalent), averaging four electrons per atom, and is an intrinsic semiconductor. Dope with more Ga or As to obtain n-type or p-type extrinsic GaAs. Or dope GaAs with a hexavalent element (sulphur, tellurium) to form n-type, or with a divalent element (zinc) to form p-type. Cadmium sulphide (light-dependent resistors) consists of Cd (divalent) and S (hexavalent).

A.5 pn junction

Adjacent regions of silicon are doped to make them *n-type* and *p-type*. Excess electrons from n-type cross junction and are captured by holes in p-type.

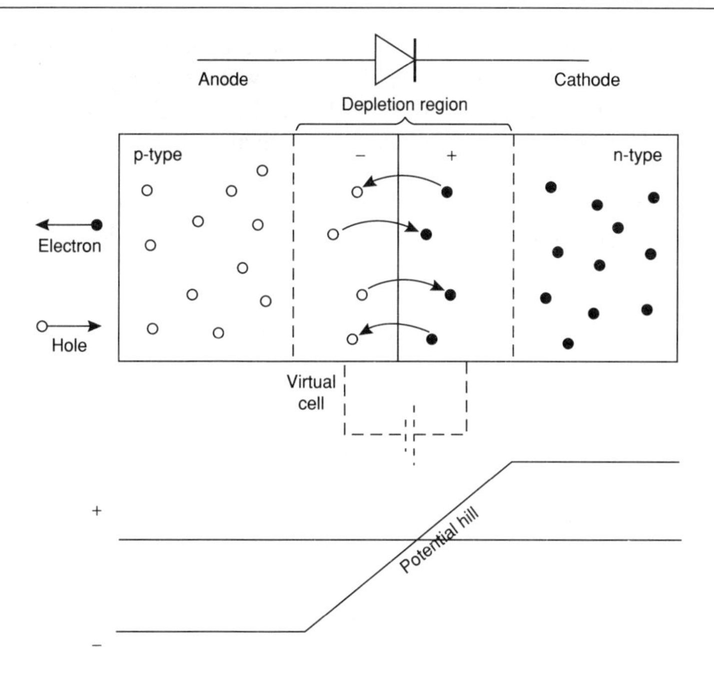

Figure A.9

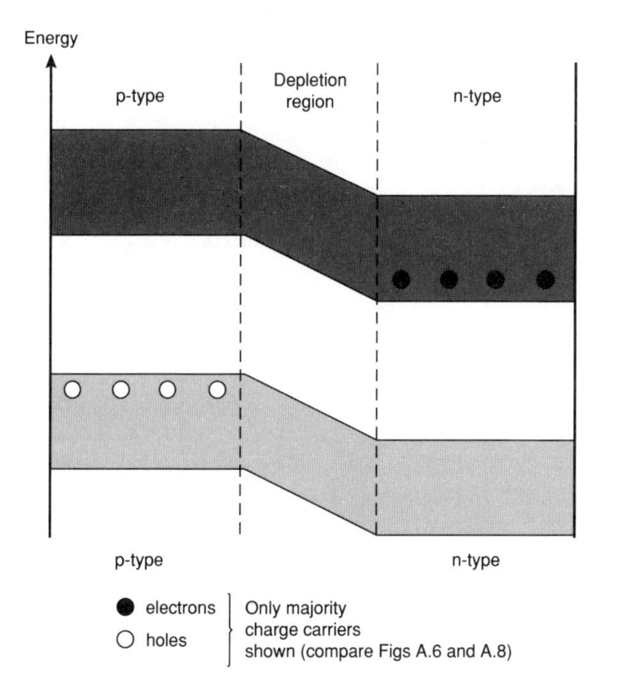

Figure A.10

Excess holes from p-type cross junction and recombine with electrons in n-type. Result is a *depletion region* (Fig. A.9) with no charge carriers, about 0.5 μm wide, narrower if more heavily doped. Loss of charge carriers leaves n-type positive of p-type, a potential hill, equivalent to *virtual cell*, 0.7 V in Si, 0.3 V in Ge. In Fig. A.10 the energy bands are displaced upward on the p-type (more negative) side of the junction and downward on the n-type side (electron energy is more negative toward the top of the page). Majority carriers cannot cross depletion region.

If junction is *forward biased* by external source, depletion region narrower or eliminated, virtual cell overcome, majority carriers cross junction, current flows (Fig. A.11).

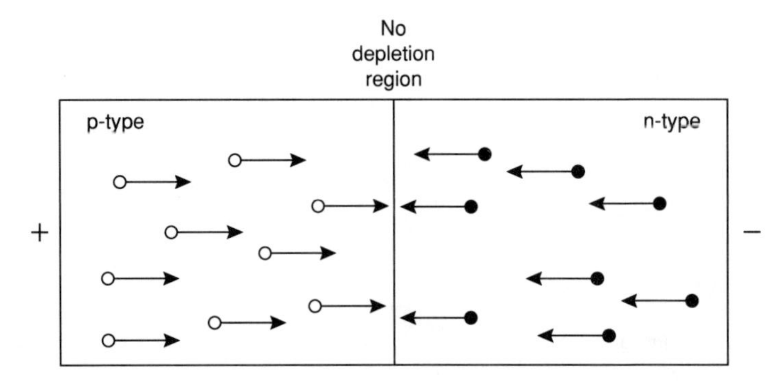

Figure A.11

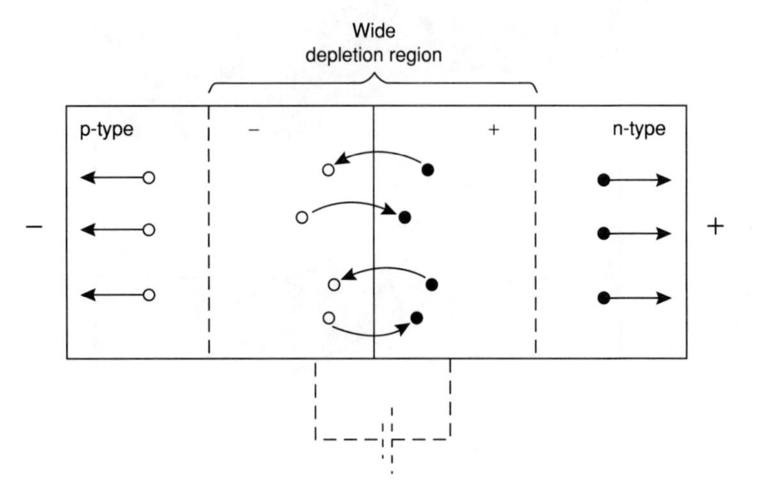

Figure A.12

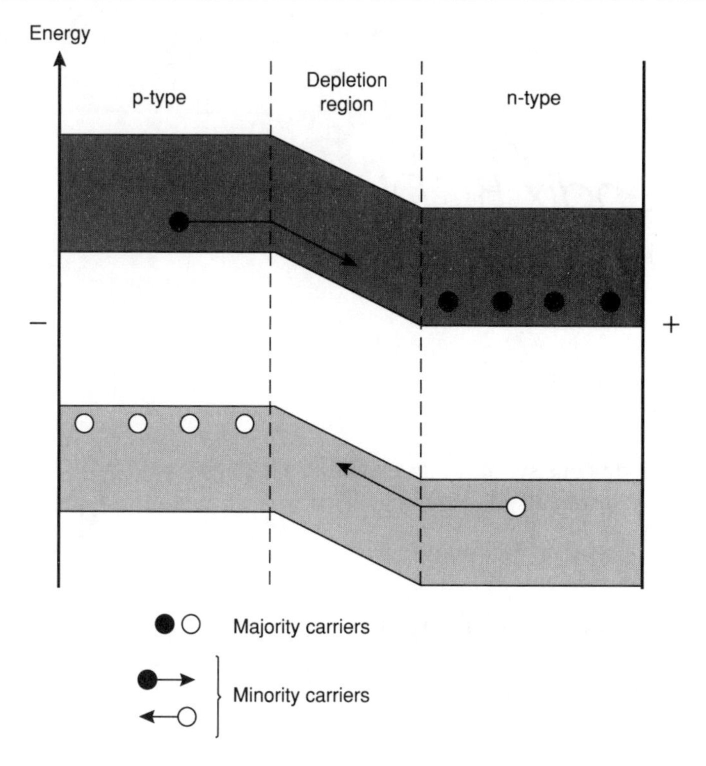

Figure A.13

If junction is *reverse biased* by external source, depletion region wider, virtual cell reinforced, majority carriers cannot cross junction, no current flows (Fig. A.12) except the reverse saturation current I_s, a few nanoamps in silicon, a few microamps in germanium (Fig. A.13). This is due to minority carriers and is more or less constant when the reverse voltage exceeds about 0.1 V. Minority carriers are produced only by the creation of electron-hole pairs so I_s increases with increasing temperature. Figure A.13 shows that minority carriers have enough energy to cross reverse-biased junction.

Appendix B Diodes

A typical diode consists of a block of Si or Ge doped as in A.5 to produce a pn junction. Terminal wires are connected to each region: n-type = cathode, p-type = anode.

B.1 Pd-current relationship

Current across forward-biased junction is i_D:

$$i_D = I_S(e^{v_D/nV_T} - 1)$$

Where:
i_D = instantaneous junction current (diffusion current) (A)
I_S = reverse saturation current (A), temperature dependent (A.1)
v_D = pd applied across the junction (V)
$n = 1$ for silicon devices and 2 for germanium devices
$V_T = kT/q$, in volts (k is Boltzmann's constant, 1.38×10^{-23} J/K, T is the junction temperature in K, and q is the electron charge, 1.6×10^{-19} C)
At 300 K(27°C), $V_T = 25.875$ mV. Given that $I_S = 1\,\mu$A for a junction in silicon:

$$i_D = 1 \times 10^{-6}(e^{v_D/0.025875} - 1) = (e^{38.65v_{D-1}})\mu A$$

If $v_D > 0.2$ V, then $e^{38.65v_D} \gg 1$ and:

$$i_D \approx e^{38.65v_D}$$

An exponential relationship (Fig. B.1), with turn-on voltage $V_\gamma \approx 0.6$ V for Si. Plotting the same curve on a larger scale for a negative voltage range (Fig. B.2) shows I_S. This is greater for germanium diodes $V_\gamma \approx 0.3$ V. If reverse bias is increased beyond the *peak inverse voltage* (PIV), there is a sudden increase

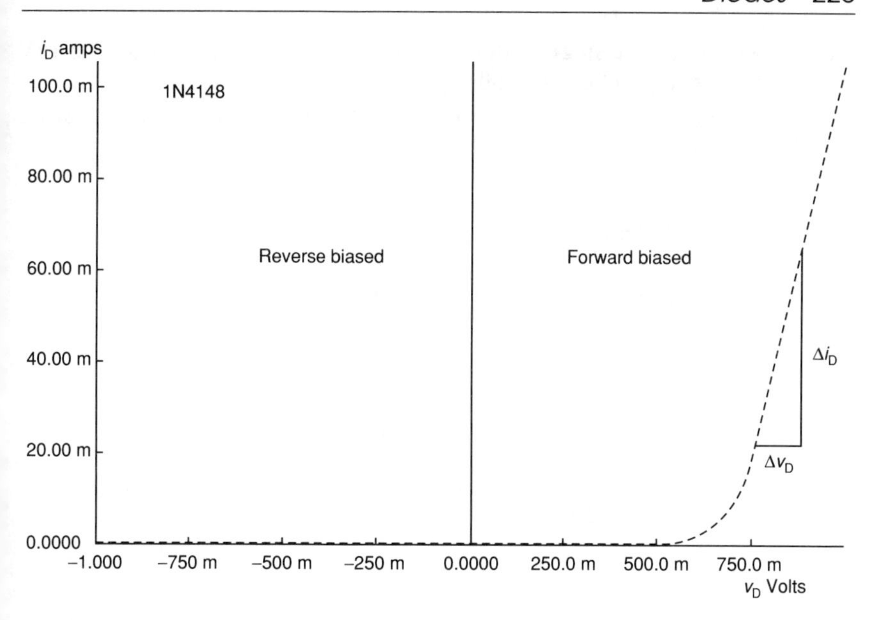

Figure B.1

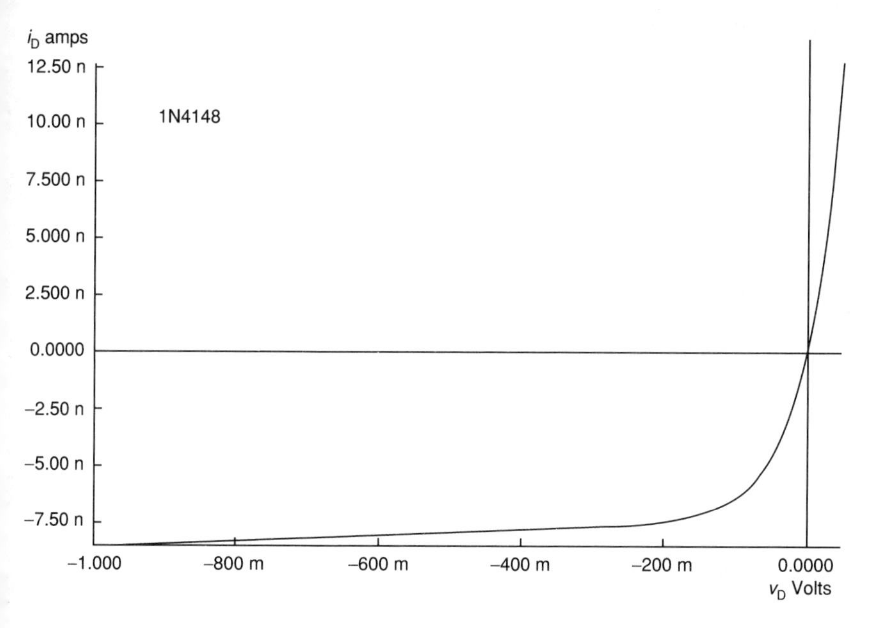

Figure B.2

in reverse current, causing the diode to overheat and destroying it. The PIV is usually several hundreds of volts.

These relationships apply to any pn junction whether in a diode, a transistor or other device.

B.2 Diode resistance

The ratio v_D/i_D is not constant (B.1), so resistance R cannot be found. Instead, find *dynamic resistance*, r.

In Fig. B.1 : $$r = \Delta v_D / \Delta i_D$$

In the limit : $$r = \mathrm{d}v_D / \mathrm{d}i_D$$

These quantities have volts divided by amps so their unit is ohms. Begin with the junction equation (B.1):

$$i_D = I_S(\mathrm{e}^{v_D/nV_T} - 1)$$

Assume $n = 1$, insert values of the q and k, standardize at 300K. Then $V_T = 0.0259$ and $1/V_T \approx 40$.
Substituting:

$$i_D = I_S(\mathrm{e}^{40v_D} - 1)$$

If $v_D > 0.2$, then $\mathrm{e}^{40v_D} \gg 1$ and:

$$i_D = I_S \mathrm{e}^{40v_D}$$

Differentiating:

$$\mathrm{d}i_D/\mathrm{d}v_D = 40 I_S \mathrm{e}^{40v_D} = 40 i_D$$

$$\Rightarrow \quad r = \mathrm{d}v_D/\mathrm{d}i_D = 1/40 i_D = 0.025/i_D$$

Dynamic resistance is inversely proportional to current and, if i_D is in milliamps:

$$r = \frac{25\,\Omega}{i_D}$$

B.3 Diode capacitance

Junction capacitance C_J in reverse-biased diode: n-type and p-type act as capacitor plates with non-conductive depletion region acting as dielectric. C_J depends on reverse voltage (greater reverse voltage → wider region → lower capacitance).

Varactor diode has an abrupt junction, giving a thin depletion region, and relatively high capacitance. Capacitance is controlled by varying the reverse voltage. Used in VHF modulators (8.3).

Example: Decreasing the reverse bias from −10 V to −1 V increases capacitance from 9 pf to 170 pf.

PIN diodes have layer of intrinsic semiconductor (i-type) between n-type and p-type. Gives wider depletion region and hence lower capacitance. Used to minimize capacitance effects in high frequency circuits. This effect is absent in a Schottky barrier diode (B.4.3).

Diffusion capacitance C_D in a forward-biased diode. Due to charge stored in the junction because of holes from p-type crossing the junction but not immediately combining with electrons. In effect there is a store of positive charge in the vicinity of the junction, equivalent to charge stored on a capacitor of several hundred picofarads.

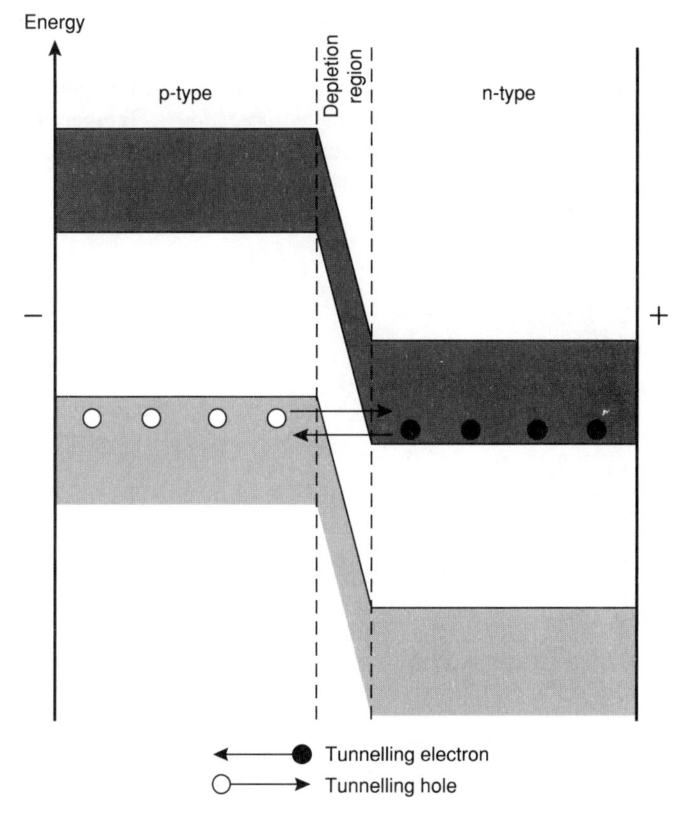

Figure B.3

B.4 Other diodes

B.4.1 Avalanche diode

Strong reverse-bias → larger I_S → excessive heat → more electron-hole pairs → electrons accelerated in strong field → strike other electrons → more electron-hole pairs → cumulative effect → avalanche → larger current → diode destroyed. Avalanche diodes show this effect at a sharply defined reverse voltage (5 V to 1000 V), without being destroyed. Used in voltage-limiting and voltage-regulating circuits.

B.4.2 Zener diode

Heavily doped → narrow depletion region → steep potential gradient (Fig. B.3). Majority carriers on both sides are brought to the same energy level. They are able to tunnel through the short distance when a reverse pd of the right amount (the *Zener voltage*, 1 V to 5 V) is applied. There is a sharp 'knee' on the pd–current curve at the Zener voltage V_Z (Fig. B.4). Below the knee, I_S increases steeply with a *dynamic slope resistance*, $\Delta v/\Delta i$.

In practical Zener diodes in the range 3 V to 8 V, breakdown is a combination of *avalanche breakdown* and *Zener breakdown*. Temperature has an opposite effect on these two processes. At higher temperature, avalanche

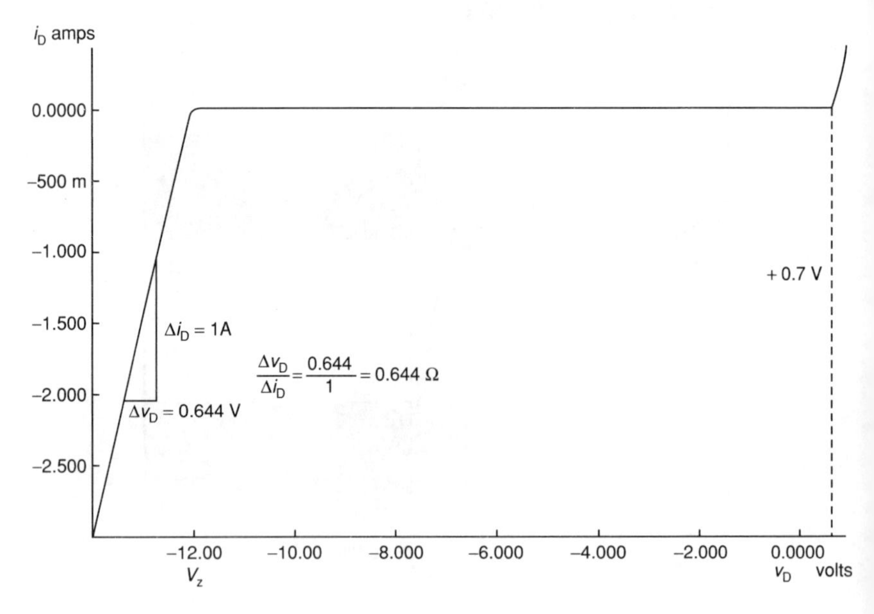

Figure B.4

breakdown occurs at a greater reverse pd. Conversely, a higher temperature provides the carriers with additional energy to initiate tunnelling and a lower pd is required for the Zener effect. In diodes rated to break down at about 5 V, the effects of temperature on avalanche and Zener effects are approximately opposite and equal. The tempco of a voltage regulator based on a 5 V Zener diode is almost zero.

B.4.3 Schottky barrier diode

Based on the rectifying action of a metal/semiconductor junction. Usually the semiconductor is n-type because the mobility of electrons is higher than that of holes and higher switching speeds are obtainable. Only majority charge carriers present so there are no charge storage effects (B.3) to produce diffusion capacitance. Schottky diodes are used for high-speed switching.

Appendix C Transistors

C.1 Transistor types

Bipolar junction transistor (BJT):	Silicon-substrate:	npn (most commonly used of BJTs) pnp	
	Germanium-substrate:	npn pnp	
Field effect transistor (FET):	MOSFET (IGFET):	Enhancement:	n-channel (NMOS, most commonly used of FETs) p-channel (PMOS)
		Depletion:	n-channel (rarely used) p-channel (rarely used)
	JFET:	Enhancement:	(not possible)
		Depletion:	n-channel (commonly used) p-channel
Insulated gate bipolar transistors (IGBT):		npn	

Only the commonest types are considered here: silicon npn BJT, enhancement NMOS, n-channel JFET, referred to for brevity as BJT, NMOS and JFET respectively. P-channel types are complementary, operating with reversed polarities and with slightly different characteristics mainly due to the lower mobility of holes.

C.2 Bipolar junction transistor

C.2.1 Structure

Conduction by electrons and holes (bipolar); based on two pn junctions (Fig. C.1a).

- emitter: n-type, heavily doped, of medium extent;
- base: p-type, lightly doped, exceedingly thin;
- collector: n-type, lightly doped, large extent.

C.2.2 Action

Equivalent to two back-to-back diodes (Fig. C.1b), with two depletion regions (Fig. C.2); electrons cannot flow between emitter and collector in the unconnected transistor. In normal operation (Fig. C.3), $v_{CE} > v_{BE}$ so base-emitter junction is forward biased and base-collector junction is reverse biased.

In Fig. C.3 the emitter terminal is common to both meshes of the network → the *common-emitter* (CE) connection (C.2.3). All subsequent discussion refers to CE connection. Holes flow into the base from the external supply (the base current i_B, Fig. C.3) while large numbers of electrons (from i_E) are emitted by the emitter and cross the base-emitter junction (Fig. C.4). In the base, a small proportion of the electrons (as few as 1 in 100, often fewer) combine with holes.

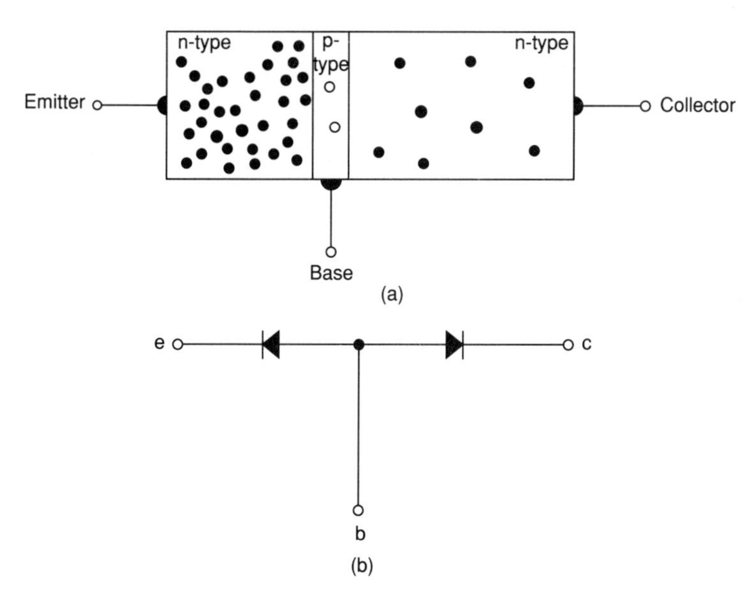

Figure C.1

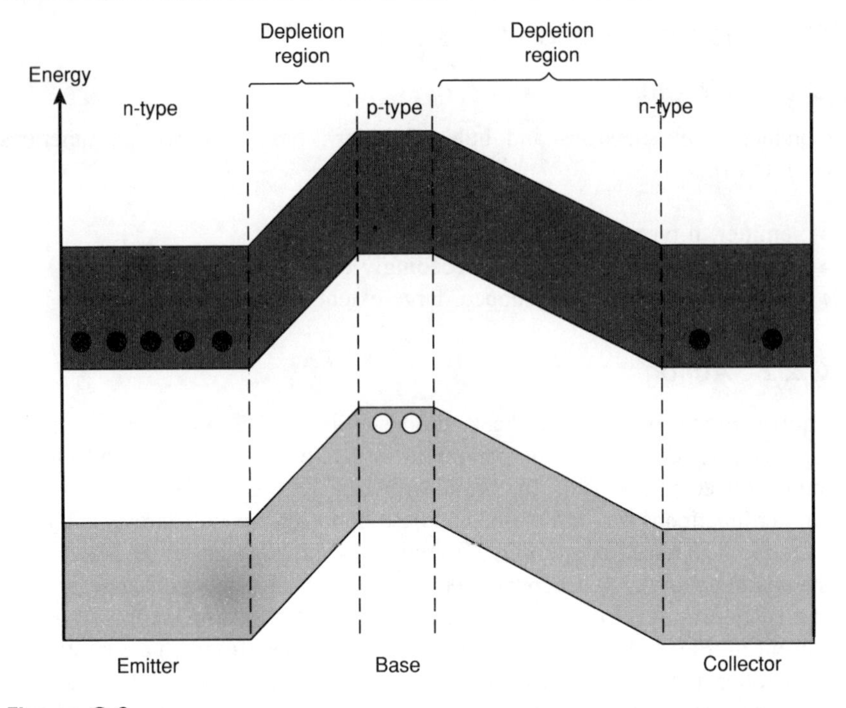

Figure C.2

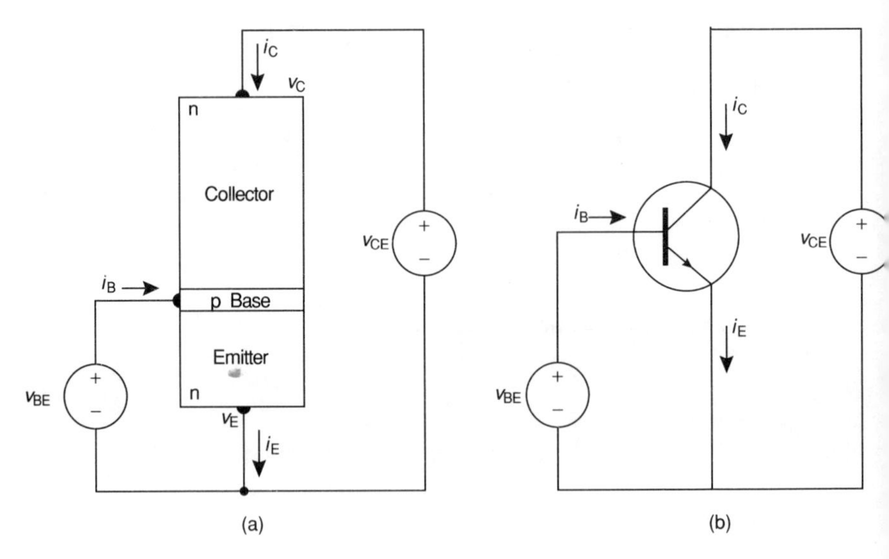

Figure C.3

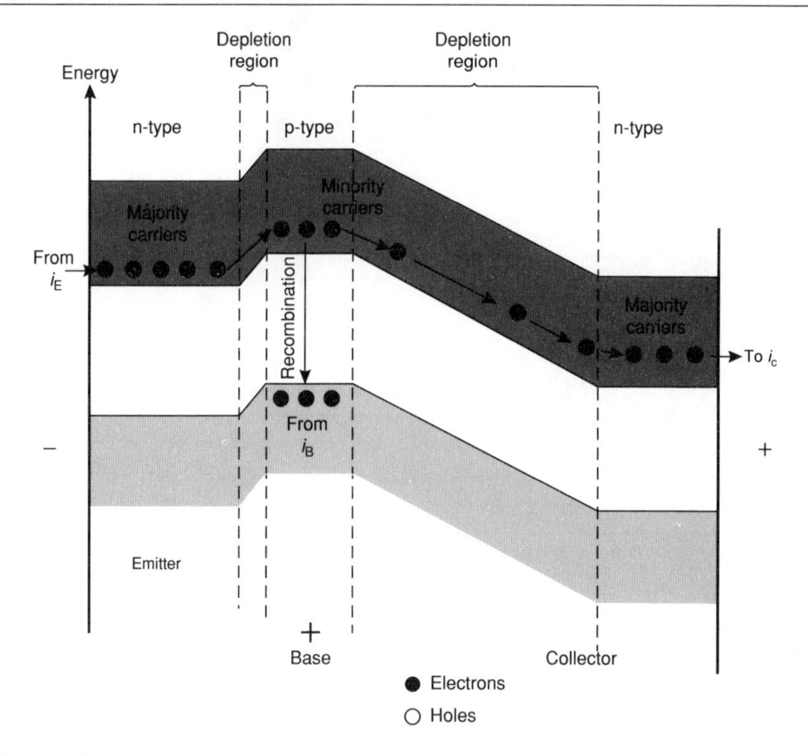

Figure C.4

This is because there are relatively few holes in the lightly doped base and the base is so thin that few electrons meet a hole during their journey across it. The key to the operation of the transistor is that, in the emitter, the electrons are majority carriers but, in the base, they count as minority carriers. This means that they ($\approx$ 99% of i_E) are able to continue across the base-collector depletion region (compare Fig. A.13) into the collector and are attracted toward the more positive potential at the collector terminal.

A BJT is driven by current. i_B supplies holes to the base. As the holes capture electrons (Fig. C.4) and become neutralized, base potential falls, attracting more holes from the external circuit. But relatively few electrons are captured and most pass through to the collector, to become i_C. In this way the small i_B controls the size of a much larger i_C. The BJT is a current amplifying device.

C.2.3 BJT output characteristics

In Fig. C.5. i_C is plotted against v_{CE} for a range of values of i_B. In the *saturation region*, i_C is limited by the small value of v_{CE}. In the *linear region*,

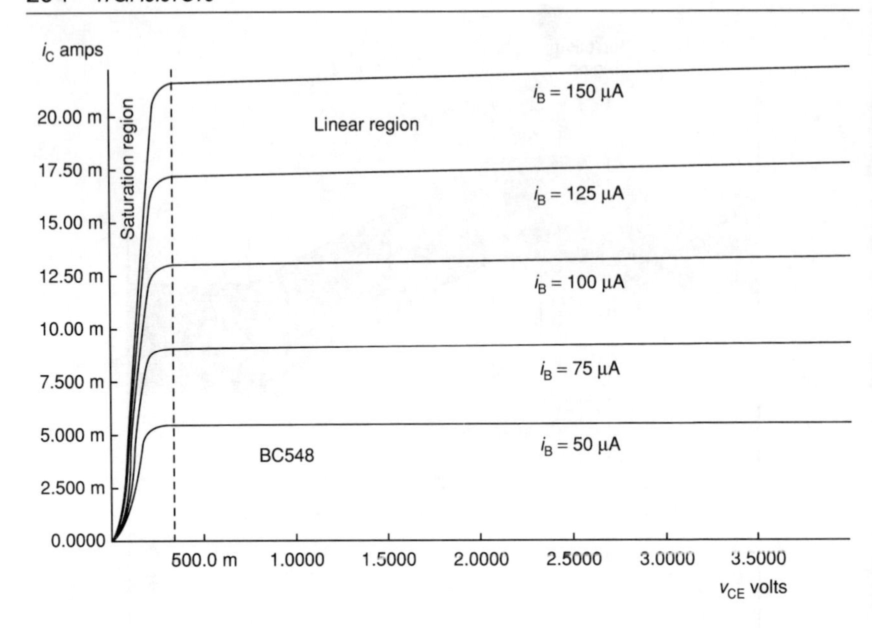

Figure C.5

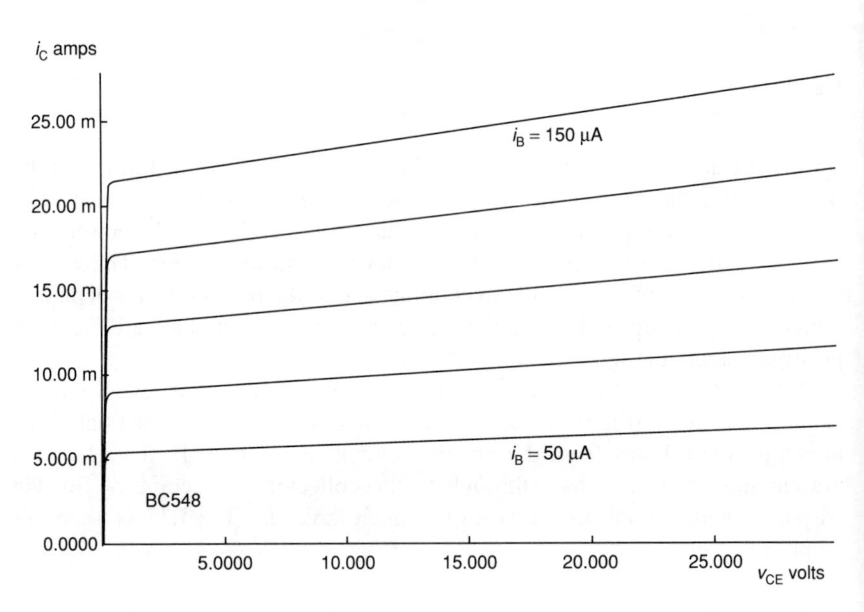

Figure C.6

$i_C \propto i_B$, and is relatively unaffected by v_{CE}. In the linear region, large signal gain h_{FE} ($= i_C/i_B$) increases with increasing i_C (linear portions of upper curves more widely spaced) and with v_{CE} (linear portions of curves slope slightly upward to right). Also, linear portions of curves are not exactly parallel with each other or with the x-axis. If they are projected backward to the left, they converge on a single point on the axis, the *Early voltage* (commonly about -100 V). This shows more clearly in Fig. C.6 which has an extended voltage range. Early effect is explained by increasing $v_{CB} \rightarrow$ base-collector depletion region wider $\rightarrow$ reduces effective width of base region $\rightarrow$ reduced recombination in base region $\rightarrow$ fewer electrons recombine with holes $\rightarrow$ slightly increased collector current = *Early effect.*

C.2.4 BJT transfer characteristic

C.2.4.1 Current gain

In Fig. C.3:

$$i_C = i_B + i_E$$

Because i_B is much smaller than the other two:

$$i_C \approx i_E$$

The *large signal current gain* of a transistor in the common-emitter connection is:

$$h_{FE} = i_C/i_B$$

In Fig. C.7, i_C (plotted in mA) is about $100 \times i_B$ (plotted in μA) indicating that $h_{FE} \approx 100$. Current amplification is practically linear.

At low i_B, h_{FE} is appreciably less than 100 (line less steep, see figure). Low $i_B \rightarrow$ fewer positively charged holes in the base region $\rightarrow$ electrons less strongly attracted $\rightarrow$ higher proportion of electrons enter base by diffusion and combine with the holes $\rightarrow$ fewer pass on to the collector $\rightarrow$ transistor action is less efficient $\rightarrow$ reduced h_{FE}.

When i_B is very large $\rightarrow$ surplus of holes $\rightarrow$ many flow directly to the base-emitter junction $\rightarrow$ combine there with electrons $\rightarrow$ do not contribute to the transistor action $\rightarrow$ reduced h_{FE}.

For small signals, it is more accurate to use the *small signal common-emitter current gain*:

$$h_{fe} = \Delta i_C/\Delta i_B$$

This is the gradient of Fig. C.7 over a limited part of its range, centred on the working currents of a given circuit. In practice, $h_{FE} \approx h_{fe}$. Manufacturers quote h_{fe} on data sheets and, since the value of h_{fe} is usually widely spread, even between transistors of the same manufacturer's type number, the difference

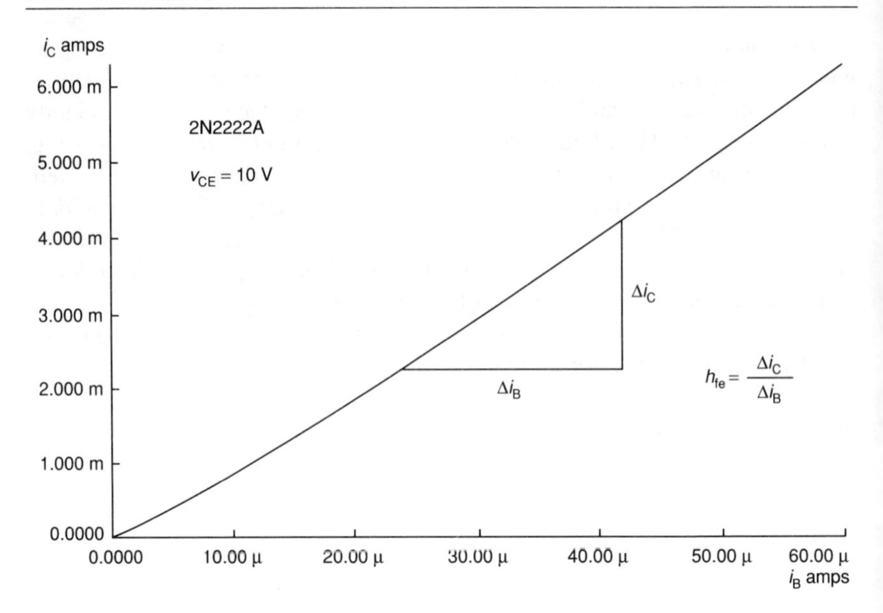

Figure C.7

between h_{FE} and h_{fe} is largely academic. The wide spread of h_{fe} between transistors of the same type causes difficulties in circuit design but these can be avoided (4.1.1).

C.2.4.2 Transconductance

Examine the relationship between i_C and v_{BE}. From B.1 and F.2.3 but with the transistor in the active mode ($v_{BE} > 0$, $v_{BC} < 0$), $i_c = i_{cc}$ and:

$$i_C = I_S(e^{v_{BE}/nv_{T-1}})$$

$i_C \gg I_S$, so ignore the -1. Also take $n = 1$:

$$i_C = I_S e^{v_{BE}/v_T}$$

$$\Rightarrow \qquad v_{BE} = v_T \ln(i_C/I_S)$$

At constant temperature v_T and I_S are constant, so:

$$v_{BE} \propto \ln i_c$$

Figure C.8 shows this linear relationship for several temperatures between 0°C and 30°C. The graph covers a wide range of i_c, a hundred-thousand fold increase. The gradients show that an increase in v_{BE} of about 60 mV results in a tenfold increase of collector current. This dependence applies to all npn

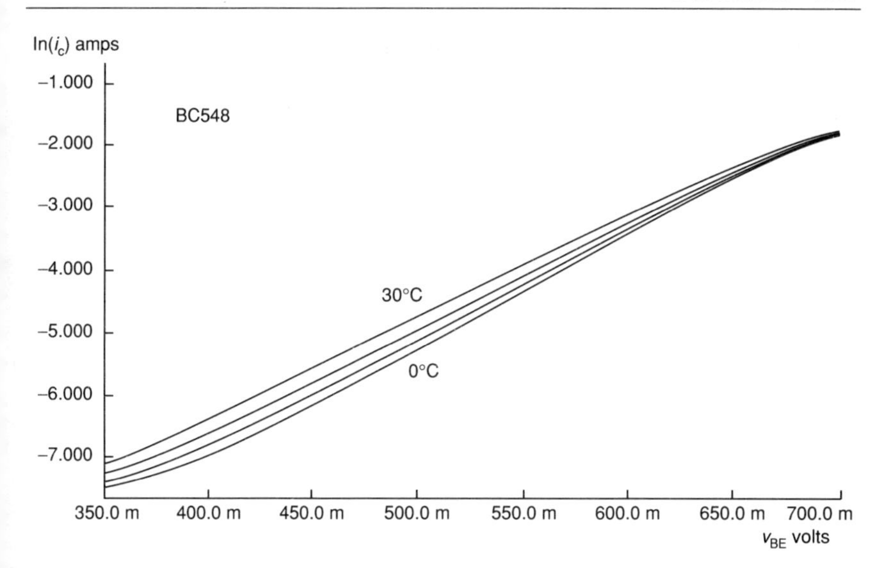

Figure C.8

BJTs, since it results directly from the Ebers–Moll equations (F.2.3) and h_{fe} has no influence on it.

If we take v_{BE} as the input (instead of i_B) and i_C as the output, the 'gain' is i_C/v_{BE}. Its unit is i/v, the same as a conductance. It is the *transconductance* of the transistor, symbol g_m, with siemens as its unit.

Evaluate g_m using the simplified h-parameter model (F.2.2):

$$i_B = v_{BE}/h_{ie} \text{(by Ohm's law)}$$

On the output side:

$$i_C = h_{fe} i_B = h_{fe} v_{BE}/h_{ie}$$

From the definition of transconductance:

$$g_m = i_C/v_{BE} = h_{fe}/h_{ie} \qquad \text{(C.2.1)}$$

The value of h_{ie} is the resistance encountered by the base current and consists of resistance r_b at the base due to the resistance of the base material and other factors, in series with similar resistance at the emitter r_e. The emitter resistance is a dynamic resistance, determined in the same way as the dynamic resistance of a diode (B.2). If i_E is in milliamps:

$$r_e = 25/i_E$$

But for a current flowing through from base to emitter the effective value of r_e is much greater. A current leaving the base to enter the emitter has to overcome the resistance offered to a current $h_{fe} + 1$ times as great. The total resistance encountered is:

$$h_{ie} = r_b + h_{fe} \times \frac{25}{i_E}$$

The value of r_b is several hundred ohms but, provided that i_E is not more than 1 mA, r_b can be ignored:

$$h_{ie} = \frac{25h_{fe}}{i_E}$$

$$\Rightarrow \qquad \frac{h_{fe}}{h_{ie}} = \frac{i_E}{25}$$

Substituting in (C.2.1) above:

$$g_m = \frac{i_E}{25} \approx \frac{i_C}{25}$$

Although there have been several approximations in the argument above, the result is a useful one:

$$g_m \approx 40 i_C \text{ mA/V}$$

This parameter applies to all types of BJT and depends on no other factor but the collector current. The effect of i_C on g_m is a source of distortion of large signals.

C.2.5 BJT connections

Common-emitter (Fig. C.9a): (C.2.3–C.2.4). Probably the most often used connection as it has the greatest h_{fe} (*c*. 20–1000). High Z_{in}, low Z_{out} . Used in switching, amplifiers, oscillators, etc. Common-base (Fig. C.9b): $h_{fe} \approx 1$. Low Z_{in}, low Z_{out}. Used in power regulators. Common-collector (Fig. C.9c): $h_{fe} \approx 1$. High Z_{in}, low Z_{out}. Used in impedance matching, high-frequency amplifiers.

C.2.6 Effect of temperature

The pd across the base-emitter junction, v_{BE}, is usually taken to be 0.6 V in calculations. In practice, it falls by about 2.1 mV/K at temperatures around room temperature.

I_{CBO} is the leakage through the reverse-biased base-collector junction (compare I_S, (B.2)). Has the same effect as the external base current i_B and is amplified to give a collector-emitter current $I_{CEO} = h_{fe} I_{CBO}$. If temperature increases → increase of I_{CBO} → increase of I_{CEO} (but much larger current) →

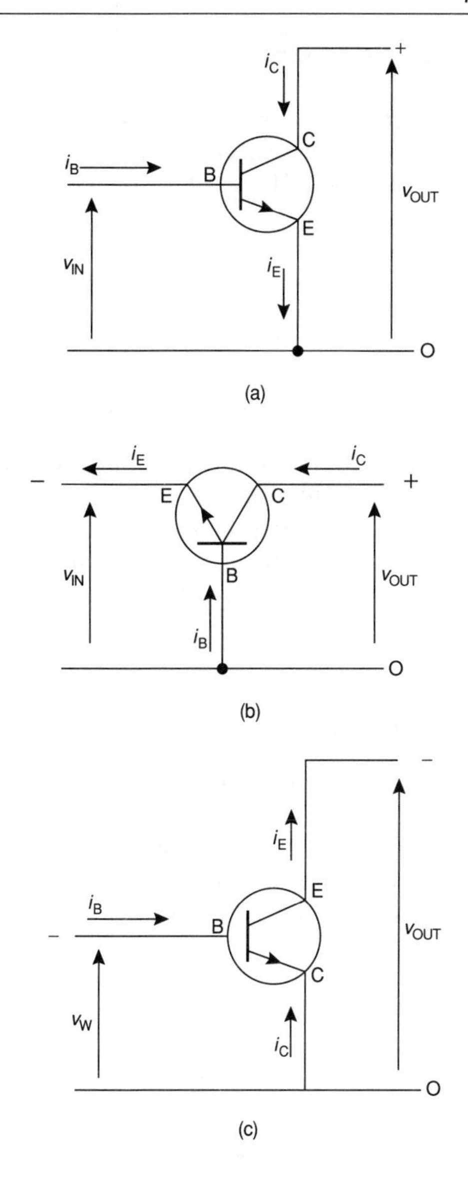

Figure C.9

I_{CEO} as large as normal collector currents → degrades signal. In more serious cases, the transistor overheats → increase in I_{CBO} → increase in I_{CEO} → cumulative effect → *thermal runaway* → transistor destroyed. A problem in Ge transistors but Si transistors are relatively immune because they have smaller I_{CBO} and are also able to withstand higher temperatures.

C.3 Field effect transistors

Unipolar: use either electrons (n-channel) or holes (p-channel); action depends on electric field → voltage driven; gate is insulated → very high input impedance (> $10^{10}\ \Omega$ for MOSFETs, > $10^{8}\ \Omega$ for JFETs).

C.3.1 Structure

MOSFETs (metal-oxide-silicon FETs) have gate insulated by SiO_2 layer → gate current ≈ 10^{-12} A. Substrate is usually connected to the source, or has a separate terminal. Figure C.10 shows n-channel enhancement mode (NMOS). Depletion mode MOSFETs have an n-type channel linking the n-type wells (Fig. C.11).

JFETs (junction FETs) use reverse-biased pn junction to insulate the gate (Fig. C.12). In an n-channel JFET, the body is a bar of n-type, with source and drain contacts at either end. Heavily doped regions on either side of the bar form the gate. The gate is usually made negative of the source, the reverse bias creates a depletion region to insulate the gate.

C.3.2 Action

In NMOS the gate is made positive of source → electrons attracted toward the region between the n-type wells → n-channel created → electrons flow from source to drain. Gate made more positive → more electrons attracted → conductivity enhanced → increased i_D.

In JFETs the gate is made negative of the source → electrons flow from source to drain in channel between depletion regions. Gate made more negative → depletion regions wider → channel narrower → reduced flow of electrons

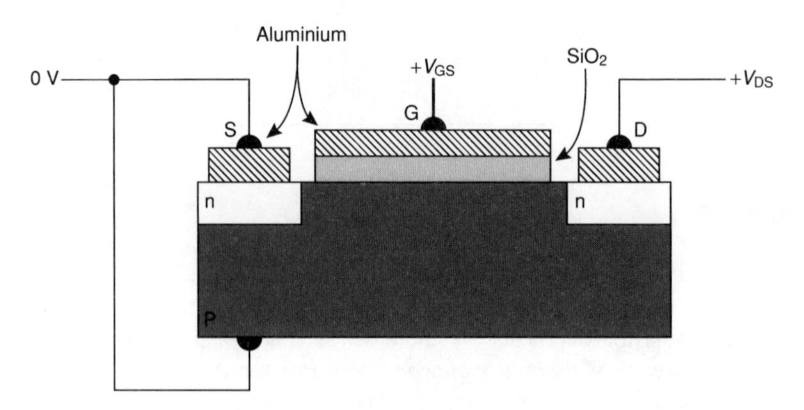

Figure C.10

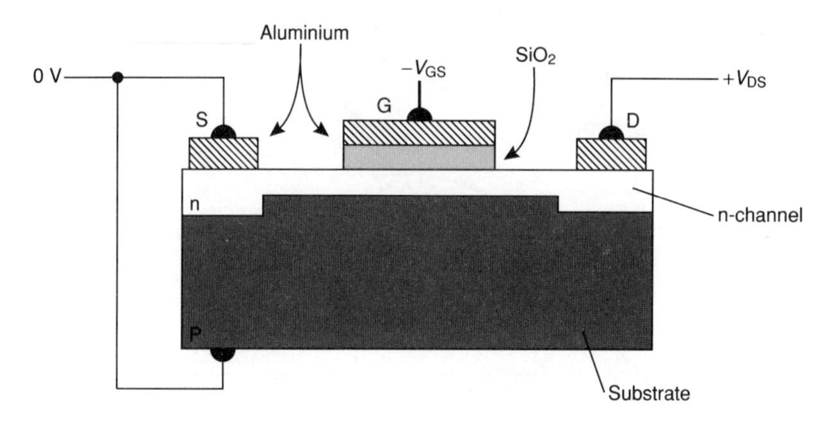

Figure C.11

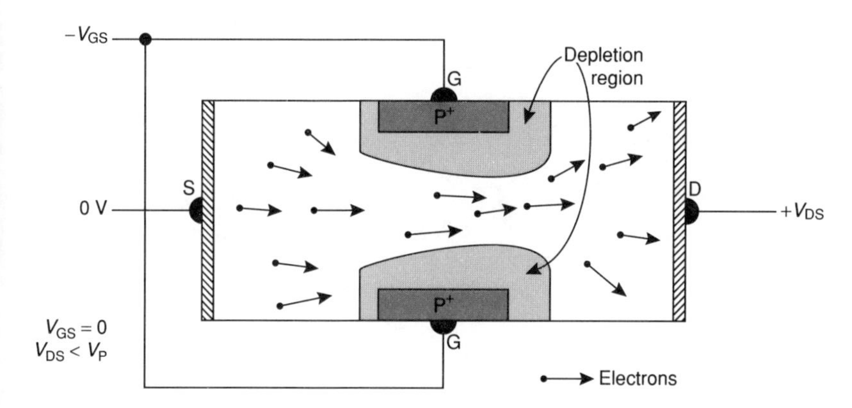

Figure C.12

→ reduced i_D. Gate can be made up to ≈ 0.5 V positive of the source → depletion region still present but thinner → increased i_D (this mode rarely used).

C.3.3 FET output characteristics

Figure C.13 shows typical output characteristics for NMOS with i_D plotted against v_{DS} at different values of v_{GS} (compare Fig. C.5). In the resistor (or *ohmic*) region i_D increases linearly (approximately) with v_{DS} and the transistor acts as a resistor. When the transistor is turned fully on, the resistance is only a few ohms and only a few hundredths of ohms in many power NMOS. The gradients of the curves in this *active region* are proportional to $v_{GS} - V_T$,

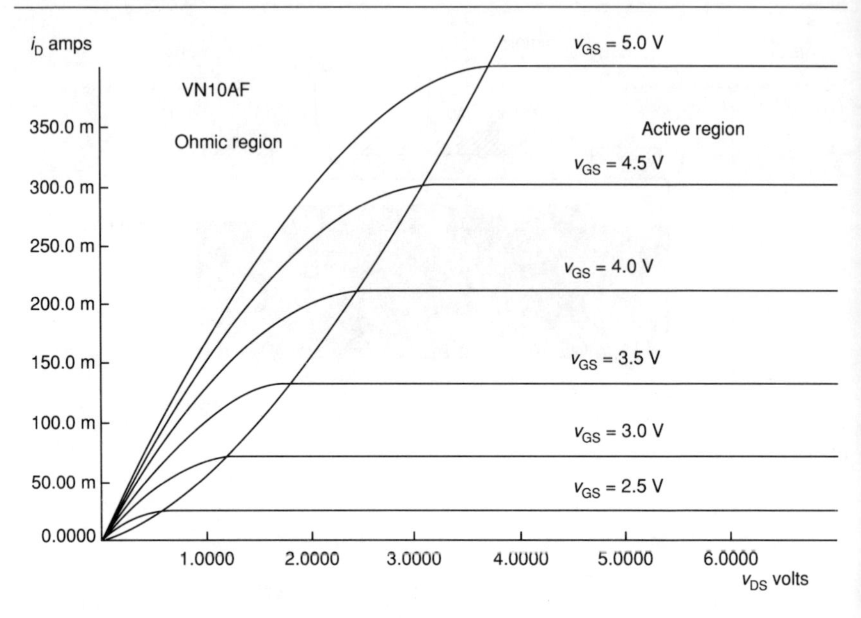

Figure C.13

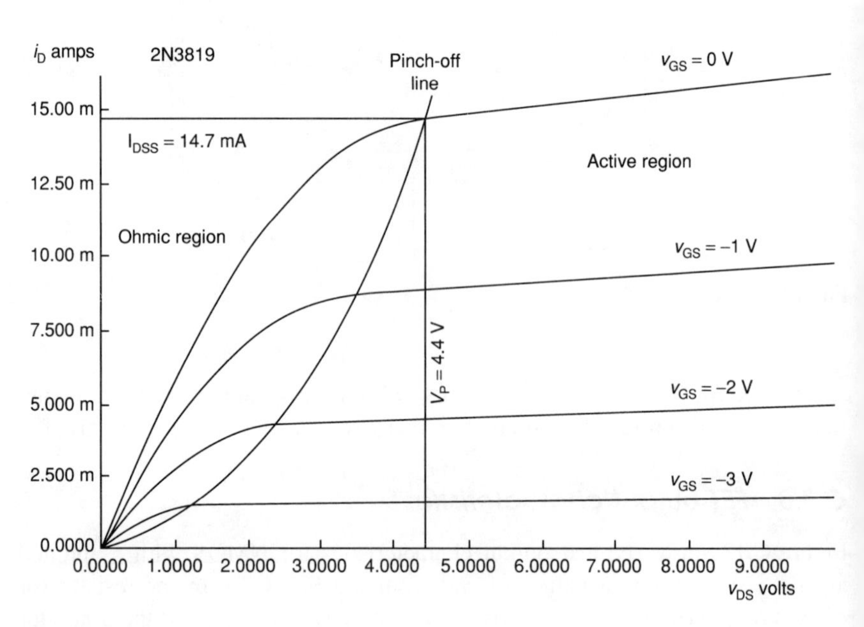

Figure C.14

where V_T is the threshold gate voltage at which i_D approaches zero. In the active region $i_D \propto v_{GS}$, but not linearly (curves not equally spaced for equal increments of v_{GS}). There is no Early effect. If v_{DS} is further increased, the transistor enters the breakdown region, with a sharp increase of i_D.

JFETs have similar characteristics (Fig. C.14), except that v_{GS} is negative (usually, see C.3.2). i_D depends inversely on the channel width, which

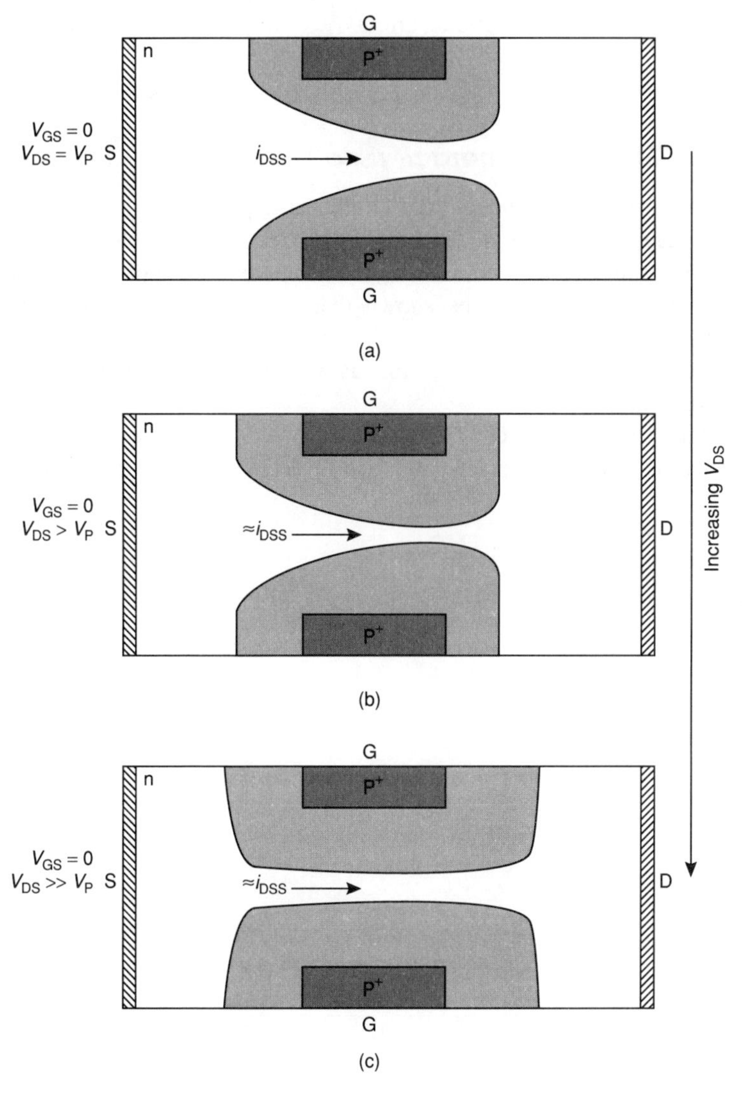

Figure C.15

is narrower toward the drain end because of stronger reverse-bias at that end (Fig. C.12). When $v_{DS} = V_P$, the pinch-off voltage, the channel reaches minimum width at the end nearer the drain (Fig. C.15a). Beyond the pinch-off line, the narrow section extends back toward the negative (source) end of the channel as v_{DS} increases (Fig. C.15b/c). The increasing v_{DS} causes a stronger electric field along the bar (which accelerates electrons), but this is compensated for by the increased length of the pinched-off part of the channel (which impedes electrons). Result: i_D increases only slightly.

The drain-source saturation current I_{DSS}, often quoted in manufacturers' data sheets, is the value of i_D at pinch-off when $v_{GS} = 0\,V$ (gate and source shorted together). In Fig. C.14, I_{DSS} is 14.7 mA when v_{DS} is 4.4 V.

C.3.4 FET transfer characteristic

Gate input current is virtually zero → h_{fe} not applicable as a measure of gain → use transconductance (C.2.4.2). For NMOS, Fig. C.16 shows the relation between i_D and v_{GS} with constant v_{DS}. The curve is parabolic, indicating that g_m is not constant. It can be shown that $g_m \propto \sqrt{i_D}$, which is a source of distortion of large signals.

The output characteristic for a JFET is similar (Fig. C.17) but shifted toward negative values of v_{GS}.

Usually, the gate current and hence the power input is exceedingly small, giving an FET a large power gain. Gate leakage is usually only a few pA at

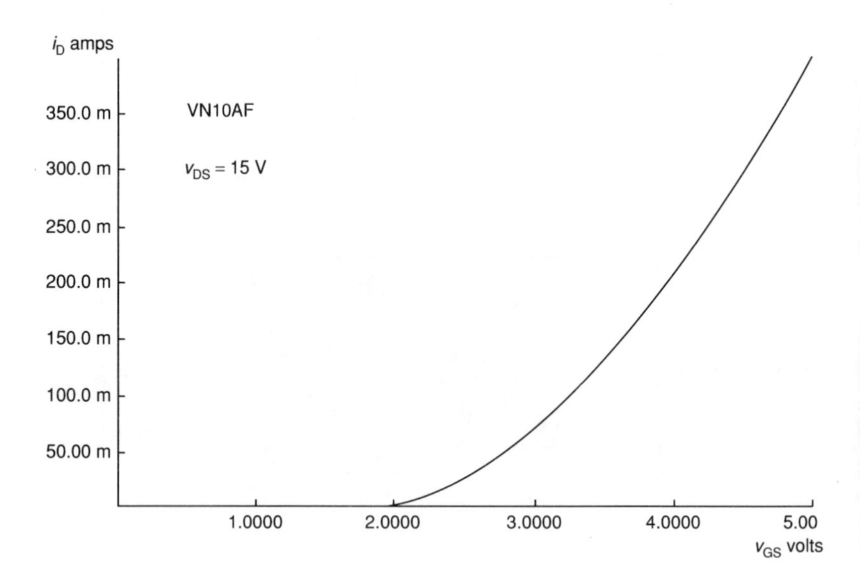

Figure C.16

room temperatures but doubles for every 10°C rise in temperature. In addition (in JFETs, especially n-channel) impact-ionization current may raise gate current to several microamps when v_{DS} and i_D are excessive. In JFETs and MOSFETs, gate capacitance at higher frequencies has the effect of increasing input impedance appreciably.

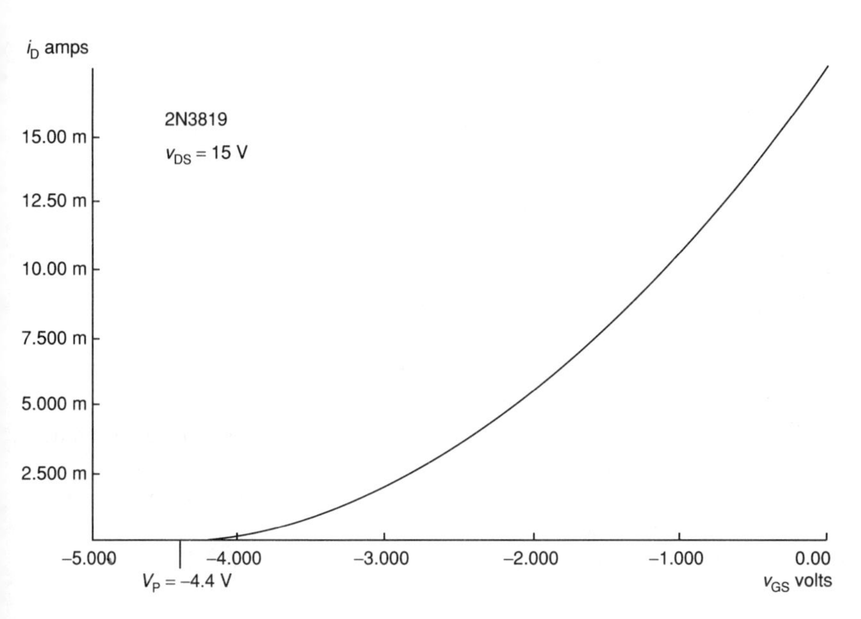

Figure C.17

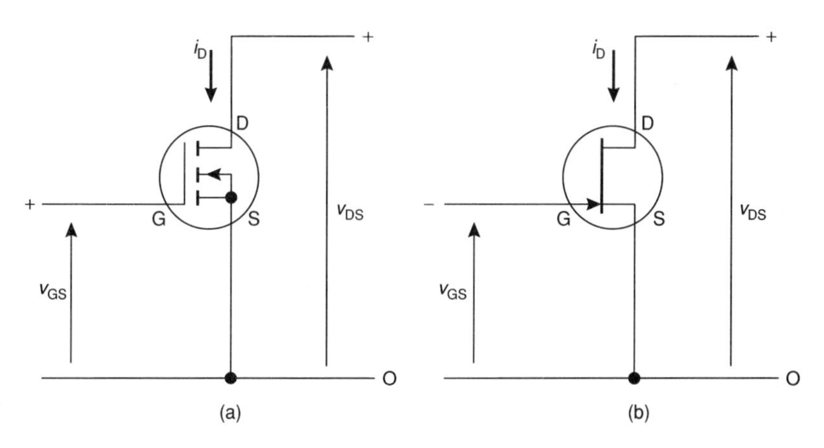

Figure C.18

C.3.5 FET connections

Common-source, CS (Fig. C.18): Most often used; $g_m = 0.05\,\text{mS}$ to $10\,\text{mS}$ for JFETs, 50 mS to 700 mS for small-signal MOSFETs, up to 30 S for power MOSFETs. High Z. Used for switching, amplifiers, oscillators, etc. Common-gate, CG: Rarely used. Common-drain, CD: High Z_{in}, low Z_{out}. Used for impedance matching.

C.3.6 FET temperature effects

When i_D large, increasing temperature $\rightarrow$ decreasing I_{DSS} $\rightarrow$ decreasing i_D $\rightarrow$ no thermal runaway (C.2.6). When i_D small, increasing temperature $\rightarrow$ increasing i_D. There is an intermediate i_D with zero tempco. Temperature effects can usually be ignored.

C.4 Comparison of BJTs and FETs

(1) FETs have much higher input impedances than BJTs.
(2) FETs have lower transconductance than BJTs.
(3) FETs have wider variation in parameters due to manufacturing tolerances.
(4) FETs are not subject to thermal runaway.
(5) Rise in temperature causes FET gate leakage current to increase exponentially; BJT base currents decrease slightly.
(6) NMOS has low drain-source resistance when turned fully on.

Appendix D Operational amplifiers

Op amps are versatile building-blocks for electronic circuits, especially analog circuits. One or more op amps are fabricated on a single chip as an integrated circuit (ic) with common power supply terminals. A typical single op amp has the terminals and ic pin-out shown in Fig. D.1. Inputs may be to bipolar, JFET or MOSFET stages.

D.1 Op amp terms and characteristics

Supply voltage: Most operate on dual supplies ($+V$ and $-V$). Supply voltage ± 1 V to ± 20 V, depending on type. Some accept v_{IN} as low as $-V$, and are then considered as single-supply devices.
Supply current: Usually low, typically 1 mA to 4 mA, as little as 10 μA in some CMOS devices.
Output voltage swing: v_{OUT} usually swings between 2 V below $+V$ and 2 V above $-V$, but can swing up to $+V$ and down to $-V$ in some types (e.g. CMOS).
Input bias current, I_{BIAS}: The amount of current required to bias input stages into operation. Defined as the mean value for the two input terminals. A few nanoamps for bipolar inputs. FET input stages take no more than a few picoamps of leakage current (C.3.4). Temperature dependent (typically 10 nA/°C for bipolar op amps).
Input impedance: High, 2 MΩ in bipolar-input op amps, up to 10^{12} Ω in MOSFET op amps. But the input impedance of an amplifier circuit based on an op amp may be much less than this.
Input offset current, I_{IN}: Differences of bias currents drawn by the two inputs when supplied by sources of equal output impedance. Typically 50 nA in

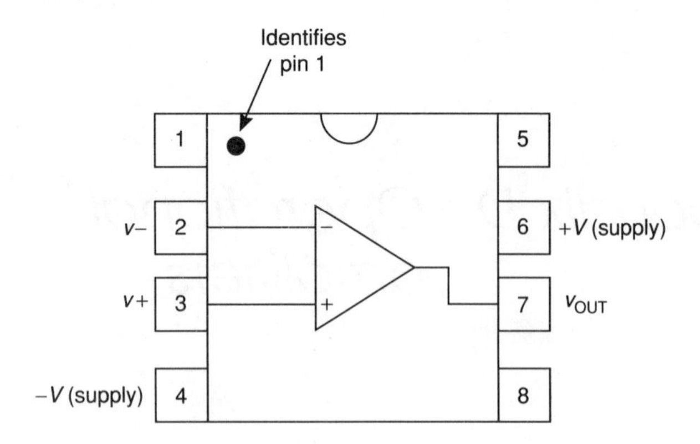

Figure D.1

bipolar op amps, 50 pA in JFET, 1 pA in MOSFET. Temperature dependent (typically $-2\,\text{nA}/^\circ\text{C}$ for bipolar).

Input offset voltage: Difference of input voltages when output is 0 V. Input offset is 10 µV for precision op amps, up to 1 mV in others. Many op amps have offset null terminals to which to connect a variable resistor for compensating for this. Temperature dependent (typically 1 to $10\,\mu\text{V}/^\circ\text{C}$).

Common-mode input range: Input restricted to voltages in a range usually less than $-V$ to $+V$.

Differential input range: The maximum allowable difference between v_+ and v_-. Usually almost as great as $-V$ to $+V$, but restricted in some bipolar op amps.

Open loop voltage gain G_o: 70 dB to 140 dB (×3000 to ×10 000 000), typically 105 dB (×200 000), but feedback in the amplifier circuit usually reduces this considerably.

Common mode gain G_{cm}: The gain when input terminals are connected together and input voltage v_{IN} varied. Ideally v_{OUT} should remain at zero, but there is in practice an increase $G_{cm} = v_{OUT}/v_{IN}$.

Common mode rejection ratio, CMMR: $\text{CMMR} = |G_o| / |G_{cm}|$. In decibels, $\text{CMMR} = 20 \log_{10}(|G_o| / |G_{cm}|)$. Typically 80 dB to 100 dB.

Power supply rejection ratio, PSRR: PSRR = (Change in v_{OUT})/(Total change in $+V$ and $-V$), in µV/V or decibels. Typically 30 µV/V.

Output impedance: Typically 75 Ω.

Slew Rate: Maximum rate at which v_{OUT} can change, measured when closed-loop gain = 1. Typically 1 V/µs to 10 V/µs, but up to 600 V/µs for high-frequency op amps. Limits the amplitude of sinusoidal outputs at high frequencies → waveform distorted → tends toward triangular. Maximum frequency for no distortion = (slew rate)/$2\pi v_{MAX}$, where v_{MAX} = maximum undistorted output voltage.

Gain-bandwidth product: Gain falls off with increased frequency. Typically it starts falling above 10 kHz (the full-power bandwidth) and reaches unity at f_T, often $\approx$ 1 MHz. This is a GBP of 1 MHz. Fall-off in gain is accompanied by gradual increase in output phase lag, approaching $-180°$ as gain approaches unity.
Noise figure, e_n (7.5): Typically 200 nV/$\sqrt{}$Hz for bipolar, 15 nV/$\sqrt{}$Hz for JFET, 50 V/$\sqrt{}$Hz for MOSFET.

D.2 Basic op amp circuits

D.2.1 Inverting amplifier

In an ideal op amp:

1 The amplifier reaches a stable state when the voltage difference at the inputs is zero.
2 Zero current enters each input.

In Fig. D.2, $v_+ = 0$ (no current in R_B or to (+) input); if v_{IN} is positive, v_{OUT} swings negative until v_- is zero. Current flows from input, through R_A and R_F (no current goes to (−) input) and into output. Current through R_A is:

$$v_{IN}/R_A$$

Current through R_F is:

$$-v_{OUT}/R_F$$

The currents are equal, so:

$$\text{voltage gain, } G_v = v_{OUT}/v_{IN} = -R_F/R_A$$

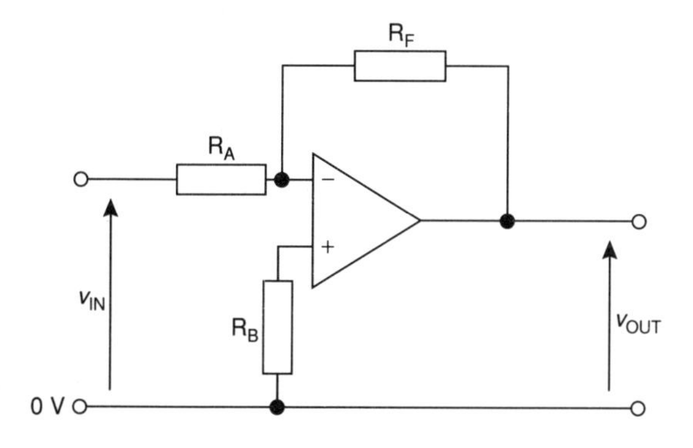

Figure D.2

Gain depends only on ratio of R_F and R_A, not on open-loop gain G_o of op amp. Gain is negative; an inverting amplifier. $Z_{in} = R_A$; a disadvantage when R_A has to be made small to give high gain.

R_B is needed with bipolar op amps because assumption of no current entering inputs is not true. I_{BIAS} causes voltage drop across resistors. Correct this by making R_B equal to R_A and R_F in parallel. R_B not required with FET op amps.

D.2.2 Non-inverting amplifier

In Fig. D.3a voltage at (+) input is v_{IN}; voltage at (−) input is same as at centre of potential divider:

$$v_{OUT} \times \frac{R_A}{R_A + R_F}$$

These are equal:

$$v_{IN} = v_{OUT} \times \frac{R_A}{R_A + R_F}$$

⇒

$$G_v = \frac{v_{OUT}}{v_{IN}} = \frac{R_A + R_F}{R_A}$$

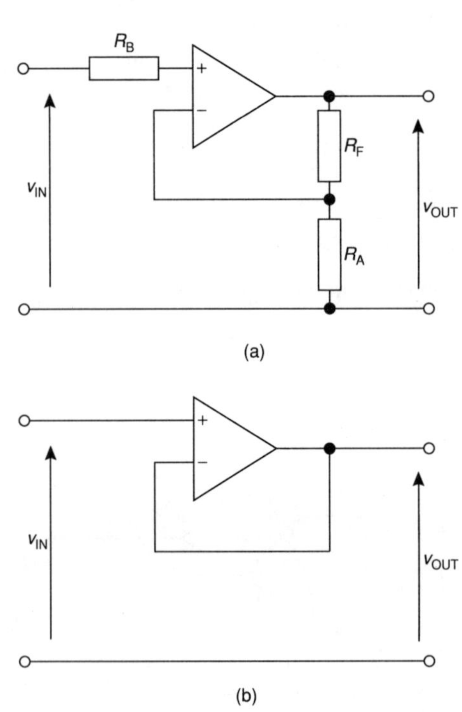

Figure D.3

Gain is dependent only on R_F and R_A, not on open-loop gain G_o of op amp. Gain is positive; a non-inverting amplifier. Z_{in} is the input impedance of the (+) terminal, 2 MΩ for bipolar, or more for FET op amps. With bipolar op amps, include $R_B = R_A \| R_F$.

Figure D.3b is a version of D.3a with $R_F = 0$. Then:

$$G_v = \frac{R_A + 0}{R_A} = 1$$

A voltage follower with high Z_{in}; ideal as a buffer.

D.2.3 Comparator

In Fig. D.4, if $v_{INA} < v_{INB}$ then $v_{OUT} > 0$. If $v_{INA} > v_{INB}$ then $v_{OUT} < 0$. For small differences between inputs, output could be close to zero but in practice output usually swings as far as it can go toward one supply rail or the other. Z_{in} is high at both inputs. Use an op amp with low input offset voltage.

D.2.4 Differential amplifier

In Fig. D.5:

$$v_{OUT} = v_{INB} \times \left(\frac{R_A + R_F}{R_A}\right)\left(\frac{R_C}{R_B + R_C}\right) - v_{INA} \times \frac{R_F}{R_A}$$

To balance bias currents make:

$$R_B \| R_C = R_A \| R_F$$

If $R_A = R_B$, and $R_C = R_F$, this simplifies to:

$$v_{OUT} = \frac{R_F \times (v_{INB} - v_{INA})}{R_A}$$

A simple differential amplifier. If all four resistors are equal:

$$v_{OUT} = v_{INB} - v_{INA}$$

This version is also known as a *subtractor*.

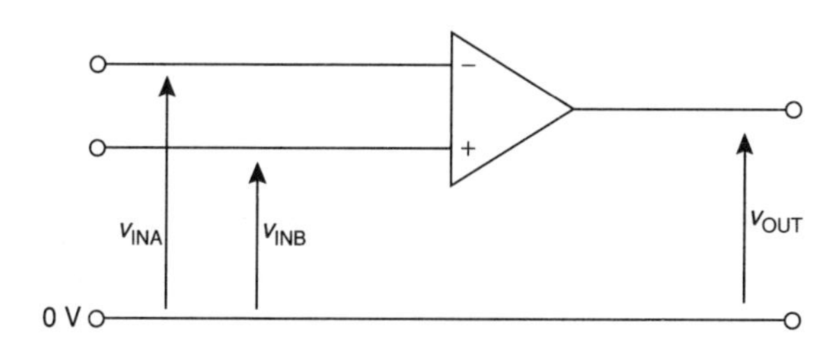

Figure D.4

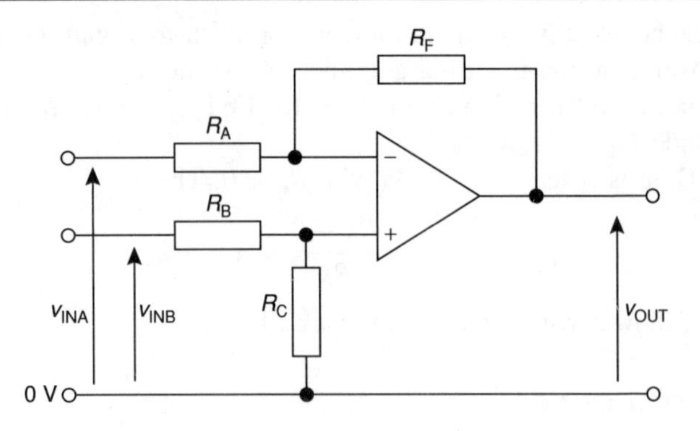

Figure D.5

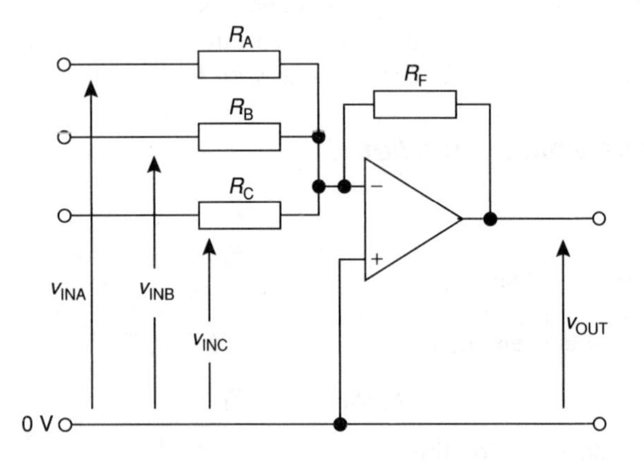

Figure D.6

D.2.5 Summing amplifier

In Fig. D.6 because (−) input is at 0 V (virtual earth), total input current flowing to (−) is:

$$v_{INA}/R_A + v_{INB}/R_B + v_{INC}/R_C$$

Current flowing through R_F is:

$$-v_{OUT}/R_F$$

This is equal to the total input current:

$$v_{OUT} = -R_F(v_{INA}/R_A + v_{INB}/R_B + v_{INC}/R_C)$$

v_{OUT} is the sum of the input voltages weighted by the reciprocals of the input resistor values. If all resistors are equal:

$$v_{OUT} = -(v_{INA} + v_{INB} + v_{INC})$$

Appendix E Circuit analysis

A summary of techniques useful for 'pen-and-paper' analysis.

E.1 Kirchhoff's laws

E.1.1 Kirchhoff's current law (KCL)

At any instant, the sum of currents arriving at and leaving a node is zero (Fig. E.1). Alternatively expressed, the total current entering a node equals the total current leaving it.

E.1.2 Kirchhoff's voltage law (KVL)

At any instant, the sum of the pds across the branches of a loop of a network is zero. A *loop* is a closed path through a network. There are three loops in Fig. E.2: ABCD, BEFC, and ABEFCD. Figure E.2 shows the KVL equation for the loop ABCD. Equations for the other loops are:

$$v_2 - v_3 - v_4 = 0 \quad \text{and} \quad v - v_1 - v_3 - v_4 = 0$$

E.2 Nodal analysis

(1) Assign symbolic values to unknown node voltages.
(2) Write equations for branch currents.
(3) Apply KCL.
(4) Solve the equations so produced to find the currents.

Example (Fig. E.3a):

(1) Assign the value 0 V to node 0, and the value V to node 1.
(2) By Ohm's law, the branch currents are:

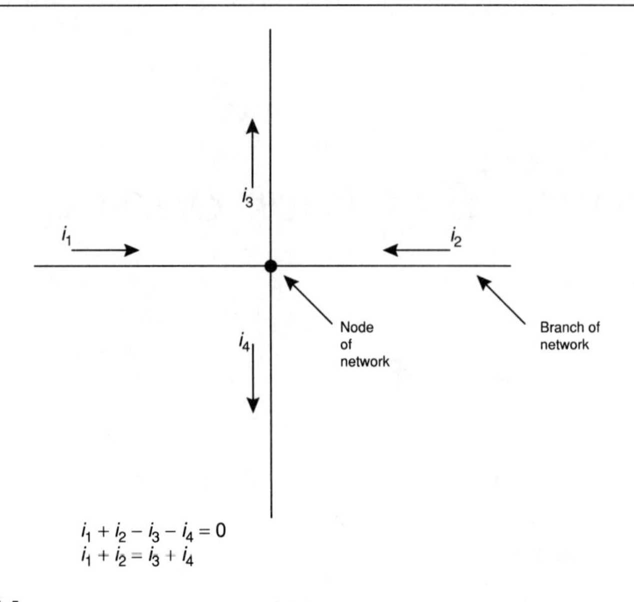

Figure E.1

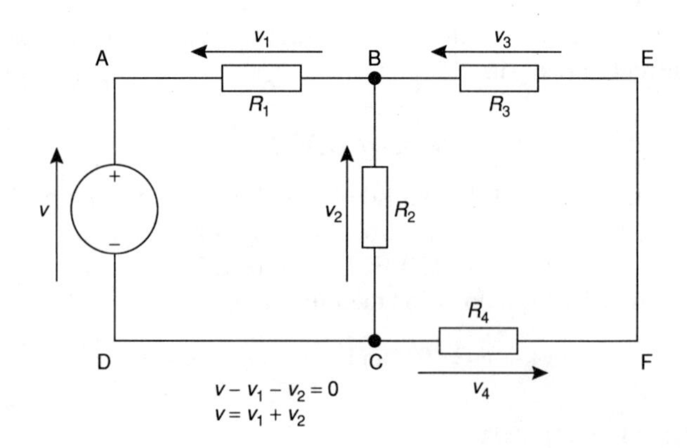

Figure E.2

$$i_1 = (10 - V)/3 \quad i_2 = V/11 \quad i_3 = V/4$$

(3) Applying KCL:

$$i_1 = i_2 + i_3$$

$$\Rightarrow \quad (10 - V)/3 = V/11 + V/4$$

$$\Rightarrow \quad V = 4.94\,\text{V}$$

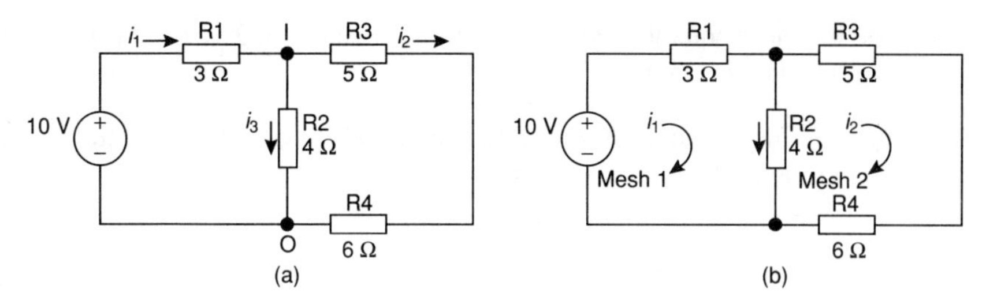

Figure E.3

(4) Substitute V in the current equations:

$$i_1 = 1.685\ \text{A} \quad i_2 = 0.449 \quad i_3 = 1.236$$

E.3 Mesh analysis

(1) Assign currents i_1, i_2, i_3, ... to each mesh. A *mesh* is a loop which does not have branches linking different parts of it. In Fig. E.2, ABCD and BEFC are meshes, but ABEFCD is not. By convention, currents circulate clockwise.
(2) Proceeding around each mesh in the direction of the current, and using KVL, write a voltage equation for each mesh.
(3) Rewrite the equations as simultaneous equations.
(4) Solve the simultaneous equations for currents.

Example (Fig. E.3b):

(1) i_1 and i_2 assigned as shown.
(2) In mesh 1, the voltage of the source is positive (potential rises as we pass through the source in a clockwise direction), but each resistor causes a voltage drop. By KVL:

$$10 - (3 + 4)i_1 - 4(-i_2) = 0$$

i_2 flows through R2 in the opposite direction to i_1, so is negative. The same applies to i_1 in mesh 2:

$$4(-i_1) + (5 + 6)i_2 = 0$$

(3) Rewriting these equations as simultaneous equations:

$$7i_1 - 4i_2 = 10$$
$$-4i_1 + 15i_2 = 0$$

(4) Solving these equations gives $i_1 = 1.687$ A and $i_2 = 0.449$ A. From these two results we calculate i_3, using KCL.

We are then able to calculate the potentials at all nodes, obtaining the same results as in the nodal analysis.

E.4 Superposition theorem

If there are two or more voltage or current sources in a network, the action of each source is superposed on that of the other sources.

(1) Eliminate all except one source by replacing current sources by open circuit and by replacing voltage sources by short circuits.
(2) Investigate the potentials and currents with the one remaining source active.
(3) and (4), (5) and (6), etc. If necessary, repeat with a different source eliminated.

Finally, sum the component currents and voltages.

Example (Fig. E.4a):
To find the values of v and i.

(1) Eliminate the current source (Fig. E.4b). The network becomes 20 Ω and 30 Ω in series across the 12 V source.
(2) Calculate $i' = 12/50 = 0.24$ A. $v' = 12 \times 20/50 = 4.8$ V.
(3) Eliminate the voltage source (Fig. E.4c). The network becomes 4 Ω in series with 20 Ω and 30 Ω in parallel. Total resistance across the source is $4\,\Omega + 12\,\Omega = 16\,\Omega$.
(4) Pd across source is $16 \times 0.15 = 2.4$ V. Pd across the 20 Ω in parallel with 30 Ω is $v'' = 2.4 \times 12/16 = 1.8$ V. But this is negative because its polarity is opposite in direction to the arrow. $i'' = 1.8/30 = 0.06$ A. This flows in the same direction as the arrow.
(5)

$$v = v' + v'' = 4.8 - 1.8 = 3\text{ V}$$

$$i = i' + i'' = 0.24 + 0.06 = 0.3\text{ A}$$

A similar procedure can be used to find other voltages and currents.

E.5 Thévenin's theorem

Any 2-terminal network consisting of one or more voltage or current sources and one or more impedances may be represented by a single voltage source

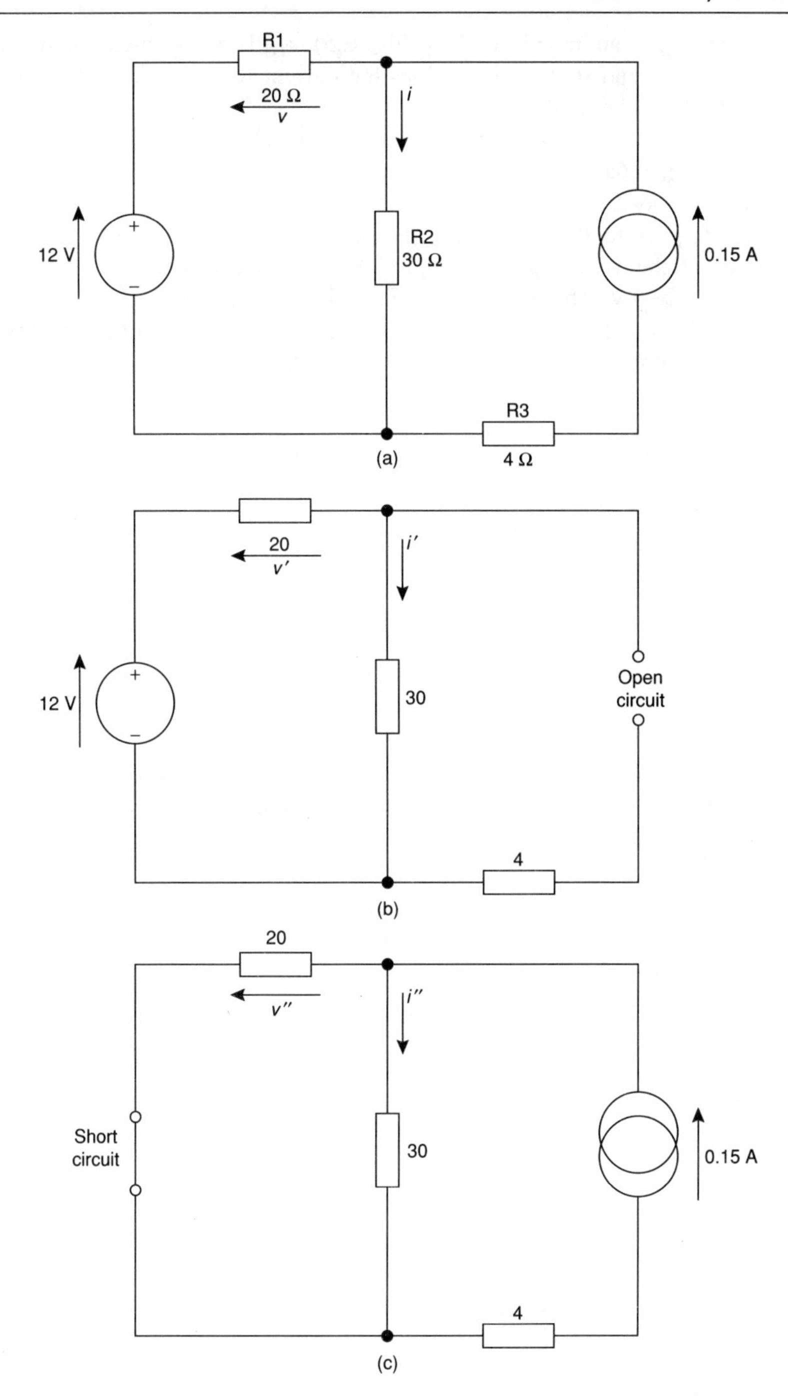
R1
20 Ω
v
i
12 V
R2
30 Ω
0.15 A
R3
4 Ω
(a)
20
v′
i′
12 V
30
Open circuit
4
(b)
20
v″
i″
Short circuit
30
0.15 A
4
(c)

Figure E.4

v_{TH} in series with an impedance R_{TH} (Fig. E.5). v_{TH} is the open-circuit voltage v_{OC} between A and B. R_{TH} determines the current i_{SC} from A to B when A and B are short-circuited.

Example (Fig. E.6a):
Use the superposition method for both calculations:
Finding V_{TH}: A to B is oc. Make the current source oc. Then $v'_{OC} = 3\,\text{V}$. Short-circuit the voltage source: $v''_{OC} = 0.2 \times 10 = 2\,\text{V}$. Summing voltages, $v_{OC} = 3 + 2 = 5\,\text{V}$. This is the Thévenin voltage: $v_{TH} = 5\,\text{V}$.
Finding R_{TH}: A to B is sc and we need to find the current i_{SC} through it. Make the current source oc. $i'_{SC} = 3/40 = 0.075\,\text{A}$. Short-circuit the voltage source.

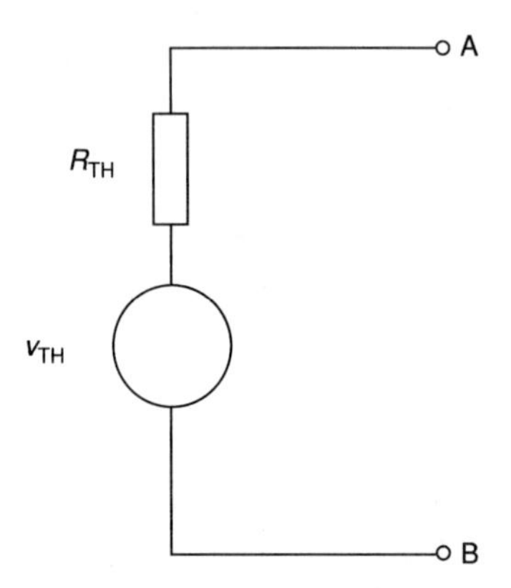

Figure E.5

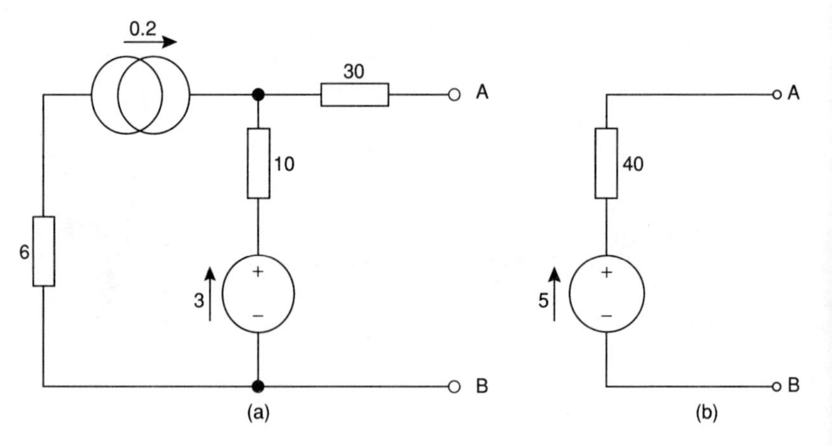

Figure E.6

By the current division rule: $i''_{SC} = 0.2 \times 10/40 = 0.05$ A.
Summing currents: $i_{SC} = 0.075 + 0.05 = 0.125$ A.

Given that $v_{TH} = 5$ V and $i_{SC} = 0.125$ A:

$$R_{TH} = v_{TH}/i_{SC} = 5/0.125 = 40\,\Omega$$

Figure E.6b is the *Thévenin equivalent* of Fig. E.6a. When various other networks are connected to the terminals A and B the behaviour of Figs. E.6a and E.6b are indistinguishable. In a larger network we identify a section such as Fig. E.6a and replace it by its Thévenin equivalent. This simplifies subsequent calculations.

E.6 Norton's theorem

Any 2-terminal network can be replaced by a current source i_N in parallel with a conductance G_N (Fig. E.7). When the *Norton equivalent* is sc, all the current from the Norton current source i_N passes through the sc. When the terminals of the Norton equivalent are oc, the current flows through the conductance and produces an open-circuit pd across it. Thus, to find the Norton equivalent of a network we need to know i_{sc} and v_{oc} (same quantities as are needed for Thévenin equivalent).

Example (Fig. E.6a):
Find i_{sc}, as before: $i_N = i_{SC} = 0.125$ A. $v_{OC} = 5$ V. Then:

$$G_N = i_N/v_{OC} = 0.025\,\text{S}$$

This equivalent can be substituted for the network of Fig. E.6a to simplify subsequent calculations.

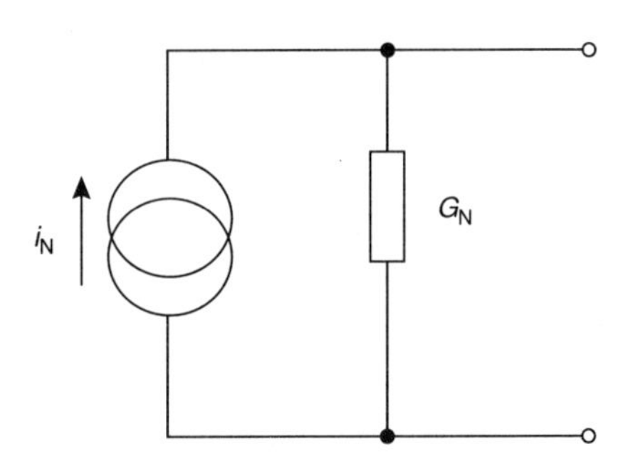

Figure E.7

E.7 Reactive networks

E.7.1 RC network

Figure E.8 includes capacitance C and a sinusoidal source:

$$\mathbf{v} = V_0 \sin(\omega t + \phi)$$

where V_0 is the amplitude, ω is angular frequency (in rad/s), t is elapsed time, ϕ is phase of signal when $t = 0$. If frequency in hertz is f, then $\omega = 2\pi f$.

The reactance of capacitance C is $X_C = 1/\omega C$ (Ω).

The pd across a resistance is in phase with the current through it, but the phase of the pd across a capacitance lags behind the phase of the current by $90^\circ (\pi/2)$. For Fig. E.8 the pds across the voltage source ($\mathbf{v}$), the resistor ($\mathbf{v_R}$), and the capacitor ($\mathbf{v_C}$), as obtained by a circuit simulator, are plotted in Fig. E.9. The fourth curve (i_R) is the current through the resistor. By the definition of resistance, $\mathbf{i_R}$ is in phase with $\mathbf{v_R}$. $\mathbf{v_C}$ lags $\mathbf{i_R}$ by $\pi/2 \rightarrow \mathbf{v_C}$ lags $\mathbf{v_R}$ by $\pi/2$. At any instant, $\mathbf{v}$ is the sum of $\mathbf{v_C}$ and $\mathbf{v_R}$ (check Fig. E.9) so $\mathbf{v}$ lags $\mathbf{v_R}$ by an amount intermediate between 0 and $\pi/2$. The curve relationships, complicated because $\mathbf{v_C}$ and $\mathbf{v_R}$ are not in phase with each other, can be investigated by drawing a phasor diagram.

E.7.2 Phasor diagram

Represents voltage or current amplitudes, or magnitudes of impedances, by lengths of phasors; represents phase relationships by angles of phasors. This need not be drawn to scale as it is solved geometrically. By convention, the horizontal direction to the right represents zero phase angle. Phasors do not represent frequency: all signals in a diagram must have the same frequency.

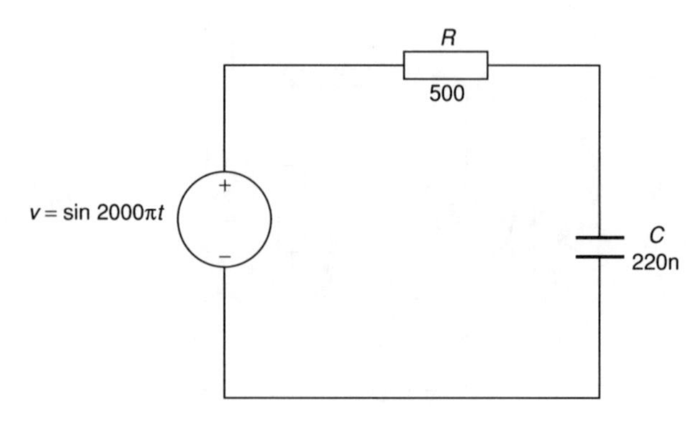

Figure E.8

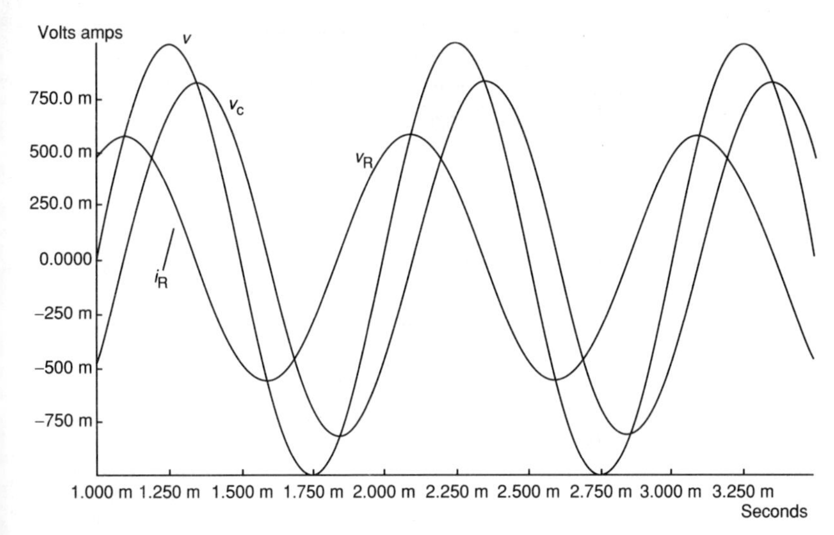

Figure E.9

Example (Fig. E.8):
The signal has 1 V amplitude and 1 kHz frequency.

(1) Calculate reactance: $\omega = 2000\pi$, so $\mathbf{X_C} = 1/\omega C = 1/(2000 \times \pi \times 220 \times 10^{-9}) = 723\,\Omega$.

(2) Draw phase diagram (Fig. E.10). Zero direction is phase of $\mathbf{i_R}$ and $\mathbf{v_R}$. $\mathbf{v_R}$ and $\mathbf{v_C}$ are perpendicular ($\pi/2$ phase lead of $\mathbf{v_R}$ with respect to $\mathbf{v_C}$). Since the source, capacitor and resistor are in series, the same current flows through all and thus the lengths of the phasors are proportional both to voltage amplitudes and to the impedances:

$$\mathbf{X_C} = 1/\omega C = 723\,\Omega \quad \mathbf{R} = 500\,\Omega \quad \mathbf{Z_T} = \mathbf{X_C} + \mathbf{R}$$

where $\mathbf{Z_T}$ is the total circuit impedance and is the vector sum of $\mathbf{X_C}$ and $\mathbf{R}$, not the arithmetic sum.

(3) Sum the phasors by using Pythagoras' theorem:

$$\mathbf{Z_T} = \sqrt{(\mathbf{X_C^2} + \mathbf{R}^2)} = 879\,\Omega$$

(4) Find the phase lag of $\mathbf{v_r}$ by trigonometry:

$$\phi = \tan^{-1}(723/500) = 55.3^\circ$$

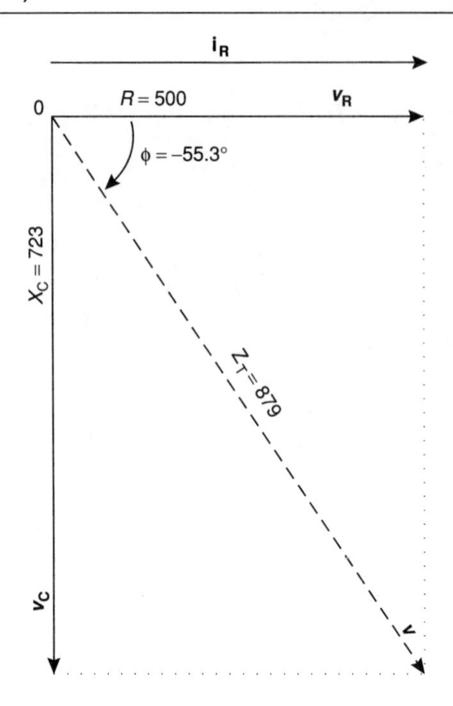

Figure E.10

(5) Calculate the other angle:
There is 90° between $\mathbf{v_C}$ and $\mathbf{v_R}$ → the phase lead of $\mathbf{v_C}$ is $90 - 55.3 = 34.7°$.

(6) Find the impedance–amplitude ratio to relate impedances to signal amplitudes:
$\mathbf{Z_T} = 879\,\Omega$ and amplitude of $\mathbf{v} = 1$ V. Ratio = 1/879.

(7) Calculate other amplitudes:
Amplitude of $\mathbf{v_C} = 723/879 = 0.82$ V
Amplitude of $\mathbf{v_R} = 500/879 = 0.57$ V

(8) Write the equations of the voltage signals:

$$\mathbf{v} = \sin(2000\pi t - 55.3°)$$

$$\mathbf{v_C} = 0.82\sin(2000\pi t - 90°)$$

and

$$\mathbf{v_R} = 0.57\sin 2000\pi t$$

But it is often more useful to take the phase of the source as the reference phase. Add 55.3° to each phase angle:

$$\mathbf{v} = \sin 2000\pi t$$

$$\mathbf{v_C} = 0.82\sin(2000\pi t - 34.7^\circ)$$

and
$$\mathbf{v_R} = 0.57\sin(2000\pi t + 55.3^\circ)$$

Confirm these by reference to Fig. E.9.

(9) Calculate $\mathbf{i_R}$, $\mathbf{I_0} = \mathbf{V_0}/\mathbf{Z_T} = 1/879 = 1.14\,\text{mA}$, plotted $\times$ 250 in Fig. E.9. It is in phase with $\mathbf{v_R}$:

$$\mathbf{i_R} = 0.00114\sin(2000\pi t + 55.3^\circ)$$

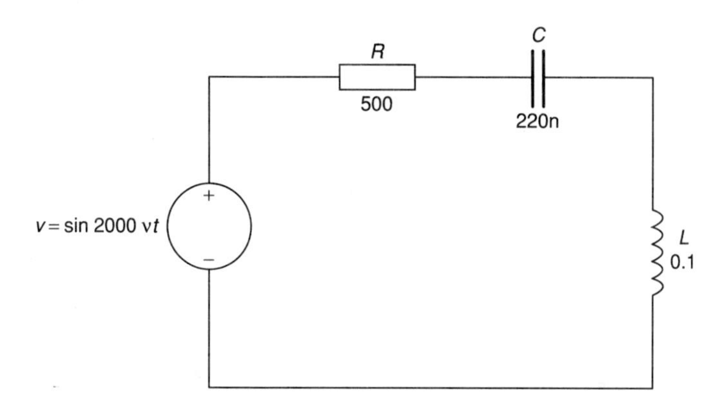

Figure E.11

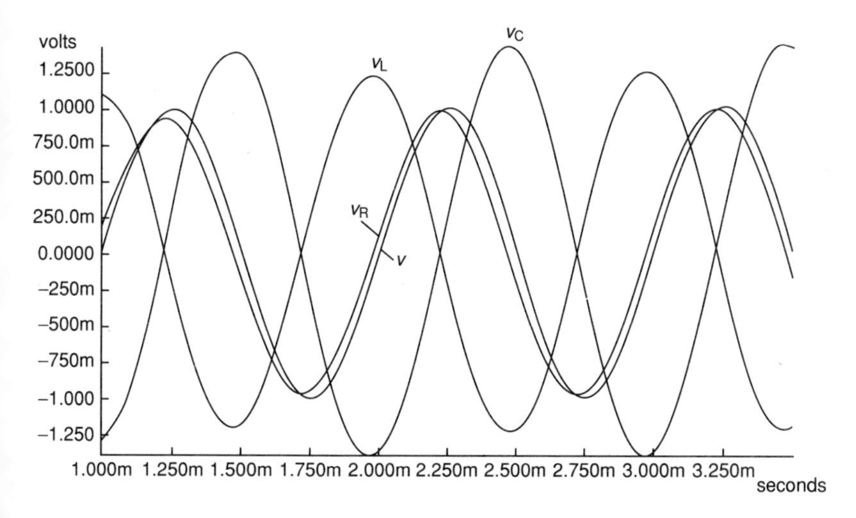

Figure E.12

E.7.3 LCR network

The reactance of an inductance L is $\mathbf{X_L} = \omega L$ (Ω)
The phase of the pd across an inductance leads the phase of the current by $90^\circ(\pi/2)$.

Analysing the LCR network of Fig. E.11 on a simulator gives Fig. E.12. The voltage signals are represented by the phasors of Fig. E.13. At 1 kHz, impedances are:

$$\mathbf{R} = 500\,\Omega \quad \mathbf{X_C} = 723\,\Omega \quad \mathbf{X_L} = 0.1\omega = 628\,\Omega$$

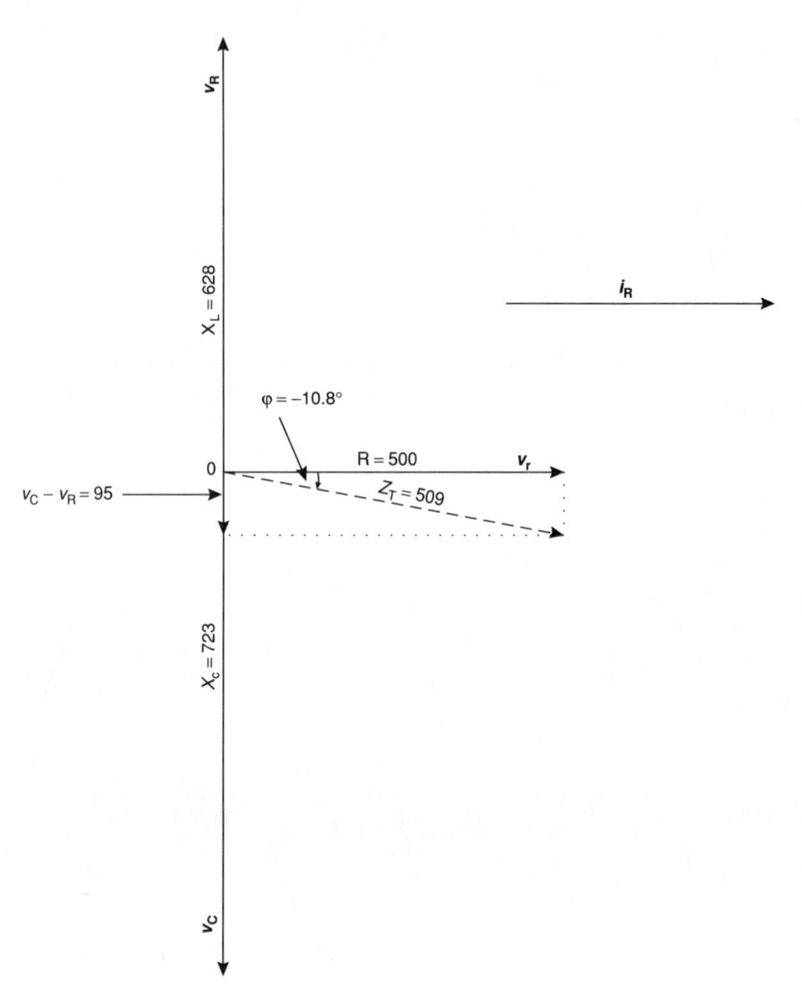

Figure E.13

The phasors $\mathbf{v_C}$ and $\mathbf{v_L}$ are both perpendicular to $\mathbf{v_R}$, but directed oppositely since $\mathbf{v_l}$ leads by $\pi/2$ while $\mathbf{v_C}$ lags by $\pi/2$. The vector sum of $\mathbf{v_C}$ and $\mathbf{v_L}$ is $723 - 628 = 95\,\Omega$ in the same direction as $\mathbf{v_C}$. Working as in E.7.2, calculate:

$$\mathbf{Z_T} = \sqrt{(95^2 + 500^2)} = 509\,\Omega$$

$$\phi = \tan^{-1}(95/500) = 10.8^\circ$$

From which, working as in E.7.2, derive amplitudes and phases with reference to phase of the source:

$$\mathbf{v} = \sin 2000\pi t$$

$$\mathbf{v_C} = 1.42\sin(2000\pi t - 79.2^\circ)$$

$$\mathbf{v_R} = 0.98\sin(2000\pi t + 10.8^\circ)$$

and

$$\mathbf{v_L} = 1.26\sin(2000\pi t + 100.8^\circ)$$

E.8 Complex impedances

E.8.1 Representing phasors by complex numbers

Addition of phasors is easier using complex numbers. Figure E.14 shows the phasors of Fig. E.10 plotted on the complex plane. $\mathbf{v_R}$ (or $\mathbf{R}$) is represented by the point 500 + j0. $\mathbf{v_C}$ (or $\mathbf{X_C}$ at the given frequency) is represented by $0 - \mathrm{j}723$. The complex representation of the phasors contains both amplitude information (the real part) and phase information (the imaginary part). The vector sum of the phasors is the sum of the complex numbers. In rectangular form:

$$\mathbf{Z_T} = (500 + \mathrm{j}0) + (0 - \mathrm{j}723) = 500 - \mathrm{j}723$$

To recover the amplitude and phase information, convert it to polar form (which is equivalent to steps 3 and 4 of E.7.2):

$$\mathbf{Z_T} = 500 - \mathrm{j}723 \rightarrow 879\,\Omega\underline{/-55.3^\circ}$$

E.8.2 Working directly with complex impedances.

Complex expression of impedances is:

$$\mathbf{X_C} = \frac{-\mathrm{j}}{\omega C} = \frac{1}{\mathrm{j}\omega C} \quad \text{(the forms are equivalent)}$$

$$\mathbf{X_L} = \mathrm{j}\omega L$$

Including j or $-$j in the expressions supplies phase information, since j is equivalent to $+\pi/2$. Expressed in these ways complex impedances may be

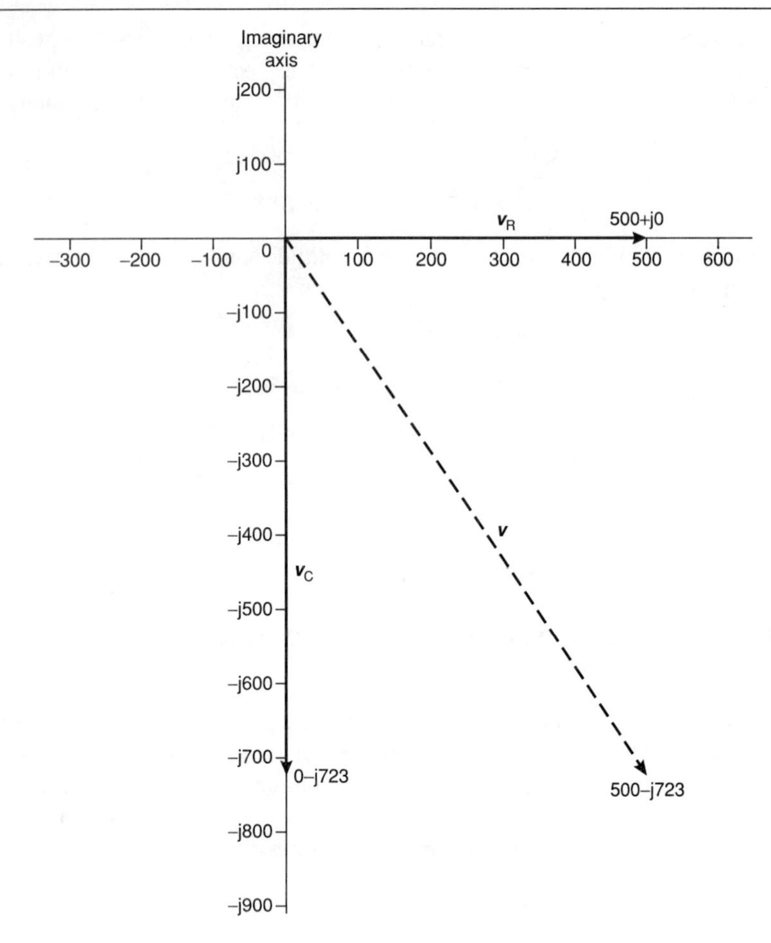

Figure E.14

written on circuit diagrams and used in analyses in the same way as real impedances.

Example (Fig. E.15a):

(1) Write the impedances in complex form (Fig. E.15b).
(2) Apply mesh analysis (E.3).

At 1 kHz, $\omega = 6283$ rad/s.
$\mathbf{X_L} = j\omega \times 300 \times 10^{-6} = 1.8849\,\Omega$
$\mathbf{X_C} = -j/(\omega \times 47 \times 10^{-6}) = -j3.386\,\Omega$

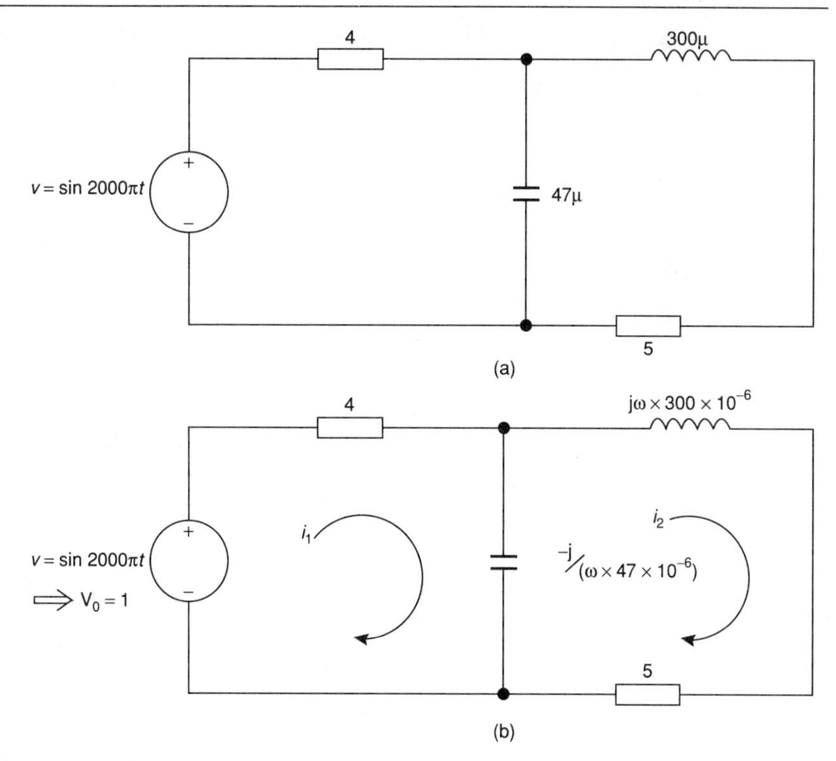

Figure E.15

Mesh equations are:

$$(4 - j3.386)i_1 - (-j3.386)i_2 = 1$$

$$-(-j3.386)i_1 + (-j3.386 + j1.8849 + 5) = 0$$

Solving these equations for i_1 and i_2 yields:

$$\mathbf{i_1} = 0.136 + j0.0614$$

$$\mathbf{i_2} = 0.0635 - j0.0731$$

Converting to polar form:

$\mathbf{i_1} = 0.149\angle 24.3^\circ$
$\mathbf{i_2} = 0.097\angle -49.0^\circ$

Current as a function of time:

$\mathbf{i_1} = 0.149\sin(2000\pi t + 24.3^\circ)$
$\mathbf{i_2} = 0.097\sin(2000\pi t - 49.0^\circ)$

These values can be used for further calculations. For example, to find the pd across the inductor, first express its inductance in polar form:

$$\mathbf{X_L} = 0 + j\omega \times 300 \times 10^{-6} = 1.8849\angle 90^\circ$$

$\Rightarrow \qquad \mathbf{v}_L = \mathbf{X}_L\mathbf{i}_2 = 1.8849\angle 90^\circ \times 0.097\angle -49.0^\circ = 0.0183\angle 41.0^\circ$

$\Rightarrow \qquad v_L = 0.0183\sin(2000pt + 41.0^\circ)$

E.9 Complex frequency variable

The expressions for complex impedance given in the previous section apply only to sinusoidal waveforms with constant amplitude and frequency. If the amplitude or frequency are changing, or if the signal is not periodic, we use the following expressions for impedance:

$$\mathbf{X}_C = 1/sC \quad \mathbf{X}_L = sL$$

where s is the complex frequency variable, as used in the Laplace transformation. Some authors use the symbol p instead of s. The variable s comprises two components, one real and one imaginary:

$$s = \sigma + j\omega$$

If the amplitude of a signal is constant, then $\sigma = 0$, and $s = j\omega$. The impedances depend only on frequency, as in the examples we have already seen and in most other network analyses.

E.10 Two-port networks

A circuit, a sub-circuit or a single component is represented by a sub-circuit with two input terminals and two output terminals. These are known as two-port or four-terminal networks.

E.10.1 Hybrid parameter network

The mesh equations of the *h-parameter network* of Fig. E.16 are:

$$v_1 = h_{11}i_1 + h_{12}v_2 \quad \text{and} \quad i_2 = h_{21}i_1 + h_{22}v_2$$

Parameters are found by pen-and-paper calculations on a schematic diagram of a network. Identify the input and output ports, then use the oc/sc technique (as in E.5 and E.6). Or use meters on an actual network or component.

(1) Open-circuit port 1, $i_1 = 0$. Apply v_1, measure v_2, i_2. Use equations on the right to calculate h_{12}, h_{22}.

$$v_1 = h_{12}v_2 \quad \Rightarrow \quad h_{12} = v_1/v_2$$

$$i_2 = h_{22}v_2 \quad \Rightarrow \quad h_{22} = i_2/v_2$$

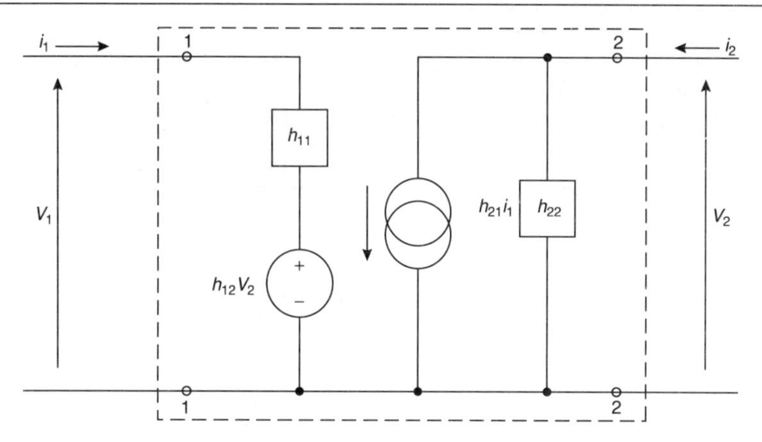

Figure E.16

(2) Short-circuit port 2, $v_2 = 0$: Apply i_1, measure v_1, i_2. Use equations on the right to calculate h_{11}, h_{21}.

$$v_1 = h_{11}i_1 \quad \Rightarrow \quad h_{11} = v_1/i_1$$

$$i_2 = h_{21}i_1 \quad \Rightarrow \quad h_{21} = i_2/i_1$$

This network is often used for analysing BJT transistor amplifiers and similar networks. In these cases the parameters receive special symbols:

h_{11} becomes h_i, the short-circuit input impedance, measured in ohms

h_{21} becomes h_f, the short-circuit forward current ratio

h_{12} becomes h_r, the open-circuit reverse voltage ratio

h_{22} becomes h_o, the open-circuit output admittance, measured in siemens

The parameters take different values according to the connection (C.4) of the amplifier and a second suffix, b, c, or e is added to indicate this. h_{fe} in this context is equivalent to h_{FE}, *the large signal current gain* (C.2.3). The *small-signal current gain* h_{fe}, as defined in C.2.4.1, has a slightly different derivation.

Example:
A common-emitter BJT amplifier circuit (4.1) is tested to find its h-parameters:

(1) Port 1 is oc (Fig. E.17a). Apply $v_1 = 100\,\mu\text{V}$, measure $v_2 = 0.25\,\text{V}$, and $i_2 = 6\,\mu\text{A}$
Calculate $h_{re} = v_1/v_2 = 4 \times 10^{-4}$
Calculate $h_{oe} = i_2/v_2 = 24\,\mu\text{S}$

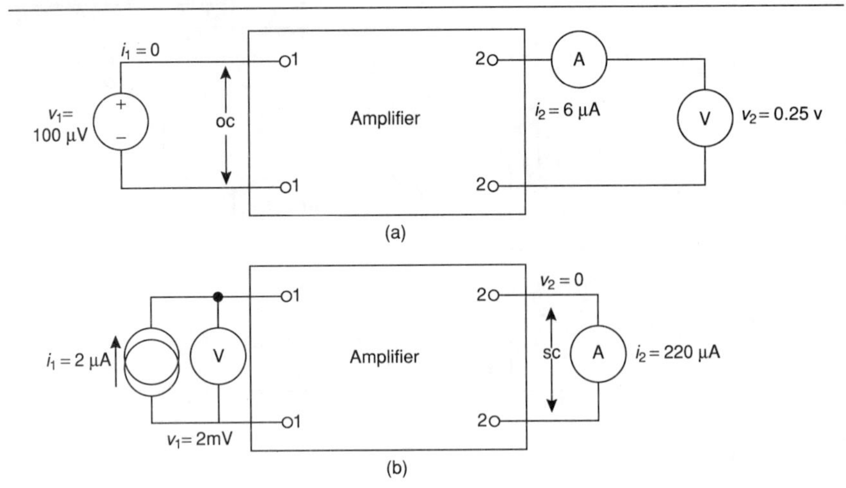

Figure E.17

(2) Port 2 is sc (Fig. E.17b). Apply $i_1 = 2\,\mu\text{A}$, measure $v_1 = 2\,\text{mV}$, $i_2 = 220\,\mu\text{A}$
Calculate $h_{\text{ie}} = v_1/i_1 = 1000\,\Omega$
Calculate $h_{\text{fe}} = i_2/i_1 = 110$

This procedure establishes values for the h-parameters of the amplifier. Using these, we calculate amplifier behaviour under other operating conditions:

Given voltage source $v_1 = 1\,\text{mV}$, and load resistor $R = 22\,\text{k}\Omega$ (Fig. E.18), find input current i_1, output voltage v_2, load current i_l and voltage amplification A_{v}. By KVL (E.1.2) on the input side:

$$v_1 - h_{\text{re}} v_2 = i_1 h_{\text{ie}}$$

$$\Rightarrow \qquad 0.001 - 4 \times 10^{-4} v_2 = 1000 i_1 \qquad \text{(E.10.1)}$$

By KCL (E.1.1) on the output side:

$$-h_{\text{fe}} i_1 = h_{\text{oe}} v_2 + v_2/R$$

$$\Rightarrow \qquad -110 i_1 = (24 \times 10^{-6} + 4.545 \times 10^{-5}) v_2 \qquad \text{(E.10.2)}$$

From (E.10.2):

$$i_1 = -6.314 \times 10^{-7} v_2$$

Substitute for i_1 in (E.10.1)

$$v_2 = -1.0314\,\text{V}$$

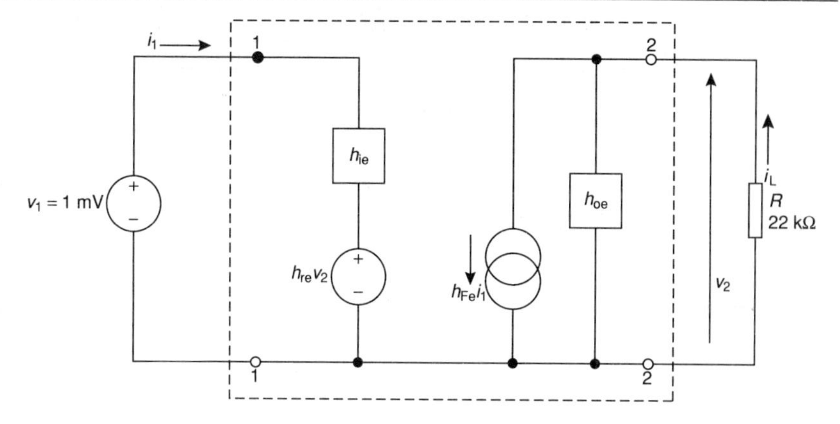

Figure E.18

Consequently:

$$i_1 = -1.0314/22\,000 = -46.88\,\mu\text{A}$$

and
$$A_v = -1.0314/0.001 = 1031$$

E.10.2 *Transmission parameters*

These are useful when a circuit consists of a number of 2-port networks chained together so that the output of one network becomes the input of the next. Also known as *ABCD parameters.*

$$v_1 = \text{A}v_2 - \text{B}i_2 \quad \text{and} \quad i_1 = \text{C}v_2 - \text{D}i_2$$

The matrix equation is:

$$\begin{bmatrix} v_1 \\ i_1 \end{bmatrix} = \begin{bmatrix} \text{A} & \text{B} \\ \text{C} & \text{D} \end{bmatrix} \cdot \begin{bmatrix} v_2 \\ -i_2 \end{bmatrix}$$

In an ABCD network (Fig. E.19) the direction of flow of i_2 is opposite to that in h-parameter networks. By making open-circuit and short-circuit tests we arrive at the following equalities:

$$1/\text{A} = v_2/v_1 \quad -1/\text{B} = i_2/v_1 \quad 1/\text{C} = v_2/i_1 \quad -1/\text{D} = i_2/i_1$$

The advantage of the ABCD parameters is that, given a chain of 2-port networks, each represented by an ABCD matrix as above, we can find the matrix of the whole chain by multiplying these matrices together.

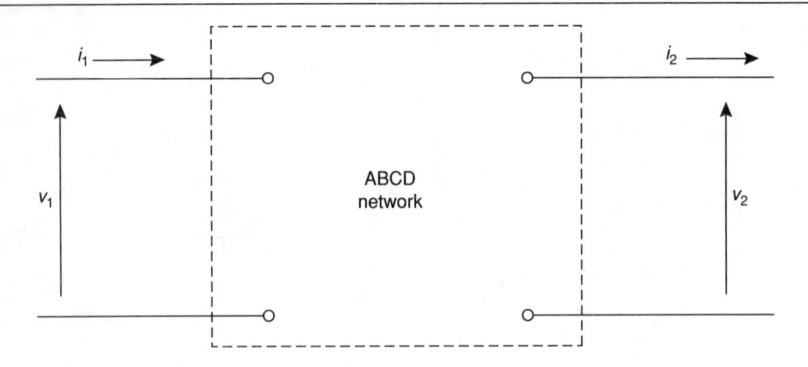

Figure E.19

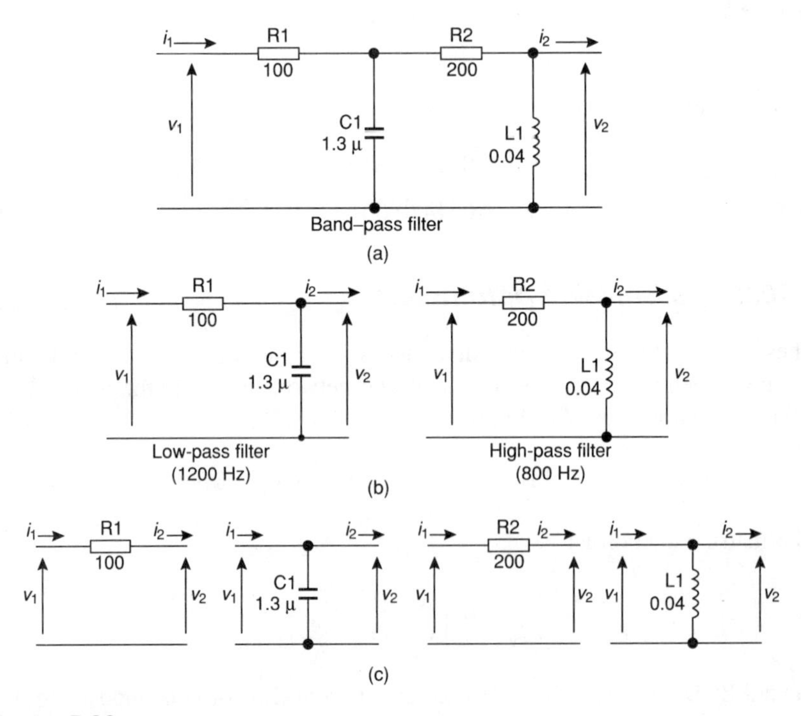

Figure E.20

Example: Band-pass filter (Fig. E.20a) (6.1.4)
Splits into two networks (Fig. E.20b), a resistor–capacitor low-pass filter, cascaded with a resistor–inductor high-pass filter.

These split into four single-element networks (Fig. E.20c). The matrix representations of these are:

$$\mathrm{R1} \Rightarrow \begin{bmatrix} 1 & 100 \\ 0 & 1 \end{bmatrix} \quad \mathrm{C1} \Rightarrow \begin{bmatrix} 1 & 0 \\ s(1.3 \times 10^{-6}) & 1 \end{bmatrix}$$

$$\mathrm{R2} \Rightarrow \begin{bmatrix} 1 & 200 \\ 0 & 1 \end{bmatrix} \quad \mathrm{L1} \Rightarrow \begin{bmatrix} 1 & 200 \\ 1/(s \times 0.4) & 1 \end{bmatrix}$$

The entries for the C parameter are the reciprocals of the capacitance and inductance because C is an admittance. For explanation of s, see E.9. R1 and C1 are cascaded to produce the low-pass filter and its ABCD matrix is obtained by multiplication:

$$\begin{bmatrix} 1 & 100 \\ 0 & 1 \end{bmatrix} \cdot \begin{bmatrix} 1 & 0 \\ s(1.3 \times 10^{-6}) & 1 \end{bmatrix} = \begin{bmatrix} 1 + s(1.3 \times 10^{-4}) & 0 + 100 \\ 0 + s(1.3 \times 10^{-6}) & 0 + 1 \end{bmatrix}$$

The transfer function of the low-pass filter is v_2/v_1, which is the reciprocal of A, the reciprocal of the forward voltage gain. We evaluate this for a sinusoid of constant amplitude of 1 V, with approximate mid-band frequency of 1 kHz. For s, substitute $j\omega = j2\pi f = 6283$. Insert this in the expression for A, then evaluate 1/A. The result is $v_2/v_1 = 1/\mathrm{A} = 0.775\angle{-39.2^\circ}$. The output is a sine wave, amplitude 0.775 V, phase lag 39.2°.

The matrix provides other information: B = 100 and this is the reciprocal of the forward transfer admittance, in this network it is the resistance of R1. $\mathrm{C} = s(1.3 \times 10^{-6})$, the reciprocal of the forward transfer impedance, in this circuit it is the reciprocal of the capacitance of C1. D = 1, the reciprocal of the forward current gain, in this network this is 1 because $i_1 = i_2$.

The characteristics of the high-pass filter are calculated in the same way, by multiplying the matrices for R2 and L1 together. This yields the matrix

$$\begin{bmatrix} 1 + 5000/s & 0 + 200 \\ 0 + 1/(s \times 0.4) & 0 + 1 \end{bmatrix}$$

Evaluating 1/A results in $v_2/v_1 = 0.783\angle{38.5^\circ}$. The output is a sine wave, amplitude 0.783 V, phase lead 38.5°. Multiplying the matrices of the two filters together gives the matrix for the whole network. If we need only the transfer function, for example, we need calculate only A. At 1 kHz this equals 1.65 – j0.3769. Evaluating the reciprocal of this gives $0.595\angle{12.8^\circ}$. Band-pass filtering reduces amplitude to 0.595 V and produces a phase lead of 12.8°.

Appendix F Models

Models are used to simplify calculations, often to represent components with non-linear characteristics (diodes, transistors). There are three main types:

- Sub-circuit models replace a component with a sub-circuit consisting of components with simpler characteristics; used in pen-and-paper analysis of networks and in computer circuit simulators.
- h-parameter models of components are based on 2-port networks (E.10.1); used in pen-and-paper analysis.
- SPICE models consist of algorithms based on sets of equations, the variables including circuit variables and parameters specific to the particular component type; used in SPICE-based computer circuit simulators.

F.1 Diode models

F.1.1 DC sub-circuit model

In Fig. F.1, the voltage source (0.7 V for Si, 0.2 V for Ge) models the virtual cell (A.5). The switch is controlled by v_D, the voltage across the diode, closed when $v_D \geq v_\gamma$, open when $v_D < v_\gamma$. R_R is the reverse-bias resistance; several megohms. The forward-bias resistance is equal to R_R and R_F in parallel (with the switch closed), a few tens of ohms. Figure F.1 does not model the exponential relationship or the effects of temperature (B.1).

F.1.2 AC sub-circuit model

In Fig. F.2, dynamic resistance r is inversely proportional to i_D (B.2). C_D and C_J (B.3) are given values typical of the working conditions.

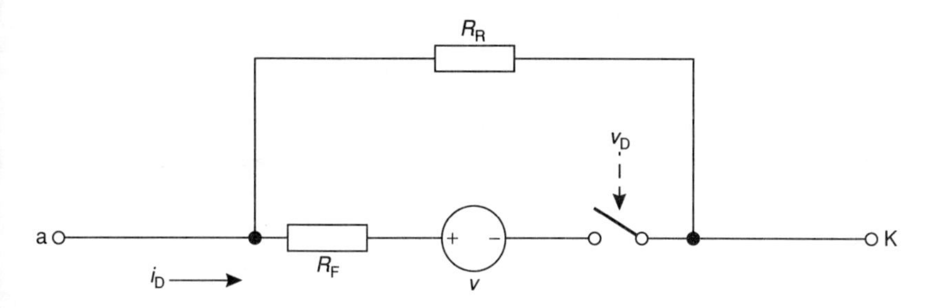

Figure F.1

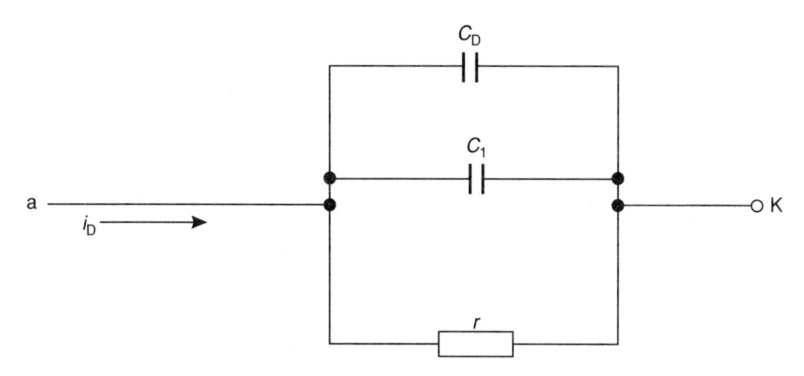

Figure F.2

F.2 BJT models

F.2.1 Sub-circuit model

Figure F.3 is based on the Ebers–Moll model (F.2.3). Input resistance at the base is modelled by R_{BB}. Capacitance at the base-collector junction is modelled by C. The diodes are, in turn, represented by models similar to Figs. F.1 or F.2. R_E senses the emitter current and controls the voltage-controlled current source (VCCS):

$$v_{RE} = 1 \times i_E$$

$$i_C = 0.99 \times v_{RE} = 0.99 \times i_E = 0.99(i_C + i_B)$$

$$\Rightarrow \quad 0.01 i_C = i_B$$

$$\Rightarrow \quad h_{fe} = i_C / i_B = 100$$

The factor 0.99 may be changed to give other values of h_{fe}. Similarly, other values for C and R_{BB} may be used in modelling specific transistor types.

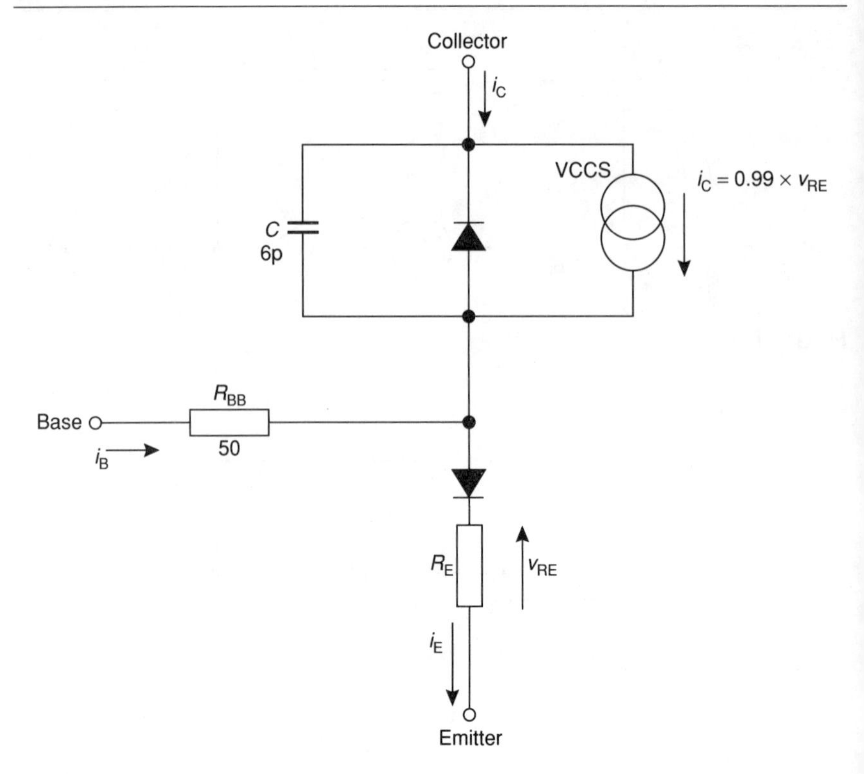

Figure F.3

F.2.2 h-parameter models

The model in Fig. F.4 has the same features as Fig. E.18, but the parameters have appropriate symbols:

parameter	description	unit	typical value
h_{ie}	input resistance	Ω	few kΩ
h_{re}	reverse voltage ratio	none	0.0002
h_{fe}	forward current gain	none	100
h_{oe}	output conductance	S	few tens of μS

Suffixes end in e to indicate a transistor in the common-emitter connection. Parameter values are obtained by measurement (D.10) or taken from manufacturers' data sheets. The model is accurate only for small DC signals. In practice, h_{re} can be ignored. The load on the transistor is in parallel with h_{oe}; the current through h_{oe} is small unless the load resistance is high. Model may be simplified to Fig. F.5.

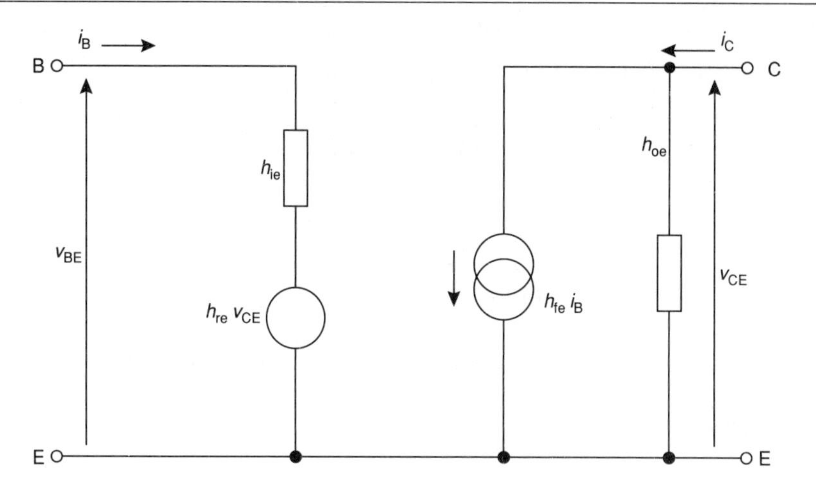

Figure F.4

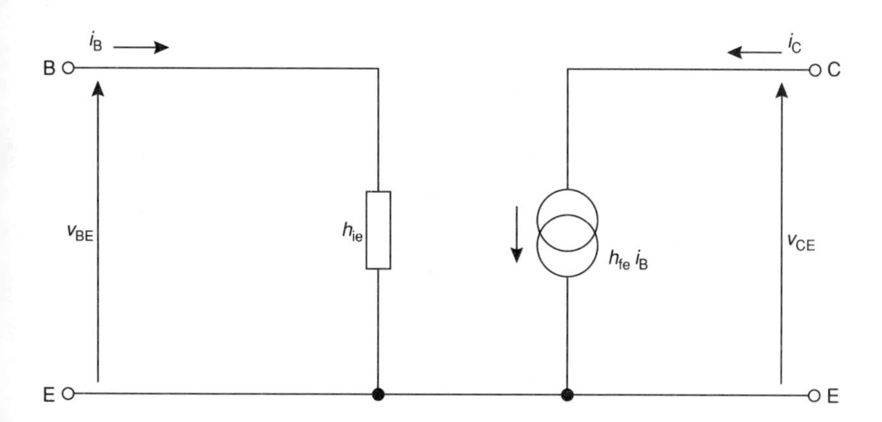

Figure F.5

Example: Calculate i_C and v_{CE} in Fig. F.6, given $i_B = 75\,\mu A$, $h_{fe} = 100$, $h_{oe} = 10\,\mu S$, $V_{CC} = 10\,V$, $R_{LOAD} = 500\,\Omega$.
The current through the generator is

$$i_G = 100 \times 75 \times 10^{-6} = 7.5 \times 10^{-3}\,A \tag{F.2.1}$$

The current through the output conductance is

$$i_H = v_{CE} \times 10 \times 10^{-6} = v_{CE} \times 10^{-5}\,A \tag{F.2.2}$$

By Ohm's Law, across R_{LOAD}:

$$v_{CE} = 10 - i_C \times 500\,V \tag{F.2.3}$$

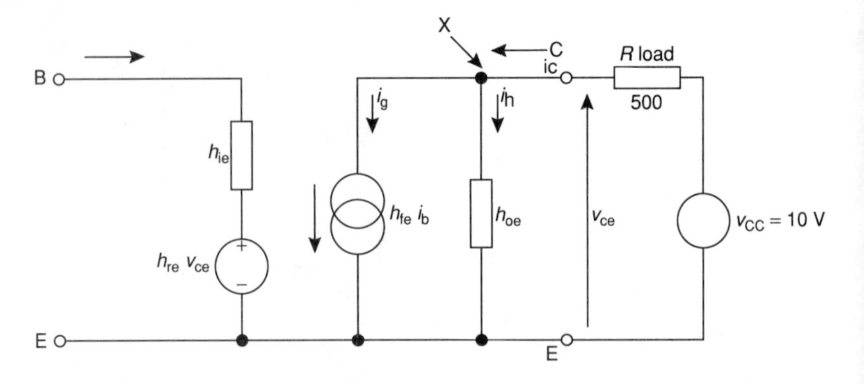

Figure F.6

By KCL, at node X, using (F.2.1) and (F.2.2):

$$i_C = i_g + i_h = 7.5 \times 10^{-3} + v_{CE} \times 10^{-5}$$

Substituting for v_{CE} from (F.2.3):

$$i_C = 7.5 \times 10^{-3} + (10 - i_C \times 500) \times 10^{-5}$$

$\Rightarrow$ $$i_C = 7.563\,\text{mA}$$

Substituting in (F.2.3) $$v_{CE} = 10 - 7.562 \times 10^{-3} \times 500 = 6.219\,\text{V}$$

F.2.3 Ebers-Moll model

The version in Fig. F.7 is the DC transport model, a variant used by SPICE (F.4), and represented by the set of equations given below. Both diodes are forward biased in Fig. F.7, but the model also works with either or both diodes reverse biased. The equations quoted below are part of the program which the computer uses for the analysis. The sequence is:

(1) Evaluate i_{CE} and i_{CC}, for the two diodes (compare B.1), inserting present circuit values of v_{BC}, v_{BE} and V_T:

$$i_{CE} = I_s(e^{v_{BC}/nv_T} - 1)$$

$$i_{CC} = I_s(e^{v_{BE}/nv_T} - 1)$$

(2) Calculate i_{CT}, the total current through the transistor, where:

$$i_{CT} = i_{CC} - i_{CE}$$

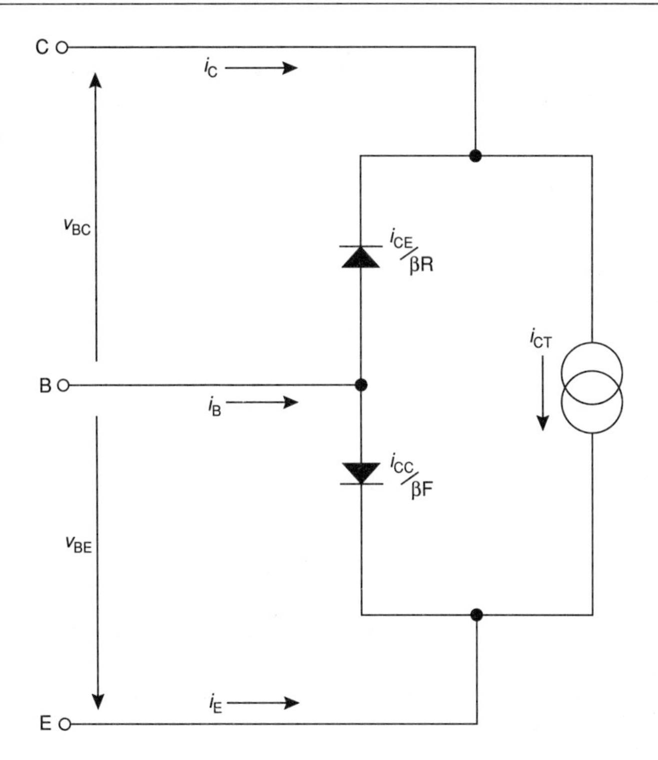

Figure F.7

(3) The expressions beside the diodes in Fig. F.7 represent the currents flowing from the base through the diodes when (or if) they are forward biased. β_R and β_F are the gains in each diode when they are forward biased. β_F is the equivalent of h_{fe}. In the model, the diode currents are reduced by dividing them by the gains, since the base currents are only small fractions of the currents flowing through the diodes. Calculate terminal currents by using KCL (D.1.1) at the three nodes:

$$i_C = i_{CT} - i_{CE}/\beta_R$$

$$i_E = -i_{CT} - i_{CC}/\beta_F$$

$$i_B = i_{CC}/\beta_F + i_{CE}/\beta_R$$

For a BJT in CE connection, the base-collector diode is reverse biased $\rightarrow$ $i_{CE} = 0 \rightarrow i_C = i_{CT} = i_{CC} = I_s(e^{v_{BE}/nv_T} - 1)$. Also $i_B = i_{CC}/\beta_F + i_{CE}/\beta_R = i_C/h_{fe}, \rightarrow i_C = h_{fe} i_B$. It can also be shown that $i_C = i_B + i_E$ and $i_C \approx i_E$, so all the theoretical relationships are modelled, including a non-linear response.

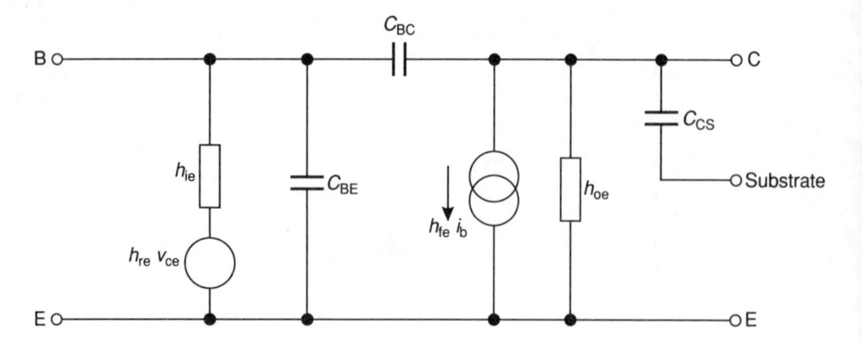

Figure F.8

F.2.4 High-frequency models

Figure F.8 is a BJT model as used by SPICE for dynamic small-signal circuits (compare with Figs. F.3 and F.4). C_{BC} is the sum of C_J and C_D (B.3) at the base-collector junction. C_J depends on the width of the depletion layer, so it is calculated from the known capacitance at zero bias, modified by the effects of v_{BC}, plus those of the virtual cell and a junction grading coefficient. All of these parameters except v_{BC} are specified in the definition of the model. The value of v_{BC} depends on circuit conditions at the time. C_{BE} is the sum of C_J and C_D at the base-emitter junction and is similarly specified, but is dependent on v_{BE}. C_{CS} is the collector-substrate capacitance. As a default, all of these capacitances have the value 2 pF, though these values may be specified for a particular type of transistor.

F.3 JFET models

A simplified h-parameter model for DC or small low-frequency signals is Fig. F.9. The figure includes the capacitors (dashed lines) for the high-frequency model.

F.4 SPICE

This is the acronym for Simulation Program with Integrated-Circuit Emphasis, first developed at the University of California at Berkeley in the late 1960s. Although intended as an aid to designing ics, it quickly became a general-purpose electronic design program. It has been through several versions, of which SPICE2 and SPICE3 have been widely adopted by software publishers as the basis for their own implementations. These invariably incorporate

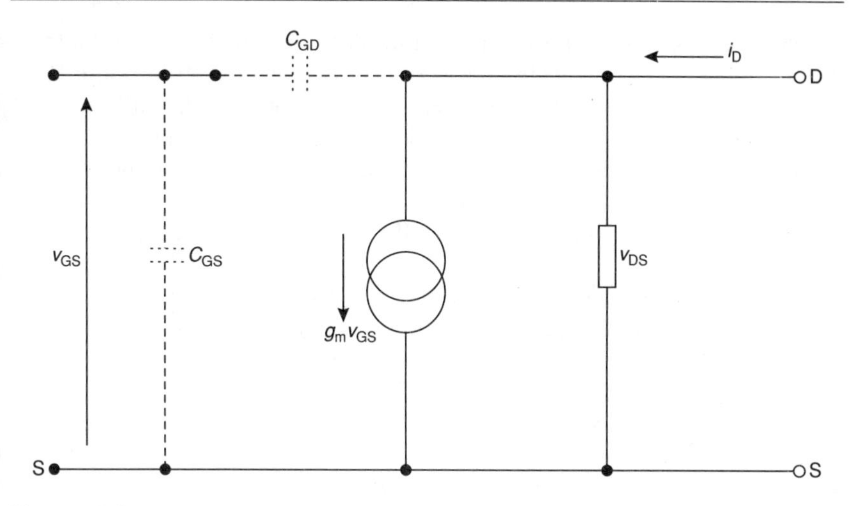

Figure F.9

user-friendly ways of entering circuit details, and a range of flexible graphics routines for displaying and printing out the results of analyses more effectively. The original versions of SPICE were run on mainframe computers and data were entered on punched cards. Today SPICE runs on a personal or laptop computer. Many publishers produce several variants of their software, ranging from inexpensive 'student' versions with limited facilities (though capable of fairly extensive analysis of small circuits) to full-scale 'professional' versions costing considerably more.

The analysis is based on a *netlist*, which is a list of all the circuit components, their values and the connections between them. Most commercial SPICE software has circuit-capture programs by which the user assembles the schematic diagram of the circuit on the computer screen, using special graphics routines and a library of component symbols. The computer then converts this to a netlist, checking for and reporting errors as it does so.

Components which may be specified in netlists include: resistors, capacitors, inductors, voltage and current sources, switches, transmission lines, diodes, BJTs (npn and pnp), JFETs, and MOSFETs. The actions of the components are described by a set of equations (F.2.3), which have a number of parameters used to match the behaviour of the component to specific manufacturers' types (many manufacturers publish lists of SPICE parameters for their products). Semiconductor devices have many parameters (e.g. BJTs have 40). Simulation software includes libraries of files defining a wide range (often several thousand) of component types.

In addition to the above, components may be represented by sub-circuits (saved as sub-circuit files for repeated use). This approach is used for modelling components (e.g. thyristors) that are not available in SPICE, and for building *generic* models of complicated devices such as transistors from more basic components (see Figs. F.3, F.7) when high precision or the modelling of certain features (such as temperature dependence) is not required. Generic models run much quicker than models with large numbers of parameters, with a considerable reduction in run time.

The analyses performed by SPICE fall under three main headings:

- DC analysis — calculates node voltages and branch currents with capacitors oc and inductors sc, and waveform generators set to their initial output values. Voltage or current sources can be swept over a range of values, as can temperature. Used for finding DC transfer functions.
- AC analysis — linear small-signal calculations. Frequency response, noise and distortion.
- Transient analysis — analysis of voltage and current changes in the time domain; Fourier analysis. Most of the simulator graphs in this book are obtained by transient analysis. Component values, frequency and temperature can be swept during the analysis.

The more recent commercial SPICE-based software allows for the simulation of digital and mixed-mode (analog plus digital) circuits. Although SPICE and its derivatives cater for a wide range of analyses, there are certain aspects of circuit design (e.g., filter design) that are better covered by software specially written for the purpose.

Like almost any mathematical algorithm, SPICE has limitations which must be understood if reliable results are to be obtained. The most common problem is convergence. SPICE works by iteration, that is, by starting off with a given set of voltages or currents and repeatedly applying the equations until it converges on a stable set of values. With certain types of circuit, it may fail to converge (2.2.1), or may converge too slowly or toward an incorrect result. The user must be aware of this kind of problem, which can usually be avoided if its cause is identified.

Any software only provides sensible answers if it is asked sensible questions. In 6.1.4 we see what can happen if the computer is not provided with all the facts. Using SPICE or other simulators is not just a matter of keying in a few figures and expecting a dependable result. The user must understand what is to be modelled, and how it is to be modelled. Which is one of the main reasons for writing this book.

Index